Appalachia's Coal-Mined Landscapes

Carl E. Zipper · Jeff Skousen
Editors

Appalachia's Coal-Mined Landscapes

Resources and Communities in a New Energy Era

Springer

Editors
Carl E. Zipper
School of Plant and Environmental Sciences
Virginia Polytechnic Institute and State University
Blacksburg, VA, USA

Jeff Skousen
Division of Plant and Soil Science
West Virginia University
Morgantown, WV, USA

ISBN 978-3-030-57779-7 ISBN 978-3-030-57780-3 (eBook)
https://doi.org/10.1007/978-3-030-57780-3

This Springer imprint is published by the registered company Springer Nature Switzerland AG
The registered company address is: Gewerbestrasse 11, 6330 Cham, Switzerland

Preface

Appalachian coal mining has been prominent during a significant period in American history. Appalachian coal was essential to the expanding American economy in the late nineteenth and twentieth centuries. Like other industries shaping America during this period, coal mining had positive and negative impacts on the land and people. Appalachian coal fueled the industrial development in the eastern USA during much of the twentieth century, including steel mills and manufacturing that became dominant producers of goods for the world. Appalachian coal also provided the low-priced electricity essential to power this industrial development and to improve the quality of life enjoyed by many. Appalachian coal is among the highest quality coals in the world and as such was shipped throughout the USA to the northeast and south into the Gulf states for electric generators, steel manufacturing, and other uses. Because of its high quality, coal from Appalachia has also been exported to Europe and Asia, bringing revenues back to the nation and region. Over more than a century, Appalachian coal has contributed to America's economic might and to industrial development in other nations. However, the use of coal is declining as low-cost gas and renewable energy replace coal as an energy source for electrical generation. As we write this in mid-2020, Appalachian coal production has declined to levels last seen in the nineteenth century, and few if any are predicting a renaissance and return to levels of the recent past.

The benefits of Appalachian coal and its mining, however, have come at a cost. Coal has dramatically transformed the bioecology and human economy of the Appalachian coalfield, a region which is among the most biodiverse in non-tropical areas of the planet. The region's humid temperate climate provides a comfortable living environment for human inhabitants and supports productive ecosystems for natural life. Spared of glaciation in its southern extents and with terrain that aided post-glacial migrations by flora and fauna, the rich biodiversity in Appalachia has remained mostly intact. But in modern history, coal mining has impacted the human society and transformed Appalachian landscapes dramatically. Due to coal mining, many mountain peaks have been removed and replaced by plateaus; numerous streams have been filled with waste rock; the mountain soils essential to the unique terrestrial ecosystems have been replaced by fractured rocks; and extensive areas have been undermined by excavated caverns, creating subsidence voids and altering hydrology. Waters draining from many mines have been degraded chemically in a manner that affects suitability as habitat for native life and use by residents. Human communities that developed in an economy supported by coal, many with inhabitants who were not prosperous even during coal's booms, are now losing that support as the industry struggles for its economic existence. Human inhabitants are less healthy, on average, in coal-mining communities than in other Appalachian regions. Not all regions of the Appalachian coalfield have been affected equally; some have experienced more intense effects than others. Everywhere coal has been mined it has left its mark.

The transformation of landscapes by human economic activity, as a generalization of the phenomenon in the Appalachian coalfield, is not a unique occurrence in world history as humans commonly alter their environment in the pursuit of economic gain. The loess plateaus of eastern China, for example, have been transformed from former rolling lands into terraced landscapes more suitable for agricultural production. People have altered urban areas in many parts of the world, changing the natural environment for housing, commercial, and industrial purposes. In parts of South America, east Asia, the American heartland, and elsewhere, broad areas have been cleared of natural vegetation and converted to agricultural production. Rarely, however, have such large areas been transformed for the extraction of a mineral resource. And rarely has such a large-scale transformation, such as that associated with coal mining in Appalachia, been as intensively studied, especially during its latter stages.

To those who walk the earth, the regional effects caused by the rise and fall of Appalachian coal has been a slow-motion occurrence, one that transcends human generations. Hence, no one generation or individual person has been able to experience fully the change from early pick-and-shovel mining to large earth-moving equipment of today, and the transformations of landscapes and related human and community impacts caused by mining. But from the standpoint of the region's natural resources and processes that formed them, the transformation has been rapid, requiring less than a century. The region's natural resources developed over eons in Appalachia's ancient landscapes through natural processes that cannot be replicated by humans, no matter how much money might be spent and even with the most advanced technologies.

We have both spent scientific careers working in the Appalachian coalfield and have seen and experienced a small microcosm of that change, including the rapid development of large-scale mining and the recent mining decline. To us, the change seems to be occurring like a slow-moving whirlwind, made all the more real by our still-increasing understanding of its past and lingering effects and the profound nature of their influence. When we both started working in the Appalachian coalfield, our scientific focus was on the effects of mining on soils, plants, and drainage. Overt effects such as vegetation loss could be easily seen and measured, and these problems were relatively easy to fix with best management practices as a means of providing short-term solutions. But with age and deeper understanding of natural processes, we see these effects of mining on soils and water, flora and fauna, ecosystem structure and function as far more subtle and problematic than we formerly perceived, with longer-term implications not well understood. Further and also with the perspective that comes with age, that set of problems also seems relatively minor compared to those stirring in the human dimension, and which are now becoming even more apparent as the economic support provided by mining to coalfield communities is falling away.

Our purpose here is to document the condition of Appalachian-mined landscapes and communities after more than a century of coal mining. This book is about the landscapes that are being left in the wake of Appalachian mining and those landscapes' inhabitants.

Countless studies of how mining has affected land, soils, water bodies, communities, and other resources have been performed, but most such studies have assessed individual resources and conditions in isolation. Our goal here is to bring that information together, to describe the resources of the Appalachian coalfield as they are today after the coal-mining whirlwind of the past two centuries. Our goal is to do this in a scientific way using science-based literature, and while not bestowing accolades of praise or pointing fingers of blame. We present this information with an intent to document regional resources' current condition with the cold eye of scientific objectivity, but also while explaining the scientific concepts that underlie what can be seen with our eyes to the extent that we and our co-authors understand them. We also endeavor to communicate what we have learned about managing the mined landscapes to reduce their environmental impact and otherwise improve their capacity to support ecological processes and human endeavors.

It is also our goal to point to the future, when we can, because the mined landscapes of Appalachia support people and their communities, many of which will remain even as local mining declines or ceases. The mined landscapes and their use by humans are ever changing as new ideas and resources are introduced and implemented. As the future unfolds, none of us can know for sure how that will happen, especially in the human societal dimension. But we think we can predict the direction of certain biological, biophysical, geochemical, and similar processes that have been set in motion by the landscape disturbances caused by Appalachian mining. With this knowledge, appropriate choices may be made on management of these new lands for ecosystem services and human use.

It is our hope that this book, and the information contained therein, can aid the management of Appalachian landscapes for the benefit of the region's human inhabitants. We make that statement while believing that the advance of scientific understanding, and its strategic application, can improve the human condition. It is also our hope that residents and managers of other world regions facing the prospect of intensive resource extraction can learn from what has occurred in the Appalachian coalfield.

Blacksburg, USA Carl E. Zipper
Morgantown, USA Jeff Skousen

Acknowledgements

We offer our sincere thanks to the many people who made this book possible.

We express gratitude to the authors, especially the lead authors who responded to our invitations to contribute chapters, answered our pleas to make deadlines, and put forth their efforts on top of many other duties. We respect their expertise and appreciate their essential contributions to this work.

We thank those who served as reviewers for the individual chapters, including Mary Beth Adams, Aysha Bodenhamer, Nicole Cavender, Charles Cravotta, W. Lee Daniels, Nick Fedorko, John Fike, Erin Haynes, Robert Hedin, Jeff Howard, Brent Means, Robert Nairn, Greg Pond, Eric Perry, David Soucek, and Jessica Turner-Skoff.

We extend sincere thanks to Patricia Donavon for the maps and geospatial analyses and to Kathryn Haering for the drawings and diagrams that appear in multiple chapters.

We also express thanks to research sponsors for contributing to our knowledge and experience in the Appalachian coalfield. These include the corporate sponsors of Powell River Project, who stayed with us even as our research priorities went beyond direct mining and reclamation effects. We also express appreciation to Virginia Tech for continuing its funding support for Powell River Project through many years. We thank the many people who have worked with us as collaborators over the years, including those with government agencies, coal companies, consulting firms, and community organizations and other residents; we thank them for providing access to properties, confiding in us, sharing ideas and friendships, and for otherwise supporting our studies of Appalachia's coal-mined landscapes.

Carl E. Zipper
Jeff Skousen

Contents

Editors and Contributors

About the Editors

Carl E. Zipper is Professor Emeritus of Environmental Sciences at Virginia Tech. He received his Ph.D from Virginia Tech in 1986, and was a member of Virginia Tech's faculty until retiring in 2019. He taught classes on environmental science to undergraduate and graduate students. He served as Director of Powell River Project, Virginia Tech's coal-mined land reclamation research and education program, for 21 years. He conducts research on environmental effects of Appalachian coal mining, and on environmental restoration of mining-influenced lands and waters.

Jeffrey G. Skousen is a Professor of Soil Science and the Land Reclamation Specialist at West Virginia University. He received his B.S. and M.S. degrees in Range Science at Brigham Young University, and Ph.D. in Agronomy at Texas A&M University in 1985. He teaches two classes: Introduction to Environmental Science and Reclamation of Disturbed Soils. He conducts research on acid soils, mined land reclamation, revegetation, reforestation, biomass for bioenergy, post-mining land use, and acid mine drainage control and treatment. He consults regularly with coal operators, consultants, regulators, and landowners on reclamation and water quality issues.

Contributors

Mary Beth Adams U.S. Forest Service, Morgantown, WV, USA

Carmen T. Agouridis University of Kentucky, Lexington, KY, USA

Christopher D. Barton University of Kentucky, Lexington, KY, USA

Elyse V. Clark Fitchburg State University, Fitchburg, MA, USA

John J. Cox University of Kentucky, Lexington, KY, USA

W. Lee Daniels Virginia Tech, Blacksburg, VA, USA

Jesse L. De La Cruz Virginia Tech, Blacksburg, VA, USA

John W. Edwards West Virginia University, Morgantown, WV, USA

Kenneth A. Eriksson Virginia Tech, Blacksburg, VA, USA

W. Mark Ford U.S. Geological Survey, Blacksburg, VA, USA

Jennifer A. Franklin University of Tennessee, Knoxville, TN, USA

Julia M. Gohlke Virginia Tech, Blacksburg, VA, USA

Sarah L. Hall Berea College, Berea, KY, USA

Nathaniel P. Hitt U.S. Geological Survey, Leetown Science Center, Kearneysville, WV, USA

Jess W. Jones U.S. Fish and Wildlife Service, Blacksburg, VA, USA

Natalie Kruse Daniels Ohio University, Athens, OH, USA

Julie A. LaBar Centenary University, Hackettstown, NJ, USA

Christopher M. Lituma West Virginia University, Morgantown, WV, USA

Louis M. McDonald West Virginia University, Morgantown, WV, USA

George T. Merovich Jr. Juniata College, Huntingdon, PA, USA

Eric R. Merriam U.S. Army Corps of Engineers, Pittsburgh, PA, USA

Lisa I. Muller University of Tennessee, Knoxville, TN, USA

George D. Santopietro Radford University, Radford, VA, USA

Kenton Sena University of Kentucky, Lexington, KY, USA

Jeff Skousen West Virginia University, Morgantown, WV, USA

David J. Soucek Illinois Natural History Survey, Champaign, IL, USA

Stephen F. Spear The Wilds, Cumberland, OH, USA

Rebecca M. Swab Restoration Ecologist, Cumberland, OH, USA

Carl E. Zipper Virginia Tech, Blacksburg, VA, USA

The Appalachian Coalfield in Historical Context

Carl E. Zipper, Mary Beth Adams, and Jeff Skousen

Abstract The Appalachian coalfield occurs within the eastern United States (US). This mountainous landscape is formed from natural dissection of sedimentary geologic strata with interbedded seams of coal and serves as headwaters for multiple rivers. The region's natural ecosystems, with a primary vegetation of mostly deciduous forest, are among the non-tropical world's most biodiverse. After first humans arrived more than 10,000 years ago, the first Europeans came to Appalachia as fur trappers and traders; agriculturalists and merchants came soon after. The region's diverse forests and rich mineral resources supported economic development as settlements expanded and populations grew. Coal mining began in the mid-1700s to supply commercial and residential users. Large-scale timber harvesting and coal mining stimulated railroad expansion in the mid-1800s, which improved transportation linkages to more populated areas and further increased coal demand. The American nation's industrial development increased usage of iron and steel, further expanding coal demands. Coal was essential to American electrification during the twentieth century. With numerous and thick coal seams accessible by both surface and underground mines, Appalachia was the US's primary coal-producing region from the 1800s through the 1970s. Appalachian coal mining has influenced the region's landscapes, forests, water, and people over more than two centuries.

Keywords Coal · Forests · Industrial development · Railroad · Steel · Timber

C. E. Zipper (✉)
Virginia Tech, Blacksburg, VA, USA
e-mail: czip@vt.edu

M. B. Adams
U.S. Forest Service, Morgantown, WV, USA

J. Skousen
West Virginia University, Morgantown, WV, USA

C. E. Zipper and J. Skousen (eds.), *Appalachia's Coal-Mined Landscapes*,
https://doi.org/10.1007/978-3-030-57780-3_1

1 Introduction

The Appalachian coalfield is a mountainous and forested region of the eastern US with biological resources that are among the world's most biodiverse (Fig. 1). The region is one of complexity, with a long history of landscape development and diversity of life forms. Within this ancient landscape, coal mining has been a major influence on human populations and environment. Many Appalachian communities developed with mining as a primary economic driver, and coal mining has affected human attitudes and culture (Bell and York 2010). Looking beyond regional boundaries, Appalachian coal has contributed to American prosperity and industrial development for more than a century.

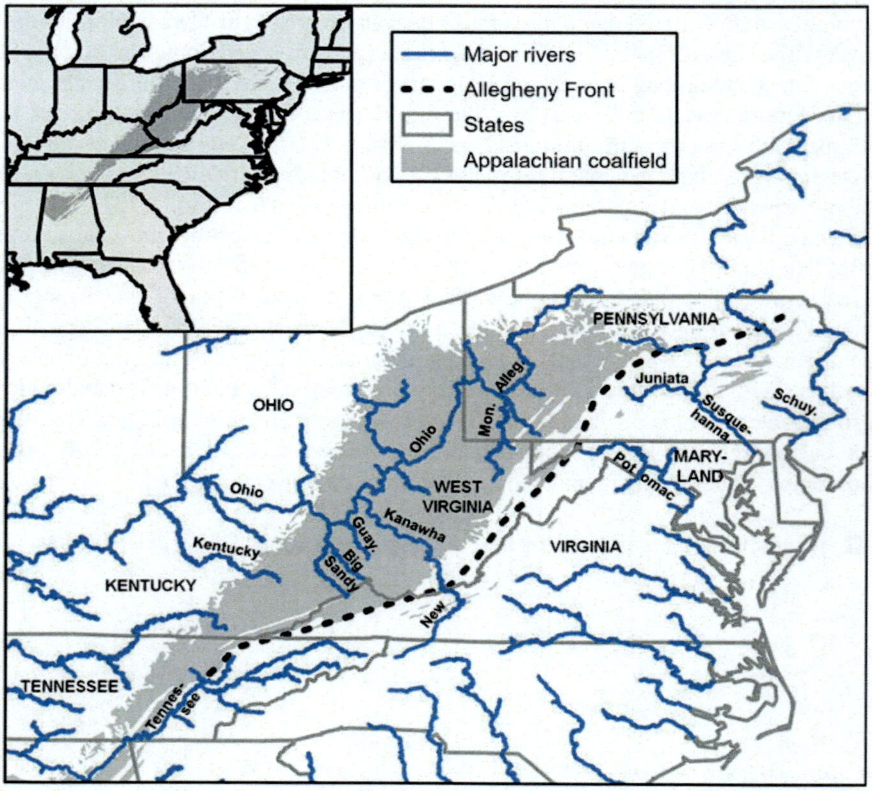

Fig. 1 The Appalachian coalfield, as it occurs in seven states of eastern U.S.A. The dark area represents surface expressions of geologic formations associated with bituminous and Pennsylvania anthracite coals. Rivers of the region include the Allegheny (Alleg.), Monongahela (Mon.), Schuylkill (Schuy.), Guayandotte (Guay), and other rivers as represented

This chapter reviews the Appalachian coalfield's natural and human history through the present day. In chapters that follow, the Appalachian coalfield's landscapes and communities are described as they have been affected by centuries of mining and removal of more than 45 billion tons of coal (Milici 1997; Zipper 2020).

2 Natural History

2.1 *Landscapes*

The Appalachian coalfield occurs within the oldest mountains on earth; their formation began early in the Paleozoic Era as thick layers of sediment and carbonate rock accumulated on the shallow sea bottom. When the seas receded, terrestrial plant communities thrived, producing organic matter that deposited with sediments. As drifting continental plates collided, they folded, faulted, and lithified former surface deposits. The mountain building was followed by deep erosion, with streams depositing layers of eroded sediments and rock debris downslope in nearby lowlands. Thus, the region's geology is sedimentary with beds of sandstone, shale, and coal, but often with overlying older rocks (Eriksson and Daniels 2021). Repeated uplift energized streams, which rapidly cut downward into the ancient bedrock. Some streams flowed along weak layers defining the folds and faults created millions of years earlier. Other streams down-cut rapidly across the resistant folded rocks of the mountain core. Then, around 220 million years ago, the prehistoric land mass known as Pangea began to break apart, and the plate tectonics that created the then-towering Appalachian Mountains were stilled. Weathering and erosion prevailed, wearing away the mountains and producing the complex and highly dissected mountainous topography found in the Appalachians today. In the eastern parts of Appalachia, terrain was formed within faulted and folded sedimentary rocks, and stream erosion has created a mountainous landscape of parallel ridges and valleys. Further west, a former plateau of generally flat-lying sedimentary geologic strata was dissected by stream drainages, forming a seemingly random pattern of narrow ridgetops, side slopes, and narrow valleys (Fenneman 1938). This mixture of youthful valleys and mature plateaus results in a complex terrain with steep slopes and deep gorges, narrow ridges, hills, and valleys prevailing in what is now known as central Appalachia, and with similar features but less steepness to the north and south.

There was a more recent geologic contribution to the development of landscapes and ecosystems in the Appalachian coalfield—continental glaciation. Rivers of ice, often thousands of meters in depth, covered the northern part of the region and contributed to the creation of current river drainages, particularly the Ohio River system. The glaciers, with their arctic air masses and icy winds, also created conditions suitable for boreal vegetation which extended as far south as present-day Georgia. As the glaciers retreated around 20,000 years ago, the climate changed again, and the boreal vegetation migrated to higher elevations and to the north even as humans began moving into the landscape.

Today, elevations in the Appalachian coalfield range from around 900 feet (275 m) in Ohio to as high as 4863 feet in West Virginia (1482 m). The region's climate is humid and temperate and varies with elevation and topography. The change in elevation, topography, and aspect can cause dramatic shifts in climate, e.g. the climate of a protected mountain valley will differ from that of an adjacent ridge top. Also, because major air masses move from west to east, terrain produces pronounced effects on precipitation, with a distinct rain shadow in some leeward areas east of the Allegheny Front divide. Mean annual precipitation ranges from 35 inches (890 mm) in valleys to 80 inches (2040 mm) on the highest peaks and is well distributed throughout the year. Because of the rugged topography, aspect exerts micro-climate effects. For example, southwest- and south-facing slopes which face the sun during mid-day and afternoon, are typically warmer and drier than adjacent northeast- and north-facing slopes. In steep terrain, west-facing slopes can receive more rainfall than east-facing slopes on the lee (drier) side of the ridges. The complex topography also drives variation in vegetation and soil, with differences in temperature and moisture often reflected by differences in forest composition and abundance, under-canopy vegetation, and productivity.

Because erosion is a major influence on these landscapes, soils are generally shallow except in depositional environments such as foot slopes, coves, and valleys, and along streams and rivers. The soils may be genetically young because of recent deposition but often with constituent sediments that are highly weathered. Due to forest vegetation, most soils are naturally acidic but high in surface organic matter. Water is generally abundant due to the ample and consistent precipitation throughout the year. Natural water quality is often dilute, with low concentrations of dissolved elements due to the highly weathered terrain. In forested watersheds, natural waters are generally of excellent quality but can be impaired after significant land-surface disturbances. Water flow in streams and rivers can be quite variable because of the high relief and shallow soils.

2.2 *Ecosystems*

Regional ecosystems have been shaped by the interactions of landscape, climate, natural and land-use history. The rich topography and dynamic climate created a complex set of varied ecosystems, with diverse flora and fauna (Butler et al. 2015). Pickering et al. (2003) reported that the Appalachian region is home to 2000 species of plants, 200 species of birds, 78 mammals, 58 reptiles, and 76 amphibians. The central and southern Appalachians are considered a "hotspot" for aquatic diversity because of the many diverse habitats created by the humid-temperate climate, the complex topography, diverse stream sections and patterns, and the relative geologic stability over evolutionary history (Chaplin et al. 2000; Flather et al. 2008).

The region is generally forested and has been since the last Ice Age. The region's extensive forests are well-described by E. Lucy Braun (1950) in her classic book *Deciduous Forests of Eastern North America.* As the Laurentide ice sheet retreated, the geographical ranges of temperate tree species migrated northward while boreal species followed the ice retreat northward and migrated to higher elevations. The generally northeast-southwest orientation of the ridges and valleys east of the Allegheny Front encouraged species migrations. Also, vertical zonation of vegetation prevails, that is, the lower elevation limits of each forest type increase toward the south, a result largely due to climate.

Like the topography, the vegetation is complex across landscapes, and indeed the two are often interrelated, with identifiable forest ecosystem types commonly found in particular elevational/temperature/moisture conditions. The most diverse are the mixed mesophytic forest types, which are largely confined to rich, moist montane soils, and may contain more than 25 tree species on a site. Some of the characteristic canopy tree species are red and white oak (*Quercus rubra* and *Q. alba*), sugar maple (*Acer saccharum*), American beech (*Fagus grandifolia*), tuliptree (*Liriodendron tulipifera*), white ash (*Fraxinus americana*), white basswood (*Tilia heterophylla*), yellow buckeye (*Aesculus octandra*), and yellow birch (*Betula alleganiensis*). Other common trees are red maple (*Acer rubrum*), shagbark and bitternut hickories (*Carya ovata* and *C. cordiformis*) and black or sweet birch (*Betula lenta*). Small understory trees and shrubs include flowering dogwood (*Cornus florida*), hophornbeam (*Ostrya virginiana*), witch-hazel (*Hamamelis virginiana*) and spicebush (*Lindera benzoin*). Perennial and annual herbs are abundant, among them herbal and medicinal plants such as American ginseng (*Panax quinquefolius*), goldenseal (*Hydrastis canadensis*), bloodroot (*Sanguinaria canadensis*) and black cohosh (*Cimicifuga racemosa*).

These same trees, shrubs, and herbs are also widely distributed in less rich, mesic forests that generally occupy cool coves, stream valleys, and flood plains throughout the region at low and intermediate elevations. At higher elevations and in the north, these mesic forests give way to less-diverse northern hardwood forests with canopies dominated by American beech, sugar maple, American basswood (*Tilia americana*), black cherry (*Prunus serotina*), and yellow birch and with far fewer species of shrubs and herbs.

Drier and rockier uplands and ridges are occupied by oak-chestnut type forests dominated by a variety of oaks (*Quercus* spp.), hickories (*Carya* spp.) and, in the past, by the American chestnut (*Castanea dentata*). The American chestnut was virtually eliminated as a canopy species by the introduced fungal chestnut blight (*Cryphonectaria parasitica*). Today, it can still be found in the understory as sapling-sized root sprouts, which succumb to the blight when they are larger. In present-day forest canopies, chestnut has been largely replaced by oaks. The richer oak forests, which often grade into mesic types in coves and on moist slopes, have dominantly white and northern red oaks, while the driest sites, particularly those in the rain shadow of the Alleghenies, are dominated by chestnut oak (*Q. montana*), or sometimes by scarlet (*Q. coccinea*) or northern red oaks, often intermixed with shortleaf, Virginia and pitch pine (*Pinus echinata, P. virginiana* and *P. rigida*). The drier oak forests generally lack the diverse small tree, shrub, and herb layers of mesic forests.

At the highest elevations, conifers dominate, mainly red spruce (*Picea rubens*), Fraser fir (*Abies fraseri*), balsam fir (*Abies balsamea*) and hemlock (*Tsuga canadensis*). As remnants of the boreal and sub-arctic vegetation which flourished during the last glaciation, these spruce-fir forests, often bathed in clouds, are home to rare and endangered species. Decimated by timber-harvesting, pollution, and an introduced invasive insect (*Adelges piceae*, the balsam woody adelgid), these high elevation ecosystems are considered endangered in the Appalachians.

Recent threats are emerging as dangers to Appalachian forest ecosystems. Such threats include serious insect and disease outbreaks, invasive exotic plants, overabundance of white-tailed deer (*Odocoileus virginianus*), and climate change. Among the most dramatic are the insect pests such as the introduced gypsy moth (*Lymantria dispar*), which infests primarily oaks, causing severe defoliation and tree mortality. The decline and likely eventual extinction of green and white ash (*Fraxinus pennsylvanica* and *americana)* is the result of the introduction of the emerald ash borer (*Agrilus planipennis*). Other pests and pathogens include the hemlock woolly adelgid (*Adelges tsugae*), which is responsible for the loss of an important conifer species in these predominantly hardwood forests; and the beech bark disease complex, which includes both a scale insect (*Cryptococcus fagisuga*) and fungal components, and affects American beech (*Fagus grandifolia*); and Dutch elm disease, caused by a fungus (*Ophistoma ulmi*) spread by elm bark beetles, which is responsible for killing millions of elm (*Ulmus*) trees. All of these insect and disease threats are creating a less diverse forest overstory. Over-browsing by deer negatively affects forest regeneration, further driving down tree species diversity. Climate change is likely to be a subtle but significant interacting effect on the forests in the coming decades (Butler et al. 2015).

3 Human Settlement and Economic History

3.1 Early Humans

The region's human history began near the end of the last Ice Age, during the early Pleistocene. Evidence of human habitation 14,000 or more years before present was found at Meadowcroft Rockshelter in Washington County, Pennsylvania (Advaosio et al. 1990). During this era of receding ice sheets, the climate was colder and supported mixed vegetation with more open canopy than in present-day forests, while the landscape was occupied by animal species adapted to savannah-like vegetation (Guilday 1982). The region's earliest humans were likely nomadic hunters of large mammals, many of which are now extinct or displaced. Primary archeological evidence of human existence during what is known as the Paleoindian era is comprised of projectile points, carved from rock and with distinctive shapes. Such "arrowheads" have been found in the Appalachian coalfield, indicating significant human dispersal during the late Paleoindian era (Fig. 2). Most Paleoindian artifacts

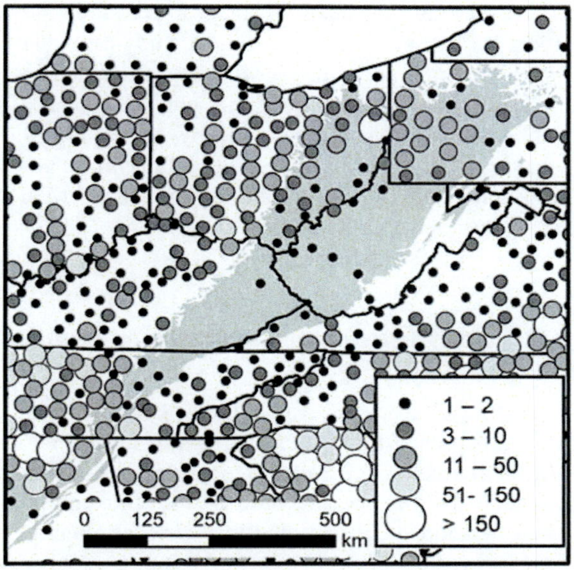

Fig. 2 Distribution of Paleoindian artifacts (arrowheads) in middle latitudes of eastern US, Numbers of artifacts reported by Anderson et al. (2019) by county are plotted by each county's geographic centroid. Counties lacking symbols have yielded no Paleoindian artifacts. Gray areas are the Appalachian coalfield

have been found in the coalfield's northern segments, where relief is less extreme than further south, and in wide river valleys such as the Ohio and Kanawha; while some artifacts have also been found in valleys of smaller rivers and in uplands (Rosencrance 2018). However, the geographic pattern of Paleoindian artifacts suggests that Appalachian coalfield regions south of present-day Ohio and Pennsylvania were not well-used by the eastern North America's earliest people, relative to the surrounding and less-mountainous terrain.

The period of Native American history known as the Archaic began approximately 9,000 years before the present as ice sheets receded further, climate warmed, and forest canopies closed. These changes were accompanied by extinctions of some megafauna that had sustained the region's Paleoindians (Maslowski 2017). Native Americans of this period tended to settle in river bottoms and establish livelihoods that included the gathering of edible plants as well as hunting of smaller game (Drake 2001). One of the most significant Archaic sites in the coalfield is at St. Albans, near the Kanawha River in southern West Virginia, which was used periodically for more than 10,000 years through 1700 A.D. (Broyles and Harding 1974).

In eastern North America, the Archaic period transitioned into what is known as the Woodland Tradition as native Americans increased reliance on cultivated plants. In Appalachia, this transition began in about 1000 B.C. (Drake 2001). During this era, indigenous peoples developed a capability to work with pottery, social structures became more complex and integrated, and some tribes constructed burial mounds for interring their dead. Archeological sites with evidence of Woodland-tradition Americans occur in the coalfield but it appears that large areas of this mountainous region remained unoccupied, even up to the time of European arrivals (Munoz et al.

2014). Resources needed for survival were found in fertile valleys near streams and rivers (Milner et al. 2001). A recent study estimated the total population of eastern North America at the dawn of European settlement, the year 1500 A.D., at about 1.2 to 1.7 million people (Milner and Chaplin 2010). However, analysis of maps published with that estimate and by Anderson (1991) suggest humans occupying what is now the Appalachian coalfield may have numbered only 100,000 or less and were most heavily concentrated in the Ohio—Monongahela, lower Kanawha, and upper Susquehanna River drainages.

3.2 Early European Explorations and Settlement

The earliest Europeans to enter the Appalachian coalfield likely were fur traders. It is known that the European fur trade with North America began in the early 1500s, but almost a century passed before fur traders penetrated inland to reach the Appalachian coalfield. During the sixteenth and seventeenth centuries, American furs were in great demand in Europe due to their high quality and scarcity from diminishing fur-bearing animals in continental trapping grounds. Hence, Europeans established trade with North American native peoples soon after their arrival, supplying them with manufactured items such as textiles, iron tools, and guns in exchange for furs.

Barriers to westward movement by early Europeans included dense woodlands, in some cases the indigenous peoples, and difficult terrain including the formidable physical barrier created by the western edge of Appalachian plateau, the Allegheny Front. As supplies of furred animals diminished along the east coast, the traders traveled further inland, often along river courses. An eastern Virginia fur trader, Abraham Woods, in alliance with the Virginia colonial government, sent several expeditions to explore western Virginia areas. Those parties came upon the westward-flowing river, which they named Woods River (now known as the New River). Woods organized one of the first recorded journeys into the central Appalachian coalfield in 1671, the Batts-Fallam expedition. With indigenous guides, the expedition followed the Woods River west to Peters Mountain, and then traveled overland to the Guyandotte River and Tug Fork of the Big Sandy, two Ohio River tributaries (Briceland 1991). Other explorations of the interior coalfield followed. In the 1690s, Albany fur traders sent an exploration party led by Arnout Viele south into the Delaware and Susquehanna River valleys. They found their way west following the Juniata River from the Susquehanna over the watershed divide and eventually reached the Allegheny and Ohio Rivers (Drake 2001).

If other seventeenth-century explorations of the coalfield region occurred, they are not as well documented; but the eighteenth century saw increasing exploration and settlement of the coalfield. In 1742, a party led by J.P. Salley followed the New River into southern West Virginia, reaching a river where he observed outcropping coal seams, now the Coal River. In 1750, a party led by Thomas Walker traveled through the southwestern Virginia coalfield and through Cumberland Gap into eastern Kentucky. The Walker party were the first Europeans to document Cumberland Gap, but they were not the first to travel that way as their journals described streams which had been previously named (Eavenson 1942). In 1751, an exploration party led by Christopher Gist entered eastern Kentucky by way of the Ohio River, then returned eastward by following the North Fork Kentucky River and then passing through Pound Gap (Eavenson 1942).

Once these exploration parties had found the way, settlers followed. Fur traders working out of Philadelphia had outposts in the Ohio Valley prior to 1750 (Drake 2001). Westward migration to the coalfield began in Maryland and Pennsylvania. Settlers reached what is now the Pittsburgh area by traveling up the Potomac River through the Allegheny Front and into the Monongahela drainage or by following the Susquehanna and Juniata Rivers (Flanders 1998). By 1770, around 10,000 families had settled in western Pennsylvania (Flanders 1998). Settlement of Pennsylvania's anthracite coalfield occurred later, as the first permanent settlement was established in 1769 (Harvey 1909). Virginia residents began migrating westward to the Kanawha Valley in the 1770s (Drake 2001). Daniel Boone led a party through Cumberland Gap to Kentucky, establishing the settlement now known as Boonesborough in 1775 (Flanders 1998). Following the Revolutionary War, settlers moved into the coalfield counties in large numbers. The first US Census, in 1790, counted approximately 200,000 people in the coalfield with the bulk of those (~85%) in Pennsylvania. By 1820, that number had increased to ~ 1,000,000 but still with the largest concentration (~85%) in the northern Appalachia area of northern West Virginia, Maryland, Pennsylvania, and Ohio (data from Forstall 1996).

3.3 *Early Agriculture and Forest Use*

Many of the region's early settlers were agriculturalists who established what can be described as "backwoods farms." They were generally located in areas with access to water and cleared the dense timber on flatter lowland landscapes for crops. They also raised livestock, often including hogs, by treating the woodlands as "open range." The form of agriculture practiced by the earliest settlers has been described as "yeomanesque," or a largely self-sustaining activity (Drake 2001). As the 1800s progressed, backwoods agriculture became more market-oriented with establishment of towns with merchants.

The first large-scale industrial activity to affect many rural coalfield areas was timber harvesting. Appalachian forests constituted an important economic resource for the new nation. Timber resources were in demand for a variety of uses: lumber for manufacturing, building construction, ship construction; bark for tanneries; conversion to charcoal for uses such as iron forging; and direct burning for heat. As demand for lumber by east-coast cities and export markets consumed coastal forests, timber-harvesting enterprises achieved greater scale as they moved west. Industrial harvesting of Appalachian forests began before the Civil War, often in areas close to rivers large enough to float logs to market. Pennsylvania was the first of the Appalachian states to experience large-scale timber harvesting, with an initial center along the Susquehanna River. After the Civil War, Pennsylvania timber harvesting expanded, driven in part by new demands created by railroad construction, and the state became the nation's leading lumber producer for a short period (Alban and Dix 2013). By the end of the century, much of Pennsylvania's woodlands had been reduced to "stumps and ashes" (Alban and Dix 2013) and Appalachian lumber producers moved further south. For example, most of West Virginia was still in old-growth forest in 1880, but most of the state's forests had been harvested by 1920 (Lewis 1998), while the timber harvesting "boom" in eastern Kentucky and southwestern Virginia peaked in the early 1900s (Eller 1982).

Expansion of timber harvest was aided by development of railroads, which moved products from harvest areas lacking rivers to markets. Railroad expansion also provided an important market for forest products, as lumber was used for fuel, rail ties, and bridge and coach production. American railroad construction began in the 1830s and accelerated in the following decades. From fewer than 3000 miles (4828 km) of rail in 1840, railroads expanded to approximately 9000 miles (14,484 km) in 1850 and more than 30,000 miles (48,280 km) in 1860 (Thompson 1925), primarily on the east coast.

The large-scale timber harvesting initiated some of the societal changes that carried over into the era of extensive coal mining. Timber harvesting of large areas interfered with what little remained of the "open-range" forest-grazing practices of early backwoods agriculturalists. Timber harvesting companies assembled large land holdings, some of which combined financial interests in both coal and timber (Eller 1982); while lumber haulage was an important revenue source for the railroads which enabled their further expansion and later transport of coal. When working large tracts in remote areas, timber companies established "man camps" that provided food and shelter for the workers, a forerunner of the "company towns" established by mining firms to house their workers. Hence, as coal mining expanded in the late nineteenth and early twentieth centuries, the model of industrial-scale natural resource extraction, along with the associated societal changes, was already underway.

4 Early Industry

4.1 Coal Mining

An early map of the upper Potomac River, dated 1736, shows the presence of "cole mines" at locations of coal outcrops (Eavenson 1942). Early commercial mining occurred at "Coal Hill" (now Mount Washington) across the Monongahela River from what is now Pittsburgh beginning in 1760 and those coals were extracted for use by soldiers at Fort Pitt. During the late 1700s, use of coal for both home-heating and commercial purposes expanded in Pittsburgh and additional mines were established. In 1800, a traveler noted that a "cloud of smoke" was visible over Pittsburgh. By the

1820s, words such as "dark veil," "gloomy," "filth," and "sulphurous vapor" were being used to describe the Pittsburgh atmosphere (Eavenson 1942).

Extraction and use of Pennsylvania's anthracite coal also began in the eighteenth century but its expansion did not occur as rapidly. Among the region's earliest settlers were blacksmiths who began using anthracite coal from the initial settlement in 1769 (Harvey 1909). Although anthracite was used locally through the rest of the century, expansion of mining beyond local needs was hindered by two factors: difficult transportation of coal out of the region, and the difficulties of burning anthracite in the open-air furnaces that were in common use at that time. Anthracite coals burn hotter and with fewer air emissions than bituminous coals but are more difficult to ignite. Anthracite mining expanded in the early nineteenth century with improvements in river-transportation and development of burning equipment designed for anthracite (Eavenson 1942).

Coal mining also began in other Appalachian areas during this period. By 1789, coal from western Maryland's upper Potomac basin was being transported to Hagerstown for use in nail making (Eavenson 1942). Late-eighteenth century land records from West Virginia describe the presence of coal as a valued commodity, but the state's first mine was not documented until 1810, in Wheeling where it supplied homes and blacksmithing (WVGES 2017). The first reported mining in Ohio dates from 1800 in Jefferson County, just north of Wheeling (Crowell 1995). In Virginia, the earliest recorded mining west of the Richmond Basin occurred in 1790, near Toms Creek at the base of Brush Mountain in Montgomery County, an area that is no longer in production; but no mining was recorded in the southwestern coalfield until much later (Hibbard 1990). By 1807, coal was being shipped from eastern Kentucky mines down the Kentucky River to Frankfort for use by blacksmiths (Eavenson 1942). The first documented production for Tennessee dates from 1839, when a state geological report described coal being hauled by wagon from Kingston to the Tennessee River for shipment on barges to markets downstream and several mines producing coal for blacksmithing (Eavenson 1942).

During that latter half of the nineteenth century, railroad extensions into and through the Appalachian coalfield aided coal-mining expansion. The Pennsylvania Railroad was extended to link Pittsburgh with Philadelphia by 1854. The Baltimore and Ohio extended a main line from the east coast up the Potomac River to reach Wheeling in 1858, and then after 1865 across the Ohio River and further west. Also following the Civil War, the Chesapeake and Ohio was constructed from Covington, Virginia, westward to the Ohio River, reaching Huntington, West Virginia, in 1873 (Thompson 1925). The Norfolk and Western Railroad was formed by merging smaller lines extending from the port at Norfolk to Bluefield, West Virginia, in the early 1880s; and was then extended through southern West Virginia to reach the Ohio River in 1892 (Lewis 1998). These rail lines provided essential transportation to markets for coals mined in areas away from the rivers.

4.2 Appalachian Coal Aided American Industrial Expansion

As the nineteenth century continued, Appalachian coal mining expanded, mining methods changed, and the geographic locus of coal mining changed within Appalachia (Fig. 3 upper). Nonetheless, Appalachia remained as the primary source of US coal up through the 1970s. (Fig. 3 lower). Readily-available and high-quality coal enabled establishment of numerous industries in the Appalachian frontier during early decades, reducing the young nation's economic dependence on other nations. Some of those industries continued expanding through the nineteenth and into the twentieth century as America became the world's leading economy.

4.2.1 Early Industry

Salt manufacturing was one of the first industries enabled by Appalachia's abundant coal. Salt springs provided a raw material source and coal provided the energy. A coal-fired salt works was established in western Pennsylvania in 1809. Wood-fired salt works in the Kanawha valley converted to coal beginning in 1817 (Eavenson

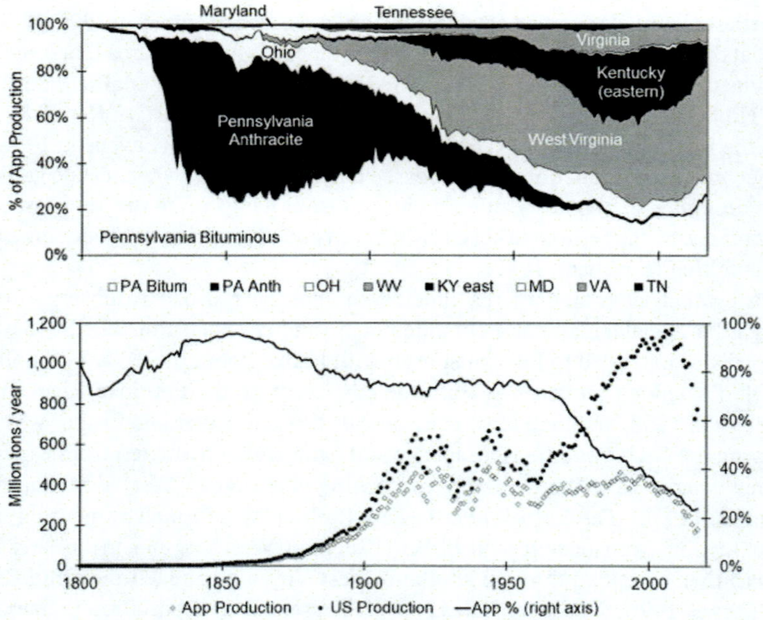

Fig. 3 Appalachian coal production, 1800–2017. (upper) Production by individual states expressed as a percentage of total Appalachian production. (lower) Annual Appalachian (App) coal production compared to total US production and expressed as a percentage of total US production (Milici 1997; EIA 2019a; Zipper 2020)

1942). By 1829, these areas were producing well over half of the new nation's salt (Lippincott 1912).

Another early American industry fueled by Appalachian coal was glass manufacturing. Following the Revolutionary War, the US had no significant capacity to produce glassware so most glass items were imported. Two glass-producing firms opened in the Pittsburgh area in 1797 using coal as fuel (Eavenson 1942). In 1800, two of the five known US glass-making factories were located in Pittsburgh (Davis 1949). By 1820, the number of US glassmakers had expanded to 33 and most of those, other than the nine west of the Alleghenies, used wood as fuel. Over the next 40 years, the wood-fired kilns suffered cost disadvantages as coal became the dominant glass-making fuel and the Pittsburgh region became the center of the US glass industry (Lamoreaux and Sokoloff 1997). This transition occurred as the industry expanded production, displacing imports to meet most of the nation's needs (Davis 1949). In the decades following 1860, natural and coal-derived gas supplanted solid coal as a glass-making fuel, but firms located in western Pennsylvania, northern West Virginia, and southern Ohio maintained leadership of the American glass industries. By the 1920s, Pittsburgh-area firms remained as dominant producers of American glass (Lamoreaux and Sokoloff 1997).

Another of earliest industries in US was iron making. Blast furnaces extract an impure form of iron, called pig iron, from the raw ore through the application of extreme heat. Multiple blast furnaces were present in the colonies even before the Revolutionary War (Hogan 1971). Following independence, iron works spread throughout the new nation. Except in the Pittsburgh area, however, many iron works relied on pig iron imported from Britain (Chandler 1972). Where available, coal was used for forging and working iron from the earliest days, but charcoal was the primary blast furnace fuel until the 1830s when a process enabling substitution of anthracite coal for charcoal was developed (Hogan 1971). This development enabled US iron production to increase from approximately 165,000 tons in 1830 to nearly 1 million tons in 1860 (Hogan 1971). The cost advantages of producing raw iron using anthracite were so profound that, during the 1840s and 1850s, much of the raw iron used by Pittsburgh's industry was being shipped from regions with better access to anthracite coal (Chandler 1972). In the late 1850s and following the Civil War, American iron manufacturers began adopting bituminous-coal and coke-based iron-making processes (Hogan 1971). By 1860, more than half of the new nation's iron-production capacity was located in Pennsylvania, and more than one-third was in Ohio, New Jersey, and New York, states with easy access to Appalachian coal (Hogan 1971).

By the late 1830s, Pennsylvania anthracite coal started to become widely available on the east coast, stimulating iron manufacturing and associated industry outside of coal-mining areas. New canals enabled anthracite shipments from production areas in central Pennsylvania, near what is now Wilkes Barre and Scranton to Philadelphia. From there, it was shipped up and down the east coast to multiple markets. Anthracite's high heat value and low price, relative to charcoal, enabled production expansion by multiple iron works in the northeast. Their expanded output enabled additional industrial activity, including expanded production of iron products by secondary manufacturers. The availability of coal for use in steam power also enabled expansion of factories for textiles and other products beyond limits imposed by reliance on water power (Chandler 1972). The northeast's rapid industrialization during this period fueled demand, enabling Pennsylvania anthracite coal production to increase from less than 250,000 tons in 1830 to more than 10 million tons in 1860 (Milici 1997).

4.2.2 Transportation

Coal was also important to pre-Civil War transportation, as it gradually supplanted wood as the power source for the steamboats that traveled the nation's rivers (Schuur and Netschert 1960). Appalachian coal was more essential to development of the young nation's network of railroads. Iron, and then steel, enabled the rail network's construction. Early in the rail industry's development, many of its iron and steel products were imported, but Pennsylvania's anthracite coal enabled expansion of the US iron industry in the decades following 1840 as the rail industry was expanding. By 1860, the domestic iron industry was fueled predominantly by anthracite coal (56% of tonnage), with some producers (13% of tonnage), primarily in the Pittsburgh area, also using bituminous coal and coke (Hogan 1971). By this time, the developing railroad industry was consuming one-fourth of domestic iron production (Fogel 1964), and domestic iron had largely supplanted imports for the production of rail (Hogan 1971). In decades following 1860 when Appalachia was the US's major coal-producing area (Fig. 4), coal displaced wood as a fuel for railroad engines and by 1921, almost 25% of US coal use was as locomotive fuel. Also during this period, shipping coal from Appalachian mines to distant markets provided the railroads with a major source of revenue.

Railroads were essential to the US's rapid industrialization during the late nineteenth and early twentieth centuries, providing low-cost transportation for the nation's widely dispersed raw materials and manufactured products (Bunker and Ciccantell 2003). Railroads have also been called a major contributor to the "take off" of the American economy, i.e., the period when a largely agrarian nation was transformed into a more industrialized economy, capable of strong economic growth (Fogel 1964). It is clear that this rapid expansion of the rail industry during the late nineteenth century was enabled by Appalachian coal.

4.2.3 Steel and Related Industries

Although small amounts of steel had been manufactured in America as early as 1810, the steel industry did not begin rapid growth until after 1860 (Hogan 1971). By 1860, Pittsburgh was already the center of the young US steel industry by virtue of its proximity to necessary raw materials, including coal. In 1867, the Bessemer process for converting pig iron to steel was established commercially, primarily for producing rail materials. By 1880, US rail production had grown seven-fold from

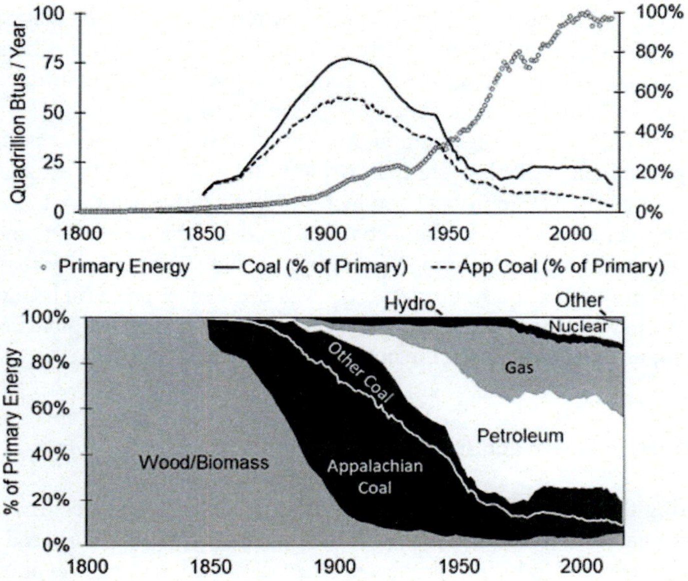

Fig. 4 US consumption of primary energy by source, 1800–2017. (Upper) Total primary energy consumption (left axis); proportion of primary energy consumption provided by coal (right axis); and estimated proportion provided by Appalachian coal (right axis). (Lower) proportions of primary energy consumption provided by various energy sources (other = geothermal, wind, solar, and miscellaneous). Data from EIA (2011, 2020b; Zipper 2020). Coal is segmented into Appalachian (App) and other sources based on sources cited for Fig. 3; hence, Appalachian coal contributions to US primary energy consumption, as a fraction of total coal contributions, are likely overestimated for recent decades since those estimates do not account for coal exports and imports

pre-Civil War levels, and steel had largely supplanted iron for rail manufacturing. By 1880, Pennsylvania facilities fueled by Appalachian coal were dominant producers of steel rail, and of iron and steel more generally. Other dominant producers were located in states with direct access to Appalachian coal, including Ohio, New York, and New Jersey (Hogan 1971).

As American industrial development continued past the era of rail expansion, the steel industry continued to play an essential role. Those contributions have become more diverse and pervasive in more recent times. Throughout most of the nineteenth and twentieth centuries, America's steel industry remained centered in mid-Atlantic areas supplied by Appalachian coal, and Appalachia's mines supplied steel producers within and well beyond Appalachia with high-quality metallurgical coals.

Another major contribution of Appalachian-supplied steel producers to American industrial development was the construction of steel-skeleton tall buildings. Pioneered in Chicago in the 1880s with steel produced by Pittsburgh's Carnegie companies (Misa 1995), these tall buildings came to dominate urban skylines. Similarly, the American military undertook a major effort to construct steel warships in the decades prior to World War I. The evolving technology required new types of steel, as guns were being developed to shoot large projectiles for long distances, along with armor-plating capable of protecting American warships from similar projectiles fired by enemies. Much of that steel was produced by the Appalachian-coal supplied firms Carnegie Steel and Bethlehem Steel (Misa 1995).

The latter half of the nineteenth and early twentieth centuries, a period during which Appalachian coal played a major role in the US economy, was a period of dynamic growth of US industrial production. During the latter half of the twentieth century, automobiles became a common mode of transportation, and automobile manufacturing became a major consumer of steel. The vast majority of American autos were domestically produced, as was the steel used in their manufacture. Throughout much of the twentieth century, Appalachian metallurgical coal was an essential input to American steel manufacturing.

4.2.4 Electric Power and Other Energy

Also, during the early-to-mid twentieth century, major changes occurred in energy sectors as coal was replaced by other fuels, including petroleum-based fuels and natural gas, for many industrial processes other than steel making. Yet another major change during this period was the electrification of the American society (Rosenberg 1998). During its early days of electrification, coal-fired steam generation was the primary means of producing electricity in areas without direct access to hydroelectric power (Fig. 5), and most American coal was Appalachian. Coal remained as the power source for more than 50% of US electrical generation for most years through the early 2000s, with Appalachia supplying more than 70% of US coal through the 1970s (Fig. 3).

5 Coal Production

5.1 Historic and Recent

Coal mining, although simple in concept, is difficult in practice, especially when prioritizing worker and environmental protections. Surface miners remove the earth materials overlying coal seams so as to extract the coal while underground miners burrow into the earth. Early mining relied on human labor, as miners removed the loose soil and rock overlying coal beds ("overburden") and then followed the coal seam into the earth for underground-mine extraction. As time passed, demands for coal increased, the excavations expanded in size, and the hazards to worker health and environment of these mining methods became more apparent. Over centuries, mining firms have innovated by developing technologies that address difficulties inherent to mining processes while remaining profitable and creating safer work spaces (Skousen and Zipper 2021).

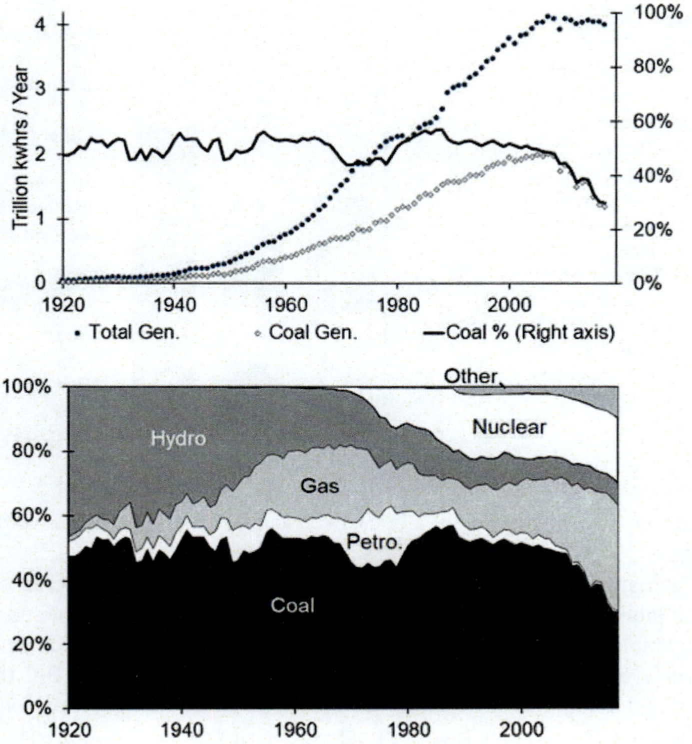

Fig. 5 Energy sources for electrical energy, 1920–2017. (Upper) Total electrical energy consumed in the US, estimated electrical energy fueled by coal, and proportion of the US-consumed electrical energy fueled by coal. (Lower) Proportions of US electrical energy consumption powered by various energy sources. Data from BOC (1975) and EIA (2020b) as compiled by Zipper (2020). "Other" energy sources include wood/biomass, wind, solar, geothermal, waste, and other miscellaneous

Appalachian coal production increased steadily through the nineteenth and early twentieth centuries, reaching 478 million tons in 1926 before declining almost to half that level in 1932 during the depths of the Great Depression (Fig. 3). Production increased leading up to World War II and then exceeded pre-depression levels, reaching 511 million tons in 1947; but declined again to <400 million tons annually during the economic transition following the War. Through the mid-twentieth century, coal was the fuel of choice for multiple uses that rely on other fuels today. During much of the twentieth century, coal producers served multiple markets, many of which were economically sensitive, resulting in dramatic fluctuations of mining activity (Fig. 6). Railroads, industrial plants, home heating, and even ocean shippers, for example, were major coal consumers in past years. American coal production, and pricing for Appalachian coal, boomed again during the late 1970s in response to the oil supply disruptions of that era which stimulated greater coal demand for

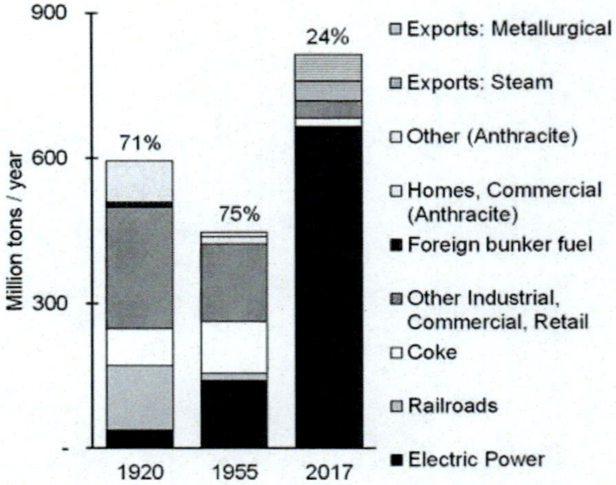

Fig. 6 Markets served by US coal in 1920, 1955, and 2017, with percentages of total production provided by Appalachian producers designated at the top of each bar. Market categories as listed by the data sources vary among years. Market segments are arranged vertically in the same order as the figure legend. Data sources: Schuur and Netschurt (1960), Table 19 for 1920 and 1955; and US EIA (2018) for 2017

electric power (Fig. 5; Black et al. 2005), even as other market sectors were reducing coal usage.

Appalachian coal production volatility declined and production increased steadily during the latter half of the twentieth century as coal producers' dependence on the relatively stable but growing electrical generation markets increased. Appalachian production averaged 428 million tons/year in the 1990s (Fig. 3). But this period also continued a major shift of American coal production to non-Appalachian areas, especially the western areas that dominate US coal production today (Fig. 5). The early twenty-first century saw a modest decline of Appalachian production to levels similar to those of the 1950s (368 million tons/year, on average, for 2002–2008), even as national production increased, but western coal producers took an increasing share of electricity markets. Over the next decade, Appalachian production declines steepened to reach production levels of <200 million tons/year by 2016 through 2019.

Although Appalachian coal has been largely displaced by other coals and fuels for electrical-generation, Appalachia remains as a primary producer of metallurgical coal for domestic coke production and for export. Metallurgical markets demand coals of specific qualities that Appalachian producers can provide. In 2018, US average prices for coking coal were more than three times those of electric power markets (US EIA, 2019a). Miners in Appalachia experience higher costs than those in other US regions due to factors such as mountainous terrain and reserve depletion that is directing production toward increasingly thinner seams. Hence, Appalachian producers' abilities to gain high prices in metallurgical-coal markets is an important driver for much of today's Appalachian mining.

Coal mining and related industries have been important contributors to Appalachian economy throughout its modern history (Thompson et al. 2001; Bowen et al. 2018). Mining's greatest local influence occurs in the central Appalachian counties of central and southern West Virginia, eastern Kentucky, and southwestern Virginia where coal production has been most intensive (Fig. 7). These areas that

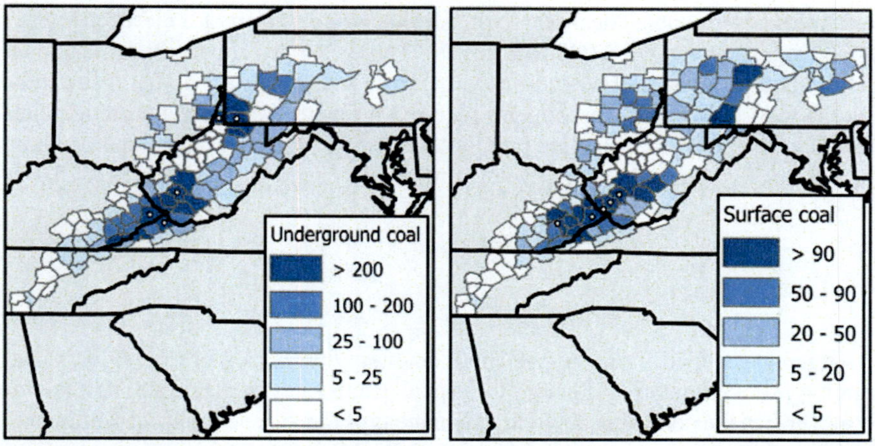

Fig. 7 Cumulative coal production (million tons), 1980–2018, by underground (left) and surface (right) mining methods, for Appalachian coalfied counties. The top five coal-production counties for each mining method are designated with small white circles. Data from US EIA (2019a) and predecessor publications

also experience high levels of poverty (Lobao et al. 2016). In 2005, coal mining was responsible for > 20% of total employment in 13 central Appalachian counties (Bowen et al. 2005). Recent declines of Appalachian coal mining have created economic hardships in these counties (Santopietro and Zipper 2021).

5.2 *Current and Future*

Appalachian coal production is declining in response to multiple factors, many of which appear unlikely to reverse (Zipper et al. 2021b), despite the ~100 billion tons of coal known to remain unmined in Appalachia (US EIA 2019a). Much of the region's unmined coal is in thin or difficult-to-access seams that are costly to mine. Such problems are exacerbated by difficult terrain, especially in central Appalachia, and increasingly strict environmental regulations including those concerning water (e.g. US EPA 2016).

Meanwhile, the nation's fleet of coal-fired electrical generators is aging and affected by increasing environmental restrictions, especially concerning air emissions (Revetz and Lienke 2016). More than 40 gigawatts of coal-fired generating capacity in primary markets for Appalachian coal, representing more than 1/3 of those states' 2011 total, was retired during 2011–2019 (US EIA 2019b, 2020c) while additional retirements are projected but no new coal-fired generation is planned (US EIA 2019c). In contrast, Appalachian natural-gas production has been expanding rapidly as prices declined in response to new extraction methods, while advancing technology also enabled declining cost and expanding capacity for electricity generation from solar and wind (US EIA 2019b). These changes are occurring as human-induced climate-change becomes more apparent, which acts as another disincentive for the long-term capital investments required for new coal-burning burning plants. Appalachian coal producers are rapidly losing their primary market, and declining production is projected to continue over future years and decades (US EIA 2020a).

6 Coal's Legacy

As mining declined in the Appalachian coalfield, the legacy of the region's coal-mining history remains. Although the Appalachian forest remains among the most biodiverse in the non-tropical world, major areas of forest have been lost (Drummond and Loveland 2010). Approximately 2.5 million acres (10,000 km^2, as compiled from federal government data by Zipper 2020), about 6.5% of the >150,000 km^2 Appalachian coalfield area, have been disturbed by surface coal mining since the late 1970s, in addition to prior disturbances. In central Appalachia, 5900 km^2 of surface-mine disturbances, approximately 7% of an 83,000 km^2 study area were evident from analysis of satellite data (Pericak et al. 2018). Many of those forests have been replaced by plant communities dominated by exotic invasive species (Sena et al. 2021) which host wildlife that also differs from those of unmined areas (Lituma et al. 2021). Soils on mined areas are transformed, as miners typically replace some or all of the natural soil with fractured rock; the resulting mine soils vary dramatically in their capabilities to support economic activities such as agriculture and development and restoration of natural ecosystems (Skousen et al. 2021; Zipper et al. 2021a). We are unaware of any estimate of the land-surface area undermined by underground coal mining. Given that underground mines have produced more than three times the tonnage extracted by surface mines (Zipper 2020), we expect such undermined areas to exceed and extend well beyond surface disturbances.

Water resources have also been affected by Appalachian mining; those effects extend beyond the mined areas themselves as mining-origin water contaminants move downstream (Clark et al. 2021; Kruse Daniels et al. 2021). Within central Appalachia, more than 1200 miles (1900 km) of headwater streams, approximately 2% of the regional stream length and 4% of first- and second-order headwater streams, were covered or otherwise lost to surface coal mining prior to 2002 (US EPA 2011), while additional losses have occurred since. Multiple studies have demonstrated extensive impacts on regional headwaters by geologic-origin contaminants in mining-influenced waters (e.g. Cormier et al. 2013; Pond et al. 2014), and that larger rivers receiving those headwaters have also been affected (e.g. Zipper et al. 2016). The region's extensive underground mining has also affected water resources. Underground mining can affect groundwater hydrology, water yields of household wells, and groundwater chemical quality while also affecting surface water quality and flows (Skousen and Zipper 2021). The geologic-origin water contaminants released from current and reclaimed Appalachian mines affect aquatic ecosystems and biota (Merovich et al. 2021).

The legacy of Appalachian coal mining, which has been so essential to the American economy, also includes a human toll. More than 150,000 people perished in US coal mines (Breslin 2010; US MSHA 2019), including more than 100,000 since the year 1900 (Fig. 8 upper), with most fatalities occurring during the era of Appalachian coal dominance. In the early twentieth century, average fatality rates exceeded 3 per 1000 workers, which means that a miner with 25-year career had nearly a 1 in 100 chance of dying in a mining-related incident. Fatality rates since then have fallen dramatically, but those rates do not include mining's other human-health effects. Among the most well-documented is a condition known as "black lung," the impairment of lung function caused by inhalation of coal particles. In the early 1970s, 25% of mine workers with >20 years of experience and participating in black-lung health assessments were diagnosed with coal worker's pneumoconiosis, a lung impairment (Fig. 8, lower left). With improving dust control in underground mines, those rates declined over subsequent decades but have been increasing recently, as have diagnoses of a related disabling lung condition, pulmonary massive fibrosis (PMF) (Fig. 8, lower right). Researchers attribute the PMF increase to exposure to air-borne particles created by excavation of rock layers above and below thin coal seams; rock particles have greater impairment effects on the lung than coal particles (Cohen et al. 2016). Underground coal-mine workers experience elevated mortality relative to the general population (Attfield and Kuempel 2008).

Appalachia's populations also reflect coal's legacy. Regional populations of coal-producing areas have been stable or declining over the twenty-first century, in contrast to national population increases. The region's population fraction of retirement age (≥65 years) is greater than that of the nation at large, and the region's working-age population is declining. Similarly, salaries and wages are lower and poverty rates higher in Appalachian coalfield counties than in the nation at large (Santopietro and Zipper 2021). Central Appalachian coalfield residents experience high rates of mortality, cancer mortality, and heart and respiratory disease relative to US population averages (Gohlke 2021), while residents of areas with close proximity to mountaintop mining, or in counties with mountaintop mining, experience elevated mortality and adverse cardiopulmonary and respiratory effects relative to residents of other areas (Broyles et al. 2017; Hendryx 2015). Whether these effects are caused by mining or are associated with mining because of socioeconomic conditions in the high-intensity mining counties is debated.

7 Conclusion

The Appalachian coalfield has changed dramatically during its natural and human history, especially over the past two centuries as nearly 50 billion tons of coal, more than half of the US's cumulative production, were extracted from Appalachian

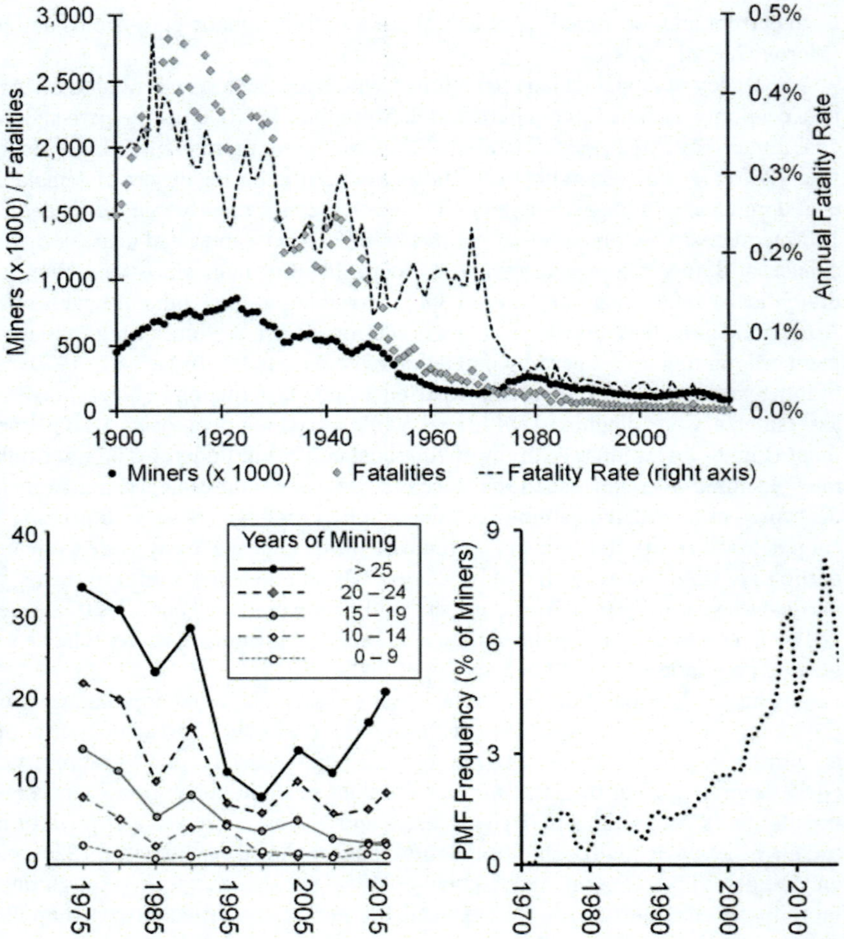

Fig. 8 (Upper) Numbers of coal miners (× 1000), numbers of fatalities, and coal-mine fatality rates by year, 1900–2016 (MSHA 2019); fatality rate is the number of fatalities divided by the number of miners for each year. (Lower) Frequencies of coal worker's pneumoconiosis (CWP) among central Appalachian coal miners, 1970–2017 five-year averages (left) and of pulmonary massive fibrosis (PMF) among US coal miners, 1970–2016 (right), participating in federal program that screens coal workers for the presence of conditions commonly known as "black lung". CWP is a condition of lung impairment caused by coal-mine dust and a significant predictor of PMF, which is totally disabling. CWP data from Blackley et al. (2018); PMF data from Almberg et al. (2018)

mines (Milici 1997; EIA 2019a). Appalachian coal was essential to the US's industrial development and remains an input to the US's electrical energy supply and its production of steel. Today, revenues gained by Appalachian coal exports aid in offsetting currency effects of imports of other products. Appalachian coal production has generated on the order of $1 trillion in sales revenues (2017$ equivalent) since the mid-1960s (Santopietro and Zipper 2021).

Appalachia is a vast and varied region, with history, culture, communities, and natural resources that extend well beyond coal. Within coal-mining areas, however, coal remains as a major influence, both economically and culturally. Those segments of Appalachian society linked with coal are facing an uncertain future while dealing with coal mining's legacy and the reality that coal's revival to production levels of past eras is unlikely.

References

Adovasio JM, Donahue J, Stuckenrath R (1990) The Meadowcroft rockshelter radiocarbon chronology 1975–1990. Am Antiq 55:348–354

Alban TM, Dix E (2013) The state of Penn's Woods. Tree Planters' Notes 56(1):4–12

Almberg KS, Halldin CN, Blackley DJ, Laney AS, Storey E et al (2018) Progressive massive fibrosis resurgence identified in U.S. coal miners filing for black lung benefits, 1970–2016. Ann Am Thorac Soc 15:1420–1426

Anderson DG (1991) Examining prehistoric settlement distribution in eastern North America. Archaeol East N Am 19:1–22

Anderson DG et al (2019) Paleoindian database of the Americas. https://pidba.tennessee.edu/ (downloaded 20 Feb. 2019)

Attfield MD, Kuempel ED (2008) Mortality among U.S. underground coal miners: a 23-year follow-up. Am J Ind Med 51:231–245

Bell SE, York R (2010) Community economic identity: the coal industry and ideological construction in West Virginia. Rural Sociol 75:111–143

Black D, McKinnish T, Sanders S (2005) The economic impact of the coal boom and bust. Econ J 115:449–476

Blackley DJ, Halldin CN, Laney AS (2018) Continued increase in prevalence of coal workers' pneumoconiosis in the United States, 1970–2017. Am J Public Health 108:1220–1222

Bowen E, Christiadi DJ, Le B (2018) An overview of the coal economy in Appalachia. Appalachian regional commission

Braun EL (1950) Deciduous forests of eastern North America. Blakiston Publishing, Philadelphia

Breslin JA (2010) One hundred years of federal mining safety and health research. US Mine Safety and Health Admin information circular 9520

Briceland AV (1991) Batts and Fallam explore the backbone of the Continent. pp 23–36, In: Mitchell RD (ed). Appalachian frontiers settlement, society and development in the preindustrial era. The University Press of Kentucky

Broyles BJ, Harding JE (1974) National register of historic places inventory—nomination form. St, Albans site

Bunker SG, Ciccantell PS (2003) Generative sectors and the new historical materialism: economic ascent and the cumulatively sequential restructuring of the world economy. Stud Comp Int Dev 37(4):3–30

Butler PR, Iverson L, Thompson FR et al (2015) Central Appalachians forest ecosystem vulnerability assessment and synthesis: a report from the central Appalachians climate change response Framework project. U.S. Forest Service, Gen. Tech. Rep. NRS-146

Chandler AD (1972) Anthracite coal and the beginnings of the industrial revolution in the United States. Bus Hist Rev 46:141–181

Chaplin S, Gerrard R, Watson H, Master L, Flack S (2000) The geography of imperilment: targeting conservation toward critical biodiversity areas. In: Stein B, Kutner L, Adams J (eds) Precious

heritage: the status of biodiversity in the United States. Oxford University Press, New York, pp 159–200
Clark EV, Zipper CE, Soucek DJ, Daniels WL (2021) Contaminants in Appalachian water resources generated by non-acid-forming coal-mining materials. Chapter 9 In: Zipper CE, Skousen JG (eds) Appalachia's coal-mined landscapes. Springer
Cohen RA, Petsonk EL, Rose C, Young B, Regier M et al (2016) Lung pathology in U.S. coal workers with rapidly progressive pneumoconiosis implicates silica and silicates. Am J Respir Crit Care Med 193:673–680
Cormier SM, Wilkes SP, Zheng L (2013) Relationship of land use and elevated ionic strength in appalachian watersheds. Environ Toxicol Chem 32:296–303
Crowell DL (1995) History of the coal-mining industry in Ohio. Ohio Department of Natural Resources, Geological Survey; bulletin 72
Davis P (1949) The development of the American glass industry. Harvard University Press
Drake RB (2001) A history of Appalachia. University of Kentucky Press, Lexington
Drummond MA, Loveland TR (2010) Land-use pressure and a transition to forest-cover loss in the eastern United States. Bioscience 60:286–298
Eavenson HN (1942) The first century and a quarter of American coal industry. Waverly Press, Baltimore
Eller R (1982) Miners, Millhands and Mountaineers. University of Tennessee Press, Knoxville.
Eriksson KA, Daniels WL (2021) Environmental implications of regional geology and coal mining in the Appalachians. Chapter 2 In: Zipper CE, Skousen JG (eds) Appalachia's coal-mined landscapes. Springer
Fenneman NM (1938) Physiography of the Eastern United States. McGraw-Hill
Flanders SA (1998) Atlas of American migrations. Facts on File, New York
Flather CH, Knowles MS, McNees J (2008) Geographic patterns of at-risk species. Fort Collins, CO: U.S. Forest Service, Gen. Tech. Rep. RMRS-GTR-211
Fogel RW (1964) Railroads and American economic growth. Johns Hopkins University Press, Baltimore
Forstall RL (1996) Population of states and counties of the United States: 1790 to 1990, U.S. Bureau of the Census, PB96–119060
Gohlke JM (2021) Human health in coalfield communities of Appalachia. Chapter 12 In: Zipper CE, Skousen JG (eds) Appalachia's coal-mined landscapes. Springer
Guilday JE (1982) Appalachia 11000–12000 years ago: a biological review. Archaeol East N Am 10:22–26
Harvey OJ (1909) A history of Wilkes-Barre Luzerne County Pennsylvania. Raeder Press, Wilkes-Barre. https://archive.org/details/historyofwilkesba01harv
Hendryx M (2015) The public health impacts of surface coal mining. Extr Ind Soc 2:820–826
Hibbard WR (1990) An abridged history of Virginia coal. Virginia Center for Coal and Energy Research, Virginia Tech, Blacksburg
Hogan WT (1971) Economic history of the iron and steel industry in the United States. Lexington Books, Lexington MA
Kruse Daniels N, LaBar J, McDonald LM (2021) Acid mine drainage in Appalachia: sources, legacy, and treatment. Chapter 8 In: Zipper CE, Skousen JG (eds) Appalachia's coal-mined landscapes. Springer.
Lamoreaux NR, Sokoloff KL (1997) Location and technological change in the American glass industry during the late nineteenth and early twentieth centuries. NBER Working Paper No. 5938
Lewis RL (1998) Transforming the Appalachian countryside: railroads, deforestation, and social change in West Virginia. University of North Carolina Press, pp 1880–1920
Lippincott I (1912) The early salt trade of the Ohio Valley. J Polit Econ 20:1029–1052
Lituma CM, Cox JJ, Spear S, Edwards JW, De La Cruz JL, Muller LI, Ford WM (2021) Terrestrial wildlife in the Appalachian post-mined landscape: status and opportunities. Chapter 6 In: Zipper CE, Skousen JG (eds) Appalachia's coal-mined landscapes. Springer

Lobao L, Zhou M, Partridge M, Betz M (2016) Poverty, place, and coal employment across Appalachia and the United States in a new economic era. Rural Sociol 81:343–386

Maslowski RF (2017) Prehistoric people. e-WV: The West Virginia encyclopedia. https://www.wvencyclopedia.org/articles/1913

Merovich GT, Hitt NP, Merriam ER, Jones JW (2021) Response of aquatic life to coal mining in Appalachia. Chapter 10 In: Zipper CE, Skousen JG (eds) Appalachia's coal-mined landscapes. Springer

Milici RC (1997) The CoalProd database: historical production data for the major coal-producing regions of the conterminous United States. USGS Open File Report pp 97–447

Milner G, Anderson D, Smith MT (2001) The distribution of eastern woodlands peoples at the prehistoric and historic interface. In: Brose DS, Cowan CW, Mainfort RC (eds) Societies in eclipse: archaeology of the Eastern woodland Indians, A.D. 1400–1700. University of Alabama Press, pp 9–18

Milner GR, Chaplin G (2010) Eastern North American population at ca. A.D. 1500. Am Anitiq 75:707–726

Misa TJ (1995) A nation of steel: The making of modern America, 1865–1925. Johns Hopkins University Press

Munoz SE, Mladenoff DJ, Schroeder S, Williams JW (2014) Defining the spatial patterns of historical land use associated with the indigenous societies of eastern North America. J Biogeog 41:2195–2210

Pericak AA, Thomas CJ, Kroodsma DA et al (2018) Mapping the yearly extent of surface coal mining in central appalachia using landsat and google earth engine. PLoS ONE 13(7):e0197758

Pickering J, Kays R, Meier A, Andrew S, Yatskievych R (2003) The Appalachians. Chapter 30 In: Mittermeier RA et al. Wilderness: Earth's Last Wild Places, Conservation International, Washington, DC

Pond GJ, Passmore ME, Pointon ND, Felbinger JK, Walker CA et al (2014) Long-term impacts on macroinvertebrates downstream of reclaimed mountaintop mining valley fills in central Appalachia. Environ Manage 54:919–933

Revetz RL, Lienke J (2016) Struggling for air: power plants and the "war on coal". Oxford University Press, New York

Rosenberg N (1998) The role of electricity in industrial development. Energy J 19:7–24

Rosencrance RL (2018) Paleoindian artifacts of West Virginia. PaleoAmerica 4:71–75

Santopietro GD, Zipper CE (2021) Economic trends in Appalachian coalfield counties. Chapter 11 In: Zipper CE, Skousen JG (eds) Appalachia's coal-mined landscapes. Springer

Schuur SH, Netschert BC (1960) Energy and the American economy, 1850–1975. Johns Hopkins University Press

Sena K, Franklin JA, Swab RM, Hall SL (2021) Plant communities on Appalachian mined lands. Chapter 5 In: Zipper CE, Skousen JG (eds) Appalachia's coal-mined landscapes. Springer

Skousen JG, Zipper CE (2021) Coal mining and reclamation in Appalachia. Chapter 3 In: Zipper CE, Skousen JG (eds) Appalachia's coal-mined landscapes. Springer

Skousen JG, Daniels WL, Zipper CE (2021) Soils on Appalachian coal-mined lands. Chapter 4 In: Zipper CE, Skousen JG (eds) Appalachia's coal-mined landscapes. Springer

Thompson S (1925) Short history of American railroads. Reprinted by Books for Libraries Press, Freeport NY, p 1971

Thompson EC, Berger M, Allen SN, Roeker JN (2001) A study on the current economic impacts of the Appalachian coal industry and its future in the region. Appalachian Regional Commission. https://www.arc.gov/research/researchreportdetails.asp?REPORT_ID=54

US Bureau of the Census (BOC) (1975) Historical statistics of the United States, colonial times to 1970. https://www.census.gov/library/publications/1975/compendia/hist_stats_colonial-1970.html

U.S. Environmental Protection Agency (EPA) (2011) The effects of mountaintop mines and valley fills on aquatic ecosystems of the central Appalachian coalfields. Office of research and development, Washington, DC. EPA/600/R-09/138F

U.S. Environmental Protection Agency (EPA) (2016) Aquatic life ambient water quality criterion for selenium—freshwater 2016. EPA 822-R-16–006

US Energy Information Administration (EIA) (2011) History of energy consumption in the United States, 1775–2009. Today in Energy, 9 Feb 2011. https://www.eia.gov/todayinenergy/detail.php?id=10

US Energy Information Administration (EIA) (2018) Quarterly coal report. 4th Quarter 2017

US Energy Information Administration (EIA) (2019a) Annual coal report. (Reference includes predecessor publications from 1980 through 2018)

US Energy Information Administration (EIA) (2019b) Electric power annual

US Energy Information Administration (EIA) (2019c) More U.S. coal-fired power plants are decommissioning as retirements continue. Today in Energy, 26 July 2019. https://www.eia.gov/todayinenergy/detail.php?id=37814

US Energy Information Administration (EIA) (2020a) Annual Energy Outlook 2019

US Energy Information Administration (EIA) (2020b) Monthly energy review (Annual data series)

US Energy Information Administration (EIA) (2020c) Preliminary monthly electric generator inventory (based on Form EIA-860M as a supplement to Form EIA-860)

US Mine Safety and Health Administration (MSHA) (2019) Coal fatalities for 1900 through 2018. https://arlweb.msha.gov/stats/centurystats/coalstats.asp

West Virginia Geologic and Economic Survey (WVGES) (2017) History of West Virginia mineral industries—coal. https://www.wvgs.wvnet.edu/www/geology/geoldvco.htm

Zipper CE (2020) Appalachian Coal Data. University Libraries, Virginia Tech. https://doi.org/10.7294/BDA6-7R15

Zipper CE, Agouridis CT, Barton CD, Skousen JG (2021a) Conversion options for mining-affected lands and waters in Appalachia. Chapter 7 In: Zipper CE, Skousen JG (eds) Appalachia's coal-mined landscapes. Springer

Zipper CE, Donovan PF, Jones JW, Li J, Price JE, Stewart RE (2016) Spatial and temporal relationships among watershed mining, water quality, and freshwater mussel status in an eastern USA river. Sci Total Environ 541:603–615

Zipper CE, Skousen JG, Barton CD (2021b) The Appalachian coalfield's energy transition and prospects. Chapter 13 In: Zipper CE, Skousen JG (eds) Appalachia's coal-mined landscapes. Springer

Environmental Implications of Regional Geology and Coal Mining in the Appalachians

Kenneth A. Eriksson and W. Lee Daniels

Abstract The history of the Appalachian Mountains commenced with breakup of a supercontinent in the Late Precambrian era, followed by passive-margin development in the Cambrian through Middle Ordovician. Collision with island arcs in the Late Ordovician and Devonian was followed by collision with Africa at the end of the Paleozoic, resulting in building of the Appalachian Mountains. Sandstones vary in composition from feldspathic to lithic to quartzose, reflecting changes in source-area compositions during evolution of the Appalachian Orogenic Belt. Coal seams of Pennsylvanian and Permian age are interspersed within sandstones and finer-grained sedimentary rocks that respond to surface weathering dependent on their composition. Feldspathic and lithic sandstones are prone to weathering via dissolution and hydrolysis of feldspar and chlorite schist-mica, respectively, whereas quartzose sandstones are durable when exposed to surface weathering. Durable sandstones are favored for mined-land fill and water conveyance structures. Physical properties of mine soils derived from rock-type mixtures are more favorable for plants than soils from pure sandstones or shales; while pre-weathered oxidized strata produce lower mine-soil rock content and more acidic conditions. The abundance of pyrite vs. neutralizers like carbonates and feldspars is related to differences in rock depositional environments and influences mine-soil chemistry and water quality.

Keywords Depositional environment · Pyrite · Durable rock · Rock type · Water quality

1 Introduction

The primary purpose of this chapter is to review the stratigraphic, structural, and plate tectonic history of the central Appalachians with an emphasis on coal and coalbed methane reservoirs in Pennsylvanian strata and prospective shale gas reserves in Ordovician and Devonian strata. Related to the subject of this book, emphasis is

K. A. Eriksson (✉) · W. L. Daniels
Virginia Tech, Blacksburg, VA, USA
e-mail: kaeson@vt.edu

C. E. Zipper and J. Skousen (eds.), *Appalachia's Coal-Mined Landscapes*,
https://doi.org/10.1007/978-3-030-57780-3_2

placed on coal-bearing Pennsylvanian and early Permian stratigraphy, depositional systems, and sandstone petrography. Furthermore, this chapter provides geologic interpretations relevant to surface coal mining operations and reclamation as they are influenced by rock type and mineralogical composition along with their influence on weathering reactions and the potential for surface- and groundwater contamination.

2 Regional Stratigraphic and Structural Framework

The Appalachian Basin extends ~1800 km from Alabama to New York and ranges from 100–500 km in width (Fig. 1). Appalachia's geologic history is a product of plate-tectonic movements and associated events that formed today's continental land masses (Table 1). Geological evolution of the central Appalachians commenced at ca. 750 million years before present (Ma) with rifting of supercontinent Rodinia and, more specifically, of the Grenville basement in the eastern United States (Read and Eriksson 2016). The geologic core of the resulting continental landmass, which has become North America, is known as Laurentia. The rifting event is recorded by the Mount Rogers Formation in southwest Virginia that consists of siliciclastic sedimentary rocks and rhyolites which are dated at 759 ± 2 Ma (U–Pb zircon; Fig. 2; Aleinikoff et al. 1995). The Mount Rogers Formation is overlain by glaciogenic sedimentary rocks of the Konnarock Formation (Rankin 1993; Miller 1994). This phase of rifting, however, did not proceed to continental separation. A second rifting event dated at 600–550 Ma (U–Pb zircon; Aleinikoff et al. 1995), fully separated Laurentia and is represented by siliciclastic sedimentary rocks of the Swift Run and Lynchburg groups (Wehr and Glover 1985), basalts and subordinate rhyolites and sedimentary rocks of the Catoctin Formation, and most of the Unicoi Formation at the base of

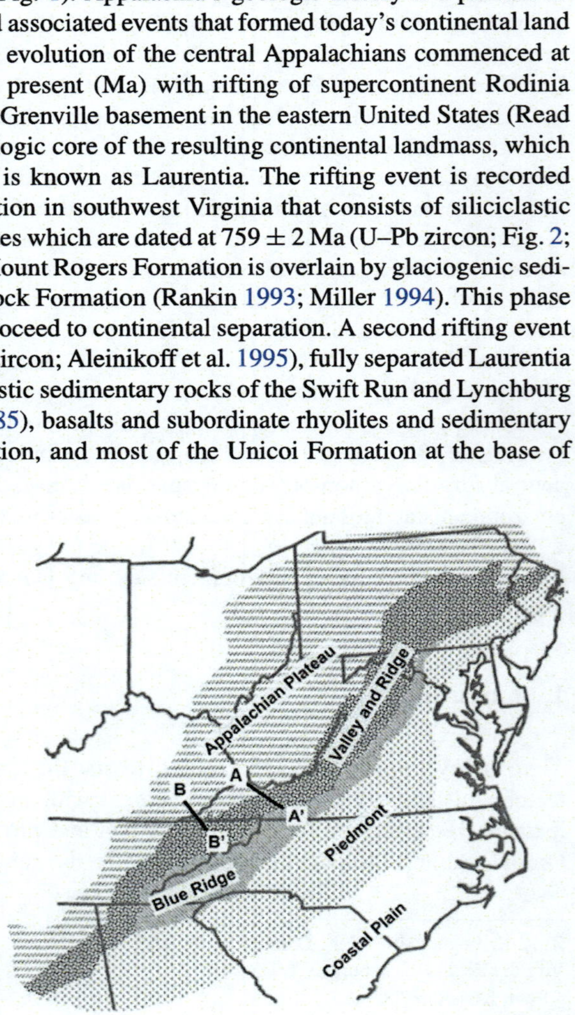

Fig. 1 Tectonic provinces of the Appalachian Mountain belt flanked to the east by the coastal plain. Location of Fig. 2 is shown by line A-A' and Fig. 3 by line B-B'. Figure by Kathryn Haering

Table 1 Summary of geologic history and events responsible for forming what is now known as the Appalachian coalfield

Eon/Era	Period	Time (Approx. Ma before present)	Significant events
Cenozoic		0–66	Many current life forms evolve. Current Appalachian topography forms as regional uplift enables topographic dissection
Mesozoic		66–252	Reptiles are dominant life forms. At approximately 175 Ma, current continental land masses form as Pangea pulls apart
Paleozoic	Permian	251–299	Mass extinctions occur in association with warming climate. Laurentia and Gondwana merge to form Pangea, causing tectonic pressures that fold and fault sedimentary rocks in what is now eastern North America, concluding the Alleghenian orogenesis
Paleozoic	Carboniferous, Pennsylvanian	299–323	Large ice masses on Gondwana undergo repeated melting and freezing cycles causing multiple changes in sea level. Large trees and primitive forests emerge as a land-cover vegetation in tropical moist areas that became Appalachia. Alleghenian orogenesis initiates. Coals form as forests are buried by rising waters and sediments within swamps, river deltas, and shallow inland seas
Paleozoic	Carboniferous, Mississipian	323–359	
Paleozoic	Devonian	359–416	Seed-bearing plants, ferns, and trees evolve. As Laurentia merges with geologic remnant, mountains rise again in what is now eastern North America (Acadian orogeny), giving rise to sedimentation and infilling of basin to the west
Paleozoic	Silurian	416–443	Vascular plants emerge on land; first jawed fishes occupy the seas

(continued)

Table 1 (continued)

Eon/Era	Period	Time (Approx. Ma before present)	Significant events
Paleozoic	Ordovician	444–485	Invertebrate animals diversify in the seas as green plants first occupy land. North American craton collides with magmatic island arcs, causing Taconic orogeny to form mountains in what is now to the east of current Appalachians; downward arching of Laurentia and infilling of the basin to the west
Paleozoic	Cambrian	490–542	Major diversification of life; first evidence of most present-day animal phyla
Neo-proterizoic	Ediacaran	541–635	Multi-celled marine animals, first skeletal animals. Second rifting of Rodinia
Neo-proterizoic	Pre-Ediacaran	635–	Life evolves >3000 Ma. 750 Ma: Initial rifting of Rodinia

the Chilhowee Group (Fig. 2). Volcanic rocks of the Catoctin Formation give ages of 570 ± 36 Ma (Rb–Sr whole rock; Badger and Sinha 1988) and 564 ± 9 Ma (U–Pb zircon; Aleinikoff et al. 1995). Quartz-rich sandstones of the upper Unicoi Formation demarcate the onset of breakup and thermal subsidence of the eastern edge of Laurentia, known as the Laurentian margin. Over geologic time, that margin became covered with sedimentary rocks that have become the Appalachian Mountains and associated physiographic features to the east and south. The overlying 3.5 km thick passive-margin deposits consist of the Hampton and Erwin Formations of the Chilhowee Group and the Shady Dolomite, Rome Shale, Elbrook Formation, and Knox Group (Fig. 2; Read 1989a; Simpson and Eriksson 1990). The Evington Group on the eastern side of the Blue Ridge (Fig. 1) is correlated with the Chilhowee Group and Shady Dolomite (Wehr and Glover 1985; Read 1989a).

Passive-margin sedimentation, an accumulation of continental-origin sediments at the continental margin in the absence of tectonic activity, was terminated in eastern North America during the Ordovician geologic era with the onset of the first of three mountain-building episodes that eventually formed the Appalachians, the Penobscotian-Taconian Orogeny (Hatcher 1989). Hinterland expressions of this orogeny are the Chopawamsic and Milton terranes in the Western Piedmont (Fig. 1), a zone of predominantly metamorphic rock that currently extends from the Danville area to north of Richmond in Virginia, and is considered to represent remnants of an extensive 450–470-Ma magmatic arc that collided with Laurentia's eastern

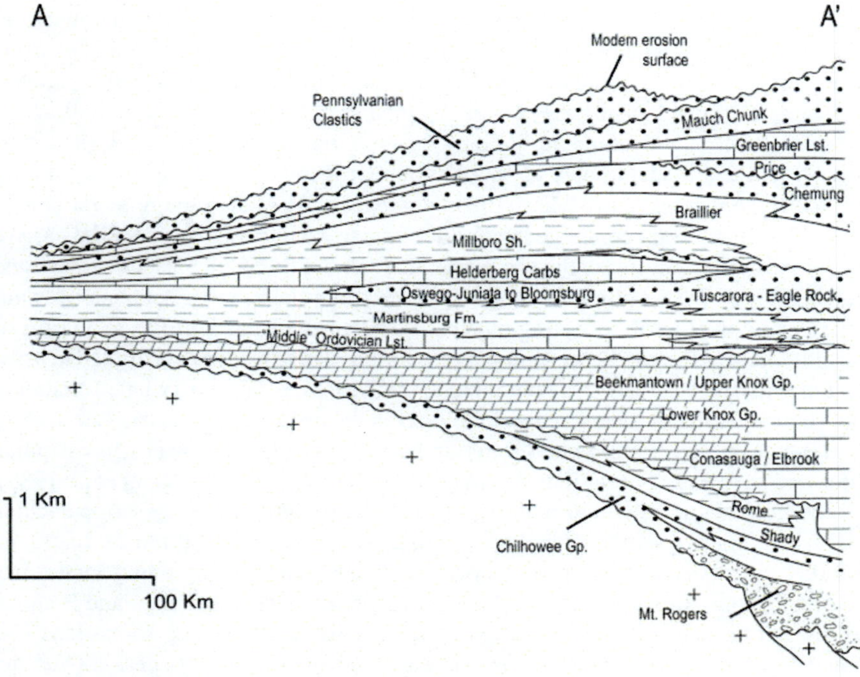

Fig. 2 Palinspastically restored, lithostratigraphic cross-section of late Neoproterozoic and Paleozoic strata of the central Appalachians (based on Colton 1970). Mineable coal seams are confined to and intercalated within Pennsylvanian clastics. Refer to Fig. 1 for approximate location of cross-section

margin starting in the Middle Ordovician (Coler et al. 2000). Intervening terranes east of the Blue Ridge, but west of the magmatic arc, are complexes known as the Smith River allochthon and Potomac terrane (Fig. 1); these geologic terranes consist of melanges, mafic and ultramafic meta-igneous rocks, and metasedimentary rocks. These terranes were amalgamated onto Laurentia and compressed between Laurentia and the approaching island arc during the Ordovician Penobscotian-Taconian Orogeny (Horton et al. 1989). In what is now the Valley and Ridge Province of the central Appalachians, the tectonic motion caused downward arching of Laurentia and a transition from passive-margin to foreland-basin sedimentation that is demarcated by the Middle Ordovician Knox unconformity that developed in response to shelf flexure (Mussman and Read 1986). Thrust-loading of the passive-margin shelf resulted in formation of an underfilled foreland basin within which the carbonaceous Utica Shale and correlative units accumulated. Further infilling of the basin by the overlying Martinsburg, Juniata, and Reedsville Formations resulted from weathering and erosion of the fold-and-thrust belt highlands to the southeast. Carbonate rocks developed on a shallow ramp on the foreland side of the basin that formed adjacent to the southeastern highlands (Read 1989b). The top of the Taconian clastic wedge is defined by an unconformity that is related to isostatic rebound that followed cessation of thrusting and resulted in erosion of the thrust load. The overlying upper Tuscarora, Rose Hill and Eagle Rock Formations (Fig. 2) reflect a phase of waning tectonism during the convergent history of the orogen (Castle 2001).

The second major mountain-building episode that formed today's Appalachian Plateau and coalfields, known as the Acadian Orogeny, occurred primarily during the late Devonian and early Mississipian geologic eras. A geologic remnant known as the Carolina terrain converged with Laurentia to produce what is now the eastern Piedmont in the Carolinas (Fig. 1; Wortman et al. 2000; Hatcher 1989; Bream and Hatcher 2002). The resulting Devonian-Mississippian clastic wedge extended toward what is now westward, reflecting renewed subsidence of the foreland and uplift of the hinterland to the southeast. Components of this clastic wedge that formed in association with the Acadian Orogeny include the Devonian Millboro Shale (including the Marcellus Shale), deposited in an underfilled, stratified and anoxic (anaerobic) foreland basin, the Brallier and Chemung Formations and the Lower Mississippian Price and Maccrady Formations that overfilled the basin (Fig. 2; Osberg et al. 1989).

The final major mountain-building episode responsible for today's Appalachian coalfield is known as the Alleghenian orogenic event and occurred as Laurentia collided with Africa to form the supercontinent of Pangea beginning in the late Paleozoic, approximately 335 Ma. Late Mississippian, Pennsylvanian, and Permian sedimentation that produced the central Appalachian coal-bearing formations was concurrent with Alleghanian orogenesis that occurred during this period (Van der Voo 1983; Hatcher 1989, 2002). Plate reconstructions indicate that Laurentia was situated in tropical equatorial latitudes favoring the development of the lush and abundant vegetation that became coal beds (Scotese 2003).

In the central Appalachians, the Laurentian craton responded to the initial crustal shortening and thrust loading by forming a broad, ~1300 km-long, 500 km-wide, and relatively shallow foreland basin (Ettensohn 2004) which was filled with sediments that currently comprise the Appalachian coalfield. This flexure generated the greatest accommodation adjacent to the thrust front, forming a SE-thickening clastic wedge of Pennsylvanian sedimentary rocks, with the thickest package preserved in southwestern Virginia. Alleghanian orogenesis advanced into a main phase of collision in the Permian (Secor et al. 1986), causing folding and faulting of sedimentary rocks immediately west of the Blue Ridge to form geologic structures that became today's Valley and Ridge terrain; these same pressures compressed the coal-forming carbonaceous materials, resulting in anthracite coal deposits in areas subjected to the greatest pressure. This orogenic event also resulted in large-scale, west-directed overthrusting of Grenvillian basement and development of a megathrust on the western side of the Blue Ridge in eastern Virginia and the Carolinas (Fig. 1; Hatcher 1972, 2002; Cook et al. 1979). Final thin-skinned deformation led to overthrusting of Pennsylvanian rocks along the Pine Mountain Thrust Fault, defining the western limit of Alleghanian deformation (Mitra 1988) (Fig. 3).

Laurentia and the now-formed North American continent separated from the remainder of Pangea around 175 Ma, ending the sequence of major tectonic events forming the geologic structure of Appalachia's coalfields, as we know it today.

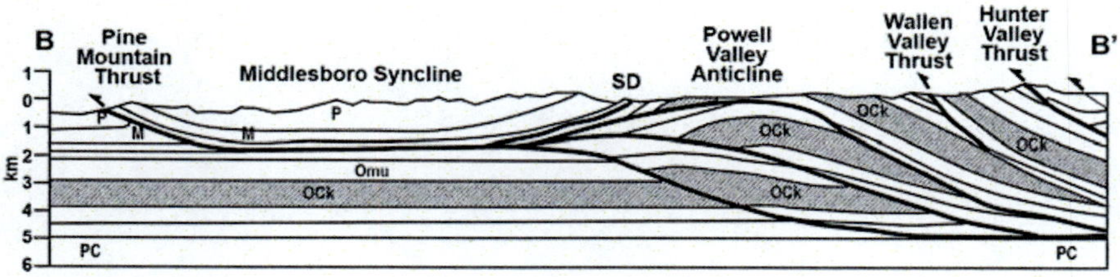

Fig. 3 Structural cross-section of the Valley and Ridge and Appalachian plateau provinces of southwestern Virginia illustrating the position of the Middlesboro syncline above the pine mountain fault (modified from Woodward 1985). The Middlesboro syncline defines the Appalachian plateau with the Valley-and-Ridge province to the east (Right). PC = Precambrian basement; OCk = Middle Ordovician through lower Cambrian mainly carbonate rocks; Omu = Upper Ordovician sedimentary rocks; SD = Silurian-Devonian sedimentary rocks; M = Mississippian strata; P = Pennsylvanian sedimentary rocks. See Fig. 1 for approximate location of transect (B–B')

3 Pennsylvanian-Permian Lithostratigraphy and Depositional Environments

The focus of this book is Pennsylvanian and Early Permian strata preserved in the coal-bearing central and northern Appalachian coalfields (hereafter referred to as the Pocahontas and Dunkard (structural) Basins,[1] respectively, (Fig. 4). At the dawn of the Pennsylvanian time period, the environment in which Appalachian coals formed was a broad plain, bounded on what is now the east by high mountains that were the source of sediments. As the period proceeded, these low-lying areas were alternately inundated by shallow oceans as global warming caused sea levels to rise, and then exposed as sea levels fell during global cooling events. The tropical climate enabled periods of dense vegetation growth; those materials were then covered by rising waters and/or deposits of sediments carried into the basin from the adjacent highlands. Repeating cycles of these events caused multiple coal beds to form.

The Pocahontas Basin extends through southern West Virginia, eastern Kentucky, and southwestern Virginia and is delineated by major bounding structures including the Appalachian Valley and Ridge to the east, the Cincinnati Arch marking the western limit in eastern Kentucky and Tennessee, and a structural hinge-line separating the Pocahontas Basin from the more northerly Dunkard Basin. The Dunkard Basin extends from the hinge line northwards into Pennsylvania and westwards into eastern Ohio to the Cincinnati Arch (Fig. 4). As discussed later in this chapter, the differences between the types of peat swamps in the Early and Middle Pennsylvanian versus the Late Pennsylvanian and Early Permian has important implications for reclamation and water quality since net acid-forming strata (pyritic-S containing strata) are much more common in the Dunkard Basin than in the Pocahontas Basin.

A long-term (25–30 Myr) change in climate during the Pennsylvanian has been inferred by Cecil et al. (1985, 2004) based on types of paleosols and coal swamp

[1]The Pocahontas and Dunkard Basins are not depositional basins but resulted from gentle folding after deposition of the sediments. Thus, they are termed structural basins.

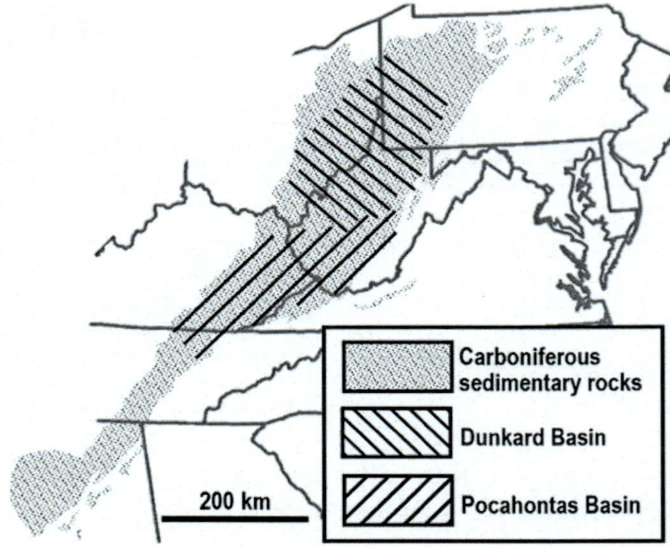

Fig. 4 Map showing the locations of the Pennsylvanian Pocahontas and Dunkard structural basins of Pennsylvania, Ohio, West Virginia, Kentucky, and Tennessee (adapted from Milici 2004; Grimm et al. 2013)

morphologies preserved in the record. Early and Middle Pennsylvanian strata lack oxidized red iron pigmentation and instead are characterized by kaolinite-rich Histosols (organic soils) attributed to "everwet" conditions that also promoted the development of domed peats (Cecil 1990; Cecil et al. 1985). In contrast, "semi-arid/seasonal" climatic conditions in the Late Pennsylvanian and Early Permian promoted the formation of abundant argillaceous red beds (including Vertisols; shrink-swell soils) and planar peats typical of conditions in which evaporation periodically exceeds rainfall. Additional evidence supporting this change in paleoclimate includes differences in floral assemblages (Kosanke and Cecil 1996) and enriched $\delta^{18}O$ signatures in Late Pennsylvanian and Early Permian paleosols interpreted to indicate evaporation associated with more arid conditions (Tabor and Montanez 2002) as evidenced by preserved petrocalcic horizons (calcrete).

The Early Pennsylvanian Breathitt Group and Lee Formation (and equivalents) unconformably overlie the Upper Mississippian Bluestone Formation and attain a maximum thickness of ~ 900 m along the SE margin of the Pocahontas Basin in Virginia and West Virginia (Fig. 5; Englund 1979; Englund and Thomas 1990; Korus et al. 2008; Blake and Beuthin 2008). Lower and Middle Pennsylvanian, coal-bearing lithostratigraphic units of the Breathitt Group include the Pocahontas, Bottom Creek, Alvy Creek, Grundy, Pikeville, Hyden, Four Corners, and Princess Formations, the lower four of which are intercalated, respectively, with the Warren Point, Sewanee, Bee Rock and Corbin quartz-rich sandstone lenses of the Lee Formation along the western margin of the Pocahontas Basin and extending into eastern

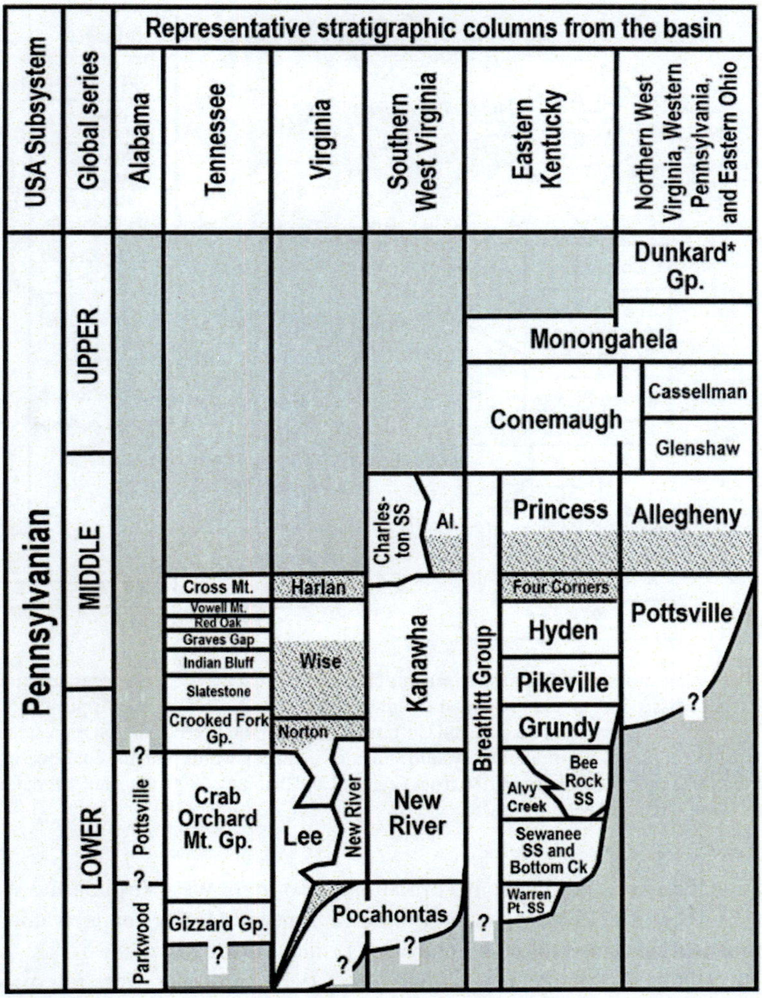

Fig. 5 Lithostratigraphy of Pennsylvanian and Permian strata in the Appalachian Plateau (modified from Greb et al. 2008). Cross-hatched strata were analyzed for leaching potentials by Clark et al. (2018a, b) and Daniels et al. (2016). For more detail, see Ruppert et al. (2014). *The top of the Dunkard group may be lower Permian

Kentucky (Fig. 5; Rice and Schwietering 1988; Chesnut 1994; Greb et al. 2008). The quartz-rich sandstone bodies onlap the Mississippian inconformity and step westwards through time (Fig. 6). Lithostratigraphic units of the Breathitt Group are subdivided by thick, marine-influenced shales, the Dark Ridge, Hensley and Dave Branch, Betsie, Kendrick, Magoffin, and Stoney Fork Members (Fig. 5; Chesnut 1994, 1996). In West Virginia, the Grundy, Pikeville, Hyden, Four Corners Formations are collectively referred to as the Kanawha Formation whereas the Princess

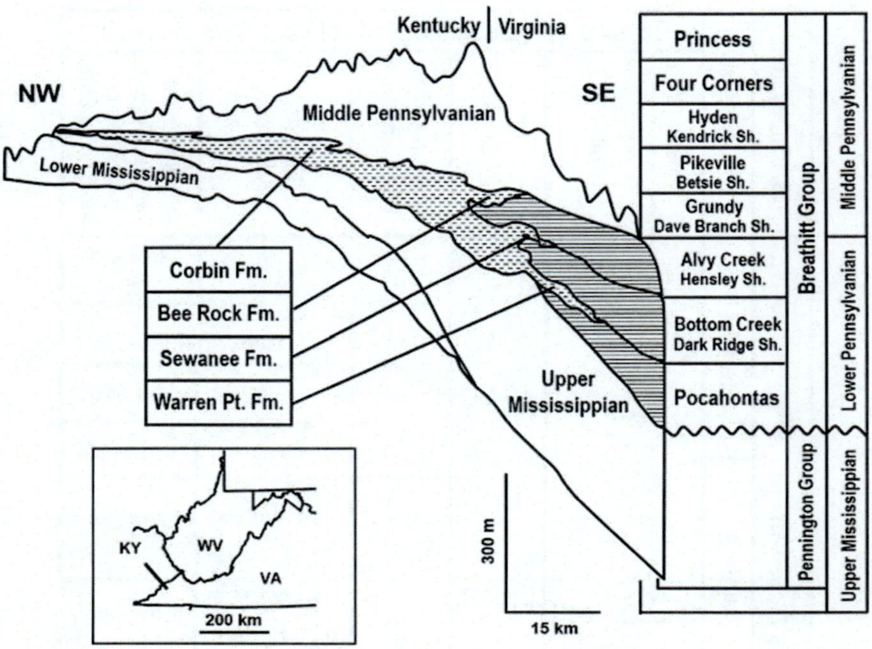

Fig. 6 Schematic cross-section and stratigraphic column showing the vertical succession and geometry of upper Mississippian and lower and middle Pennsylvanian strata in the Appalachian Basin. The Warren Point, Sewanee, Bee Rock, and Corbin Members of the Lee Formation represent quartz-rich, sandstone bodies of the trunk river system and interfinger eastwards with the coal-bearing strata of the transverse river system. (modified from Englund and Thomas 1990; Chesnut 1996; Greb and Chesnut 1996; Greb and Martino 2005)

Formation is named the Allegheny Formation. In northern West Virginia and Western Pennsylvania, Lower Pennsylvanian strata are absent and Middle Pennsylvanian units are collectively grouped into the Pottsville Formation (Fig. 5).

In the 1970′s, Lower Pennsylvanian quartz-rich sandstones were interpreted as beach deposits (Ferm 1974; Hobday and Horne 1977; Ferm and Horne 1979), but were later demonstrated to be multi-cycle, braided fluvial deposits based on detailed sedimentology and architectural element analysis (Wizevich 1992). A braided fluvial origin for this facies association is supported by the presence of thick conglomerate lags, a variety of cross-strata, channel scour-and-fill structures, downstream accreting sand bars, and compound channel macroforms, unidirectional paleocurrents for bedforms and macroforms, closely spaced erosional surfaces, fining-upward sequences, floodplain paleosols, abundant plant fossils including upright fossil tree trunks, and the absence of marine body fossils (e.g. Greb and Chesnut 1996; Grimm et al. 2013). Massive quartz-rich sandstones from these units are commonly utilized as durable rock components for active mine fill drainage structures as discussed later.

Current paleogeographic models for the Early Pennsylvanian depict the interaction of longitudinal and transverse fluvial systems draining contrasting terranes. Transverse fluvial facies of sublithic sandstones and intercalated mudstones are preserved as coal-bearing, immature siliciclastic strata of the Breathitt Group. Lithic sandstones accumulated as channeled sandstone-mudstone bodies 1-10 km wide and up to 30 m thick that filled broad alluvial valleys oriented transverse to the thrust front (Korus et al. 2008). These immature sandstones have been interpreted to be derived from low-grade metamorphic terranes uplifted during Alleghanian deformation in a fold-and-thrust belt to the southeast (Davis and Ehrlich 1974; Houseknecht 1980; Reed et al. 2005). Longitudinal fluvial facies of quartz-pebble bearing, texturally and mineralogically mature, quartz-rich sandstones with rare intercalated siltstones are preserved in strike-parallel elongate belts 17–100 km wide and up to 200 m thick along the western periphery of the basin, situated between the lithic-rich Breathitt Group and the elevated Cincinnati Arch (Rice and Survey 1984; Chesnut 1988). This mature sandstone belt is interpreted as the product of a long-lived, SW-flowing longitudinal braided river system that was fed by an extensive cratonic drainage network (Houseknecht 1980; Rice and Survey 1984; Rice and Schwietering 1988; Archer and Greb 1995; Greb and Chesnut, 1996; Eriksson et al. 2004; Grimm et al. 2013). The transverse and longitudinal fluvial deposits occupy incised valleys that developed during eustatic lowstands. Transverse fluvial systems merged into large, bedload-dominated longitudinal (axial) braided river systems draining an Amazon-scale cratonic watershed (Rice and Schwietering 1988; Greb and Chesnut 1996; Korus et al. 2008). The axial drainage dispersed sediment to the SW, toward open-marine settings in the Black Warrior Basin and Ouachita Foredeep (Graham et. al. 1975; Archer and Greb 1995; Churnet 1996). During subsequent sea-level rise, these broadly incised fluvial systems were transgressed forming extensive estuaries (Greb and Chesnut 1996; Greb and Martino 2005; Bodek 2006; Korus et al. 2008). Following inundation, small-scale, tropical, fluvial-dominated bayhead deltas prograded into the basin, infilling available accommodation (Korus et al. 2008).

Middle Pennsylvanian strata in southern West Virginia, eastern Kentucky, and southwestern Virginia lack the quartz-rich sandstone bodies typical of the Early Pennsylvanian (Figs. 5 and 6). Rather, sandstones are rich in lithic constituents such as schist and mica grains, are between 10 and 15 m thick (locally over 20 m thick), and are interpreted as incised valley fills. Marine shelf mudstones and limestones include the Kendrick, Magoffin, and Stoney Fork Members that are overlain by progradational deltaic deposits (e.g. Aitken and Flint 1995; Greb et al. 2008). Sediment was derived exclusively from the fold-and-thrust belt to the southeast and overfilled the basin beyond the Cincinnati Arch (Fig. 4). The lithic sandstones and siltstones of these collective facies are generally higher in weatherable minerals than the quartz-rich lower Pennsylvanian rocks and tend to generate higher levels of total dissolved solids (TDS) upon weathering (Orndorff et al. 2015).

Late Pennsylvanian coal-bearing strata attain their greatest thickness (up to 620 m thick) in the Dunkard Basin of central and northern West Virginia, southern Pennsylvanian, and eastern Ohio and include the Conemaugh (Glenshaw and Casselman Formations), Monongahela and Dunkard Groups (Fig. 5). The Dunkard Group was considered by Greb et al. (2008) to be partly Early Permian in age but more recent work by Schneider et al. (2013) has assigned an Early Permian age to the entire Dunkard Group based on insect biostratigraphy.

Similar depositional settings and paleogeography to those that existed in the Middle Pennsylvanian persisted into the Late Pennsylvanian Glenshaw Formation (Martino 2004; Greb et al. 2008). The Ames Limestone at the top of the Glenshaw Formation (Fig. 5) represents the last fully marine incursion into the Appalachian Basin (Lebold and Kammer 2006). The Late Pennsylvanian Monongahela and the early Permian Dunkard Groups were deposited on a low-gradient alluvial plain that passed downslope to the west into a fluvial–lacustrine deltaic plain complex (Fedorko and Skema 2013; Montanez and Cecil 2013). Lithologies are comprised of channel sandstones, floodplain fines consisting of interbedded grey, green and red mudstones including preserved Aridisols (desert soils) and Vertisols (shrink-swell soils), coals, and lacustrine limestones. Red beds are less common in the Monongahela Group and more common in the Dunkard Group whereas limestones dominate the Monongahela Group and are less abundant and thinner in the Dunkard Group. Limestones and coals thin and disappear to the south in both groups (Fedorko and Skema 2013). In the northern extent of the Dunkard Basin, lithologies are associated in repetitive cycles consisting of coal beds and underlying mudstones or claystone seat earth (highly leached paleosols), overlain by red-colored mudstones, siltstones, and sandstones and capped by fine-grained limestones (<2 m thick) that typically are brecciated and exhibit evidence of pedogenic alteration. Coals and seat earths are considered to have developed during humid interludes in an overall semi-arid setting that favored formation of red beds including Vertisols as well as Aridisols and limestones (Cecil et al. 1985; Cecil 1990; Montanez and Cecil 2013).

4 Origin of Sulfur and Other Elements in Coals

Extensive coal seams intercalated within Early and Middle Pennsylvanian as well as earliest Late Pennsylvanian strata (Conemaugh Group) developed from marshes that formed in estuaries and following abandonment of deltas Cecil et al. (1985; 1993). Cecil (1990) proposed that these coals developed in ombrogenous peat swamps that were dominated by peat-forming vegetation communities lying above groundwater saturation or tidal levels and thus dependent on rainwater for mineral nutrients. Sulfur (as well as ash) contents of these coals are low ranging from 0.7–1.2% (Neuzil 2000; Trippi et al. 2014). Compositions of these coal beds are related to the nature of the peat swamps that were not subjected to influx of detrital sediment and dissolved solids.

In contrast, coal seams associated with latest Pennsylvanian and Early Permian strata (Monongahela and Dunkard Groups) developed from marshes that formed in continental settings. In contrast with middle Pennsylvanian times, Cecil et al. (1985; 1993) and Cecil (1990) argue that late Pennsylvanian coals developed in topogenous peat swamps that are confined to topographic depressions. Sulfur contents of these coals, usually occurring as pyrite (FeS_2), is higher and ranges from 2.2–4.2% (Neuzil 2000; Trippi et al. 2014). Topogenous swamps are subject to influx of suspended sediments and dissolved solids accounting for the higher ash and pyrite contents of the resulting coal seams. The overall differences in pyritic-S and carbonate/neutralizer content between the rocks of the Early and Middle Pennsylvanian versus the Late Pennsylvanian and Early Permian are particularly important from the standpoint of resultant mine soil chemistry and potential water quality impacts. Net acid-forming strata, and often entire mining sections involving multiple coal seams, are much more common north of the hinge line between the two basins. This led to certain areas being deemed "unsuitable for surface mining" due to excessive acid-production potentials in the decades following the passage of the federal Surface Mining Control and Reclamation Act in 1977 and associated state regulations.

Coals also contain multiple impurities in addition to sulfur. At least 74 of the 94 naturally occurring elements are known to occur in coal (Schweinfurth 2009). Most notably, coals contain non-burnable components, designated as ash. These originate from multiple sources, including ash content of original plant matter, mineral impurities from soil and rock fragments that became embedded in the coal-source organic deposits, and epigenetic minerals that formed from dissolved elements in waters that infiltrated coal-source deposits subsequent to their enclosure by overlying rocks. Ash contents of Appalachian coals can vary from <1% to nearly 50% by dry weight, although Appalachian coals with >10% ash content are typically not marketable in the modern age. Primary ash components are the crustal elements, i.e. Al, Si, Fe, K, and Ca. However, coal also includes multiple trace elements at widely varying concentrations. For example, Appalachian coals contain the elements As (<1 to >200 ppm), Cd (<0.01 to >0.5 ppm), Cr (<5 to >50 ppm), Hg (< 0.01 to > 1 ppm), Ni (<4 to > 50 ppm), and Se (<1 to >10 ppm) (Data from USGS 2015) which are subject to potential emission to the atmosphere when coal is burned. Many of these elements, particularly As and Se, are commonly substituted into pyrite structures. Multiple factors contribute to this variation, including elemental content of source materials and, especially, epigenetic processes (Finkelman 1998). Similarly, additional elements of environmental significance vary widely in concentration within Appalachian coals, including elements such as Ba (<10 to >300 ppm) that are often present as mobile forms within the coal combustion products (fly ash). Appalachian coal's N contents are also of environmental significance but vary less widely, typically ranging from 0.5 to 2.5% (data from Palmer et al. 2015); most coal N originates from the coal-source organic materials, including both plant matter and aquatic algae present in the coal-forming shallow-water environments (Baxby et al. 1994). Table 2 presents a summary of basic mineralogical information on common minerals found in regional sedimentary strata.

Table 2 Common mineral forms occurring within Appalachian coal-bearing rocks

Mineral	Chemical formula[a]	Comment/Significance
Calcite	$CaCO_3$	A carbonate form that is high in calcium. More soluble and more easily weatherable than dolomite or siderite
Carbonate	MCO_3	A mineral form that weathers to produce bicarbonate, an acid neutralizer. Solubility varies with M
Chert	SiO_2	Non-crystalline Si precipitate; highly resistant to weathering
Chlorite	M (Al, Si, O, OH)	2:1:1[b] Phyllosilicates with partial substitution of Fe, Mn, Mg, and/or other metallic cations for Si; often of metamorphic origin
Dolomite	$[Ca, Mg] CO_3$	A carbonate form with both magnesium and calcium as dominant cations
Feldspar	M (Al, Si, O)	Crystalline minerals with Al and Si as major components; can occur with K, Na, and Ca as primary cations. Solubility and potential for hydrolysis varies with chemical structure and cation; weathers to consume protons and produce alkalinity as (OH^-), frequently generating kaolinite as a weathering product. Most abundant mineral form of the earth's crust
Kaolinite	$Al_2Si_2O_5(OH)_4$	A 1:1[b] phyllosilicate clay mineral, generally insoluble. Can form as a weathering product of other alumino-silicate minerals
Mica	M (Al, Si, O)	A group of 2:1[b] phyllosilicate clay minerals; releases cations such as K, Na, Ca, Mg, Fe via weathering; weathers to form other 2:1 minerals like vermiculite or Al interlayered chlorite
Muscovite	K (Al, Si, O)	A K-rich form of mica
Plagioclase	M (Al, Si, O)	A form of feldspar with minimal K content, with Na and Ca as dominant cations
Pyrite	FeS_2	An iron-rich sulfide mineral that is present in most Appalachian coal-bearing rocks. Major source of acid mine drainage; also present in association with minerals producing acid neutralizers, such as carbonates and feldspars, in rocks producing net alkaline mine drainage
Quartz	SiO_2	A crystalline primary mineral, highly resistant to weathering
Siderite	$FeCO_3$	A carbonate with iron as the primary cation. Will initially general alkalinity, but this is offset by subsequent oxidation and hydrolysis of the Fe^{2+} to produce acidity. Thus, it can cause a significant positive error estimating neutralizers in conventional acid–base-accounting procedures

(continued)

Table 2 (continued)

Mineral	Chemical formula[a]	Comment/Significance
Sulfide	MS, MS_2	A crystalline mineral form with reduced S as a primary component; can occur with various acid-soluble metallic elements (Fe, Cu, Zn, etc.) counter ions. Oxidizes and hydrolyzes to form strong acids and SO_4^{2-}

[a]For most entries, chemical formula is simplified. M = metallic element
[b]1:1, 2:1 and 2:1:1 specify the ratio of Si-tetrahedral to Al-octahedral layers in sheet phyllosilicate minerals

5 Sandstone Mineralogy

Sandstones consist of mineral grains of pre-existing rocks, primarily sand-sized, with intervening pores occupied by minerals that act as cementing agents by binding the fragments together. Because of the low geochemical reactivity of quartz, physical and chemical properties of sandstones are typically determined by non-quartz grains such as feldspar, metamorphic lithics, and micas, as well by the compositions of cement. For example, the sulfides, carbonates, and feldspars described below with reference to mine soil and water quality typically occur within the cementing agents of sandstones and associated sedimentary rocks.

Early Pennsylvanian sandstones of the Lee Formation have an average total quartz content of 92% and an average monocrystalline quartz content of 83% (Wizevich 1991; Reed et al. 2005; Clark et al. 2018a). In contrast, Breathitt Group sandstones are more lithic and feldspathic consisting of up to 30% metamorphic lithics and chert and up 20% plagioclase and potassium feldspar (Reed et al. 2005). An exception is that sandstones in the Allegheny Formation as well as the overlying Late Pennsylvanian Glenshaw Formation lack feldspar (Flanick et al. 2005). Metamorphic lithics are dominated by quartz-chlorite schist and mica grains. Sandstones of Late Pennsylvanian Casselman Formation are similar in composition to those of the Breathitt Group (Flanick et al. 2005). As discussed later, the abundance of both feldspars metamorphic lithics plays an important role in the relative cation content of mine discharge waters and influences mine spoil neutralization potentials. Quantitative petrographic data for sandstones of the Monongahela and Dunkard Groups are not available, although Davis and Ehrlich (1974) report an increase in feldspar in latest Pennsylvanian sandstones.

The predominant cements in sandstone are siderite, authigenic quartz, kaolinite, and calcite/ferruginous carbonate/dolomite. Siderite cement is rare and occurs as rims coating framework grains. Abundances of siderite cement range up to 10%. Quartz cement occurs as syntaxial overgrowths on monocrystalline quartz grains and post-dates siderite. Kaolinite cement is characterized in thin section by a distinctive vermicular texture. It typically infills remaining primary pore spaces between muscovite grains and post-dates quartz cement. Carbonate cement and replacement other than by siderite is rare. Carbonate consists mainly of ferruginous carbonate (stained blue) and dolomite (unstained) with rare calcite. The relatively low abundance of carbonates as a cement in massive sandstones and siltstones from the Pennsylvanian strata of the Pocahontas Basin is somewhat contradictory to the fact that these rocks commonly possess relatively high (>2%) net acid neutralizers (Sobek et al. 2000) and even higher apparent neutralizer levels via calcium carbonate equivalent (CCE) lab titration methods. Whether or not carbonates may exist in very fine disseminated phases that cannot be resolved via petrographic analysis and/or in forms that do not respond to petrographic stains warrants further study.

Sandstone grains including feldspars, metamorphic lithics and micas are commonly replaced by secondary minerals including clays, carbonates, and limonite-goethite. These products are most typical of medium- and coarse-grained sandstones. Plagioclase, and to a much lesser extent *K*-feldspar, display variable degrees of sericitization. Twinning in plagioclase commonly is visible through the alteration whereas, in rare instances, sericitization has completely pseudomorphed plagioclase with alignment of sericite/illite along cleavage planes. Both plagioclase and K-feldspar and rarely muscovite are either partially or completely replaced by kaolinite with accompanying formation of micro-porosity. Commonly, siderite rhombs are associated with kaolinite replacement. Locally, carbonate (ferruginous carbonate, dolomite, and calcite) replaces feldspar and quartz. Limonite and goethite are pervasive alteration products in sandstones. Limonite and goethite formed at the expense of siderite, ferruginous carbonate, and chlorite schist. The fact that feldspar alteration to kaolinite is quite evident even in non-oxidized gray rocks (Clark et al. 2018a) that have historically been presumed to be "non-weathered" is notable. This indicates that current concepts of the nature of the pre-mining weathering profile (e.g. brown vs. gray sandstones; see Sect. 7 in this chapter) are relatively simplistic and this also emphasizes the potential importance of feldspar hydrolysis in overall long- and medium-term acid neutralization reactions.

6 Rock Hardness, Particle Size, and Durability

Quartz-rich, durable sandstones are commonly employed by active coal mines to construct "blanket drains" that underlie large valleys fills or to line and protect other water conveyances (Skousen and Zipper 2021). Furthermore, the final slope grades for the faces of valley fills and highwall backfills depend on the predicted stability (safety factor) of the fill materials, which is enhanced when the spoils are primarily composed of "durable rock." In general, hard non-weathered massive sandstones and siltstones are considered durable and resistant to slaking and physical degradation over extended periods of time. More lithic sandstones as well as finer-textured, fissile and/or weakly-cemented mudrocks and shales are subject to slaking and are not durable. However, the presence of significant pyrite in any rock type influences durability due to its influence on (a) accelerating chemical weathering reactions and (b) physical weathering effects due to secondary mineral hydration and expansion which can shatter the surrounding rock matrix. As discussed below, the extent of pre-disturbance near-surface weathering reactions (oxidation and leaching) also has an overriding influence on rock durability and even high-quartz massive sandstones can become non-durable over relatively short periods of time when exposed to repeated wet-dry cycles in a mine fill.

7 Rock Type and Pre-weathering Effects on Mine Soil Physical and Chemical Properties and Plant Growth

The physical properties of mine soils created by the final reclamation process are highly dependent on the dominant strata that comprise the reconstructed surface soil (Skousen et al. 2021). In general, mixed lithologies generate superior physical properties such as loamy textures in the soil sized fraction (<2 mm) than strata that are composed entirely of coarse sandstones or mudrocks/shales (Daniels and Zipper 2018). However, the extent to which the strata have been pre-weathered via combined sulfide oxidation, neutralization, and leaching processes have an overriding influence on physical properties, particularly the ratio of rock fragments (>2 mm) to soil-sized materials. Whereas the depth of pre-mine oxidation and weathering can vary widely based on landform position and extent of stress-relief fracturing (Wyrick and Borchers 1981), the upper 5 to 10 m of strata below the ground surface are generally brown to brownish-yellow in color, commonly weathered to a saprolitic stage, and generate blasted mine spoils that are much lower in rock fragments and higher in silt + clay in the soil-sized fraction than deeper non-weathered gray to whitish colored strata (Haering et al. 2004). For example, initial rock-fragment content of spoils derived from non-weathered massive gray sandstones often exceeds 80% whereas spoils derived from pre-weathered oxidized materials are commonly <50% rock and can be as low as 20% (Daniels and Amos 1985). The combination of lower rock content and finer textures in the soil-sized fraction greatly enhances the plant-available water holding capacity of these pre-weathered spoils. Finally, it should be noted that spoils derived primarily from siltstones and mudrocks tend to develop a hard silt crust layer at their surface following rain events which can inhibit seedling germination during subsequent revegetation efforts.

The extent of pre-mine weathering also has a dominant influence on post-mine chemical properties. Oxidized and weathered strata are typically considerably lower in pH (4.5–6.0) and exchangeable nutrient cations (Ca, K and Mg) than gray non-weathered spoils that commonly range from pH 7.0–8.5. These higher pH levels are due primarily to carbonate dissolution and feldspar hydrolysis as discussed below, but are often rapidly offset to lower levels by sulfide oxidation. However, the pH of mine soils derived from low sulfur non-weathered spoils can remain above 7.0 for many years following reclamation and is particularly detrimental to the establishment and persistence of native woody tree species which are adapted to low pH native soil conditions (Angel et al. 2005). These higher pH conditions are favored for lands returned to hayland/pasture land uses, however, and can lead to improvements in surface water quality where utilized as new cover materials on formerly-mined acidic sites.

8 Importance of Sulfide Abundance and Morphology on Rate and Extent of Acid-S Weathering Processes

Sulfide oxidation reactions and processes play a dominant role in the production of acid mine drainage and in controlling post-reconstruction mine soil chemical properties (Kruse Daniels et al. 2021). In the absence of neutralizers (discussed below), mine soil pH values can drop below 3.0 during what is known as the "active sulfuricization" phase and may remain below pH 4.0 for decades in the "post active" phase (Fanning and Fanning 1989). In general, sulfur occurs in three phases in Appalachian rocks, organic-S, sulfate-S, and sulfide-S; with sulfidic forms dominating non-weathered rocks and sulfates occurring as a weathering product which leaches away over time into drainage waters, but may also persist during dry periods as acid sulfate salts. Organic-S in coal and fossil inclusions in lithic rocks weather and oxidize very slowly and are not considered to be major contributors to acid weathering reactions. However, sulfides can vary widely in reactivity. Finely divided framboidal pyrite is highly reactive over very short time periods (Geidel and Caruccio 2000) whereas larger highly crystalline (often readily visible) forms are much slower to react and, in fact, may not play a major role in short-term (e.g. years) acid–base reactions. Intermediate sized and less crystalline sulfides are presumed to be fully reactive over relatively short periods (e.g. years to decades) if the rock fabric allows water and oxygen penetration.

As discussed earlier, the pyritic-S content in Appalachian rock spoils varies widely, but is commonly quite low (<0.2%) in coarser-grained rocks found in southern WV, southwest Virginia and eastern KY (Pocahontas Basin), but can be greater than 2% in shales directly associated with coal seams and even higher in coarser-grained strata in the northern Dunkard Basin. Spoils with <~0.25% pyritic-S are seldom potentially acidic enough to be designated as "potentially toxic," which would require special handling and/or isolation away from surface and groundwater. Conventional acid–base accounting procedures also assume that previously weathered near-surface strata that are brown to brownish-yellow and high in Munsell color value (≥4.0) and chroma (>2.0) have lost all sulfides to oxidation reactions over extended periods of time and any remaining S is in organic or sulfate forms.

9 Importance of Carbonates, Other Neutralizers, and Metamorphic Lithics

The pH of young mine soils derived from non-weathered (gray) rock spoils is typically between 6.5 and 8.5 due to carbonate dissolution coupled with surface hydrolysis/weathering reactions on feldspar grains and broken edges of 2:1 clay minerals (micas, vermiculites, and chlorites). Whereas carbonates do occur as thin discrete strata (e.g. Magoffin beds in the Wise/Breathitt formation), they are most commonly found as disseminated secondary cementing agents. Siderite is also found in these strata as both nodules and in rim grain coatings, but does not contribute to long term spoil/soil neutralization reactions due to the offsetting acidity produced by Fe^{2+} hydrolysis (Skousen et al. 1997). The total acid neutralizing potential (NP) of Appalachian strata has historically been presumed to be dominantly derived from carbonates (Sobek et al. 2000), primarily because the cation complement in drainage waters is usually dominated by Ca and Mg which appear to result from direct carbonate neutralization of acidity produced by sulfide oxidation reactions. However, recent work (Clark et al. 2018a) as summarized above on a wide array ($n = 16$) of coarse- and medium-grained rocks found very little carbonate present despite the fact these same rock samples possessed substantial total NP via conventional Acid–Base-Accounting procedures and even higher analytical estimates (>3%) of calcium carbonate equivalence (CCE). Therefore, we believe that hydrolysis of feldspars and other net neutralizing reactions are much more important in these materials than previously assumed.

As noted previously, the study by Clark et al. (2018a) also revealed that metamorphic lithic fragments make up a substantial portion of the grains in many rocks of the region and commonly contain a significant component of chlorite (Mg interlayered 2:1:1 phyllosilicate). The origin of such fragments is unknown; however, they are most likely derived from remnants of the forerunners of today's Appalachians, the ancient highlands created by the Taconic and Acadian orogenies and which were source areas for the sediments that infilled the Appalachian basin during the Pennsylvanian and early Permian geologic eras. These lithics may contribute to the neutralizing ability of these rocks via hydrolysis weathering and may also be a source of Mg^{2+} release to weathering soils and leaching solutions. Higher than expected ratios of Mg/Ca have been reported in column leaching experiments from similar strata (Orndorff et al. 2015; Clark et al. 2018b) and also in receiving streams in related field studies. While enhanced Mg levels could also be due to dolomitic cementing agents, the abundance of this mineral is generally low in these same strata.

The amount and distribution of ions discharged in drainage waters from mining sites that enter headwater streams as TDS is dominated by reaction products from combined sulfide oxidation and neutralization reactions, particularly sulfate and Ca (Clark et al. 2021). Thus, the mineralogic backbone of the rock spoils, particularly the abundance and form(s) of sulfides and neutralizers, has a dominant influence on the amount and timing of TDS release by these materials. However, it also clear that the extent of pre-disturbance weathering is even more important in that pre-weathered materials will have lost their reactive sulfides, carbonates, and soluble ion load over their extended weathering history (Orndorff et al. 2015; Daniels et al. 2016).

10 Summary and Conclusions

The geologic history of the Appalachian coal region is attributed to thrust loading related to the approach of Africa from the east coupled with sea-level changes related to the waxing and waning of the Gondwanan glaciation. Major valleys eroded during periods of low sea level and backfilled with sandstones during periods of rising sea level. Deltas built into the basin from the east during periods of high sea level and consist of mudstone-sandstone successions. Terrestrial plant systems developed in estuaries during drowning of the continent and on abandoned deltas and produced the coal seams characteristic of the Pennsylvanian time period.

Rock type, grain size, degree of cementation, and overall mineralogic composition of these rocks have important impacts on the utilization and management of resultant mine spoil materials on Appalachian coal mines. Compositions of Pennsylvanian and Early Permian sandstones range from quartzose to lithic to feldspathic that have a profound influence on their response to weathering. In the weathering environment, feldspar grains are either replaced by kaolinite or dissolved to create porosity. Chlorite schist and mica grains occur in all sandstones and are also susceptible to leaching and formation of porosity. In contrast to feldspathic and lithic sandstones, quartzose sandstones, cemented mostly by authigenic quartz, are resistant to weathering.

High-quartz durable sandstones are utilized for valley fill drainage structures and overall rock durability directly influences fill stability and design. As rock spoils weather to mine soils, resultant physical and chemical properties are directly controlled by rock type and mineralogical properties. Mixed lithologies that have been pre-weathered to some extent before final placement generally produce superior mine soils for plant growth whereas non-weathered coarser-grained rocks typically generate much rockier mine soils that are too alkaline for native tree species but support hayland/pasture uses well. Pyrite abundance and morphology/reactivity have an overriding influence on local mine soil properties and water quality and vary widely across the region, with higher levels found in the Dunkard Basin of northern Appalachia. Within landscapes formed by mining, acidity produced by pyrite oxidation can be offset by neutralizing mineral components, primarily carbonates and feldspars, when present in sufficient quantities. The role of metamorphic lithics in the leachate chemistry from these mine spoils has been overlooked and deserves further study.

11 Glossary of Geologic and Soil Science Terms Used in this Chapter

Allochthon: a geological formation not formed in the region where found and moved to its present location by tectonic forces.

Aridisol: one of 12 global soil orders in USA Soil Taxonomy (USDA-NRCS 2014). Aridisols are soils formed in low rainfall desert areas of the World.

Craton: an ancient part of the Earth's continental crust which has been more or less stable since Precambrian times such as the Laurentian craton of ancestral North America.

Domed peats: a peat-forming vegetation community lying above ground water level (synonymous with ombrogenous swamp).

Epigenetic: formed later than the host rocks.

Framboidal pyrite: very fine (<20 μm) poorly crystalline and highly reactive form of pyrite (FeS_2). From the French term *Frambois* (raspberry) due to the shape of the crystal aggregates.

Foreland basin: a foreland basin is a structural basin that develops adjacent and parallel to a mountain belt and is infilled with sediment derived from the mountain belt.

Grenville Province: a continental crustal province that resulted from the mid-late Mesoproterozoic age (c. 1250–980 Ma) Grenville orogeny associated with the assembly of the supercontinent Rodinia.

Histosols: one of 12 global soil orders in USA Soil Taxonomy (USDA-NRCS 2014). Histosols are soils formed in saturated or very cold environments that lead to accumulation of thick surface layers of organic materials.

Incised valley: a valley cut by a river into its own floodplain typically associated with fall in sea level.

Laurentia: a large continental craton of Precambrian age that forms the ancient geological core of the North American continent.

Lithic sandstone: a sandstone with a significant fraction of lithic grains such as fragments of pre-existing igneous, metamorphic and sedimentary rocks.

Longitudinal (axial) fluvial system: a major river system flowing along the axis of an elongate sedimentary basin, typically a foreland basin, and deriving sediment from various sources including a craton.

Paleosol: weathered soil profile formed in an earlier geologic environment that has been buried and preserved by more recent sediments or other geologic/soil materials.

Pedogenic: feature derived from soil-forming processes that integrate the effects of climate, vegetation, and topographic position on parent materials over time.

Petrocalcic horizon: a very hard cemented subsoil layer that forms in soils via long term leaching of carbonates from the surface soil coupled with secondary precipitation at depth. Usually found in desert soils (Aridisols), but may also be preserved in sedimentary strata and paleosols.

Passive margin: a sedimentary basin located within a plate at the transition between oceanic and continental crust and developed above a rift basin.

Orogeny: a mountain-building event involving, amongst other geological processes, faulting and folding associated with collision of continents.

Paleogeography: the ancient geography of Earth's surface typically depicted on maps.

Planar peats: a peat swamp confined to topographic depression and characterized by a relatively horizontal upper surfaces (synonymous with topogenous swamp).

Structural basin: a basin-shaped feature resulting from gentle folding at some time following deposition of the sediments now contained within the basin.

Terrane: the area or surface over which a particular rock or group of rocks is prevalent.

Thermal subsidence: subsidence of the earth's crust associated with cooling.

Thrust fault: a reverse fault with a dip of 30° or less that places older rocks above younger rocks.

Thrust loading: subsidence of the crust associated with thrust faulting and typically associated with collision of continents.

Transverse fluvial system: a river flowing at right angles to the axis of an elongate sedimentary basin, typically a foreland basin, and deriving sediment from a fold-and-thrust belt along the active margin of the basin.

Vertisols: one of 12 global soil orders in USA Soil Taxonomy (USDA-NRCS 2014). Vertisols are clayey soils that shrink-swell and have cracking open to surface for significant periods of time.

Acknowledgements Figures 3 through 6 were drawn by Kathryn Haering and Fig. 2 was drawn by Jody Booze-Daniels; these figures are based on reference materials cited by the figure captions.

References

Aitken JF, Flint SS (1995) The application of high-resolution sequence stratigraphy to fluvial systems: a case study from the upper carboniferous breathitt group, eastern Kentucky, USA. Sedimentology 42:3–30

Aleinikoff JN, Zartman RE, Walter M, Rankin DW, Lyttle PT, Burton WC (1995) U-Pb ages of metarhyolites of the Catoctin and Mount Rogers formations, central and southern Appalachians: evidence for two pulses of Iapetan rifting. Am J Sci 295:428–454

Angel P, Davis V, Burger J, Graves D, Zipper C (2005) The Appalachian regional reforestation initiative. US Office of Surface Mining, Appalachian Regional Reforestation Initiative, Forest Reclamation Advisory Number, p 1

Archer AW, Greb SF (1995) An Amazon-scale drainage system in the early Pennsylvanian of central North America. J Geol 103:611–628

Badger RL, Sinha AK (1988) Age and Sr isotope signature of the Catoctin volcanic province: implications for subcrustal mantle evolution. Geology 16:692–695

Baxby M, Patience RL, Bartle KD (1994) The origin and diagenesis of sedimentary organic nitrogen. J Petrol Geol 17:211–230

Blake BM, Beuthin J (2008) Deciphering the mid-carboniferous eustatic event in the central Appalachian foreland basin, southern West Virginia, USA. In: Fielding CR, Frank TD, Isbell JL (eds) Resolving the late Paleozoic ice age in time and space. Geol Soc Am Spec. Pap vol 441, pp 249–260

Bodek RJ (2006) Sequence stratigraphic architecture of early Pennsylvanian, coal-bearing strata of the Cumberland block: a case study from Dickenson, County, Virginia. MS Thesis, Virginia Tech Blacksburg, p 55

Bream BR, Hatcher RD Jr (2002) Southern Appalachian terranes amended: timing of accretion and delimiting provenance from detrital zircon and Nd isotopic data. Geol Soc Am Abstr Program 34:41

Castle JW (2001) Appalachian basin stratigraphic response to convergent-margin structural evolution. Basin Res 13:397–418

Cecil CB (1990) Paleoclimate controls on stratigraphic repetition of chemical and siliciclastic rocks. Geology 18:533–536

Cecil CB, Dulong FT, Cobb JC, Supardi (1993) Allogenic and autogenic controls on sedimentation in the central Sumatra basin as an analogue for Pennsylvanian coal-bearing strata in the Appalachian basin. Geol Soc Am Spec Pap 286:3–22

Cecil CB, Dulong FT, West RD, Stamm R, Wardlaw B, Edgar NT (2004) Climate controls on the stratigraphy of a middle Pennsylvanian cyclothem in North America. In: Cecil CB, Edgar NT (eds) Climate controls on stratigraphy. SEPM (Society for Sedimentary Geology) Spec Pub 77:151–180

Cecil CB, Stanton RW, Neuzil SG, Dulong FT, Ruppert LF, Pierce BJ (1985) Paleoclimate controls on late Paleozoic sedimentation and peat formation in the central Appalachian basin. Int J Coal Geol 5:195–230

Chesnut DR, Jr. (1988) Stratigraphic Analysis of the Carboniferous Rocks of the Central Appalachian Basin. Ph.D. dissertation, University of Kentucky Lexington p 297

Chesnut DR, Jr. (1994) Eustatic and tectonic control of deposition of the Lower and middle Pennsylvanian strata of the central Appalachian basin. In: Dennison JM, Ettensohn FR (eds) Tectonic and eustatic controls on sedimentary cycles. SEPM Concepts in Sedimentology and Paleontology 4:51–64

Chesnut DR Jr (1996) Geologic framework for the coal-bearing rocks of the central Appalachian basin. Int J Coal Geol 31:55–66

Churnet HG (1996) Depositional environments of Lower Pennsylvanian coal-bearing siliciclastics of southeastern Tennessee, northwestern Georgia, and northeastern Alabama, USA. Int J Coal Geol 31:21–54

Clark EV, Daniels WL, Zipper CE, Eriksson KA (2018a) Mineralogical influences on water quality from weathering of surface coal mine spoils. Applied Geochem 91:97–106

Clark EV, Zipper CE, Daniels WL, Keefe MJ (2018b) Appalachian coal mine spoil elemental release patterns and depletion. Appl Geochem 98:109–120

Clark EV, Zipper CE, Soucek DJ, Daniels WL (2021) Contaminants in Appalachian water resources generated by non-acid-forming coal-mining materials. Chapter 9 in: Zipper CE, Skousen JG (eds) Appalachia's coal-mined landscapes. Springer

Coler DJ, Wortman GL, Samson SD, Hibbard JP, Stern R (2000) U-Pb geochronologic, Nd isotopic, and geochemical evidence for the correlation of the Chopawamsic and Milton Terranes, Piedmont Zone. South Appalachian Orogen J Geol 108:363–380

Colton GW (1970) The Appalachian basin—its depositional sequences and their geologic relationships. In: Pettijohn FJ, Reed JC (eds) Fisher PW. Central and Southern. Interscience Publishers New York, Studies of Appalachian Geology, pp 5–47

Cook FA, Albaugh DS, Brown LD, Kaufman S, Oliver JE, Hatcher RD Jr (1979) Thin skinned tectonics in the crystalline southern Appalachians: COCORP seismic-reflection profiling of the blue ridge and piedmont. Geology 7:563–567

Daniels WL, Amos DF (1985) Generating productive topsoil substitutes from hard rock overburden in the southern Appalachians. Environ Geochem Health 7:8–15

Daniels WL, Zipper CE (2018) Creation and management of productive mine soils. Virginia Cooperative Extension, Pub. #460–121, Virginia Tech, Blacksburg, VA. https://www.pubs.ext.vt.edu/460/460-121/460-121.html

Daniels WL, Zipper CE, Orndorff ZW, Skousen J, Barton CD, McDonald LM, Beck MA (2016) Predicting total dissolved solids release from central Appalachian coal mine spoils. Environ Pollut 216:371–379

Davis MW, Ehrlich R (1974) Late Paleozoic crustal composition and dynamics in the southeastern United States. In: Briggs G (ed) Carboniferous of the Southeastern United States. Geol Soc Am Spec Pap 148:171–185

Englund KJ (1979) The mississippian and Pennsylvanian (Carboniferous) systems in the United States—Virginia. Geol Soc Am Spec Pap 1110-C:1–21

Englund KJ, Thomas RE (1990) Late Paleozoic depositional trends in the central Appalachian basin. US Geol Surv Bull 1839-F, p. 1–19.

Eriksson KA, Campbell IH, Palin JM, Allen CM, Bock B (2004) Evidence for multiple recycling in Neoproterozoic through Pennsylvanian sandstones of the central Appalachian basin. J Geol 112:261–276

Ettensohn FR (2004) Modeling the nature and development of major Paleozoic clastic wedges in the Appalachian basin, USA. J Geodyn 37:657–681

Fanning DS, Fanning MCB (1989) Soil morphology, genesis and classification. Wiley, New York, pp 69–79

Fedorko N, Skema V (2013) A review of the stratigraphy and stratigraphic nomenclature of the Dunkard Group in West Virginia and Pennsylvania, USA. Int J Coal Geol 119:2–20

Ferm JC (1974) Carboniferous environmental models in eastern United States and their significance. In: Briggs G (ed) Carboniferous of the Southeastern United States. Geol Soc Am Spec 148:79–95

Ferm JC, Horne JC (1979) Carboniferous depositional environments in the Appalachian region. Dept Geol, Univ, South Carolina, Columbia, vi, p 760

Finkelman RB (1998) Trace and minor elements in coal. Pp. 593–607 (Chapter 28) In: Engel SH, Macko SA (eds) Organic geochemistry principles and applications. Plenum Press, New York

Flanick M, Bradford I, Dailey J, Cole R (2005) Petrography and provenance of middle Pennsylvanian Allegheny Formation and Conemaugh Group sandstones of the Alleghanian foreland basin, Southwestern Pennsylvania. Geol Soc Am Abstr Program 37:15

Geidel G, Caruccio FT (2000) Geochemical factors affecting coal mine drainage quality. In: Barnhisel RI et al (ed) Reclamation of drastically disturbed lands. Agronomy 41. American society of agronomy, Madison, WI. pp 105–130

Graham SA, Dickinson WR, Ingersoll RV (1975) Himalayan-Bengal model for flysch dispersal in Appalachian-Ouachita System. Geol Soc Am Bull 86:273–286

Greb SF, Chesnut DR Jr (1996) Lower and lower middle Pennsylvanian fluvial to estuarine deposition, central Appalachian basin: effects of eustasy, tectonics, and climate. Geol Soc Am Bull 108:303–317

Greb SF, Chesnut DR, Jr., Eble CF (2004) Temporal changes in coal-bearing depositional sequences (lower and middle Pennsylvanian) of the central Appalachian basin, USA. In: Pashin JC, Gastaldo RA (eds) Sequence stratigraphy, Paleoclimate, and tectonics of coal-bearing strata. AAPG Studies in Geology 51:89–120

Greb SF, Martino RL (2005) Fluvial-estuarine transitions in fluvial-dominated successions: examples from the lower Pennsylvanian of the central Appalachian Basin. In: Blum M, Marriott SB, LeClair S (eds) Fluvial sedimentology VII. Int Assoc Sedtol Spec Publ 35:425–452

Greb SF, Pashin JC, Martino RM, Eble CF (2008) Appalachian sedimentary cycles during the Pennsylvanian: changing influences of sea level, climate, and tectonics. In: Fielding CF, Frank TD, Isbell JL (eds) Resolving the late Paleozoic Gondwanan ice age in time and space. Geol Soc Am Spec Pub 441:235–248

Grimm RP, Eriksson KA, Carbaugh J (2013) Orogenic controls on the temporal and spatial distribution of longitudinal and transverse alluvial facies belts in an early Pennsylvanian foreland basin, Virginia, USA. Basin Res 25:450–470

Haering KC, Daniels WL, Galbraith JC (2004) Appalachian mine soil morphology and properties: effects of weathering and mining method. Soil Sci Soc 68:1315–1325. https://doi.org/10.2136/sssaj2004.1315

Hatcher RD Jr (1972) Development model for the southern Appalachians. Geol Soc Am Bull 83:2735–2760

Hatcher RD, Jr. (1989) Tectonic synthesis of the U.S. Appalachian. In: Hatcher RD, Jr, Thomas WA, Viele GW (eds) The Appalachian-Ouachita Orogen in the United States. The geology of North America. Geol Soc Am Boulder Colorado F-2:511–535

Hatcher RD, Jr. (2002) Alleghanian (Appalachian) orogeny, a product of zipper tectonics: rotational, transpressive continent–continent collision and closing of ancient oceans along irregular margins. In: Martínez Catalán JR, Hatcher RD, Jr, Arenas, R, Díaz García F (eds) Variscan-Appalachian dynamics: The building of the late Paleozoic basement. Geol Soc Am Spec Pap 364:199–208

Hobday DK, Horne JC (1977) Tidally influenced barrier island and estuarine sedimentation in upper Carboniferous of southern West-Virginia. Sed Geol 18:97–122

Horton JW, Jr., Drake AA, Jr., Rankin DW (1989) Tectonostratigraphic terranes and their Paleozoic boundaries in the central and southern Appalachians. In: Dallmeyer RD (ed) Terranes in the circum-atlantic Paleozoic Orogens. Geol Soc Am Spec Pap 230:213–246

Houseknecht DW (1980) Comparative anatomy of a Pottsville lithic arenite and quartz arenite of the Pocahontas Basin, southern West-Virginia - petrogenetic, depositional, and stratigraphic implications. J Sed Pet 50:3–20

Korus JT, Kvale EP, Eriksson KA, Joeckel RM (2008) Compound paleovalley fills in the lower Pennsylvanian new river formation, West Virginia, USA. Sediment Geol 208:15–26

Kosanke RM, Cecil CB (1996) Late Pennsylvanian climate changes and palynomorph extinctions. Review of Palaeobot Palynol 90:113–140

Kruse Daniels N, LaBar J, McDonald LM (2021) Acid mine drainage in Appalachia: Sources, legacy, and treatment. Chapter 8 In: Zipper CE, Skousen JG (eds) Appalachia's coal-mined landscapes. Springer

Lebold JG, Kammer TW (2006) Gradient analysis of faunal distributions associated with rapid transgression and low accommodation space in a late Pennsylvanian marine embayment: Biofacies of the Ames member (Glenshaw Formation (Conemaugh Group) in the northern Appalachian Basin, USA. Paleogeog Paleoclim Paleoecol 231:291–314

Martino RL (2004) Sequence stratigraphy of the glenshaw formation (Middle-late Pennsylvanian) in the central Appalachian basin. In Pashin JC, Gastaldo RA (eds) Coal-bearing strata: sequence stratigraphy, Paleoclimate, and Tectonics of coal-bearing strata. AAPG studies in geology 51:1–28

Milici RC (2004) Assessment of Appalachian basin oil and gas resources: carboniferous coal-bed gas total petroleum system. U.S. geological survey open-file report 2004–1272, p 59

Miller JMG (1994) The Neoproterozoic Konnarock Formation, southwest Virginia, USA: glaciolacustrine facies in a continental rift. In: Deynoux M, Miller JMG, Domack EW, Eyles N, Fairchild IJ, Young GM (eds) Earth's Glacial Record. Cambridge University Press, New York, pp 47–59

Mitra G (1988) 3-dimensional geometry and kinematic evolution of the Pine mountain Thrust system, Southern Appalachians. Geol Soc Am Bull 100:72–95

Montanez IP, Cecil CB (2013) Paleoenvironmental clues archived in non-marine pennsylvanian-lower permian limestones of the central Appalachian basin. Int J Coal Geol 119:41–55

Mussman WJ, Read JF (1986) Sedimentology and development of a passive-to-convergent margin unconformity: middle Ordovician Knox unconformity, Virginia Appalachians. Geol Soc Am Bull 97:282–295

Neuzil SG (2000) Summary report on the coal resources, coal production, and coal quality of the Allegheny group No. 5 Block, and the Pottsville Group Stockton and Coalburg, Winifrede/Hazard, Williamson/Amburgy, Campbell Creek/Upper Elkhorn No. 3, and Upper Elkhorn Nos. 1 and 2/Powellton Coal Zones, Central Appalachian Basin Coal Region. U.S. Geol Surv Prof Pap 1625–c:128

Orndorff Z, Daniels WL, Zipper C, Eick M, Beck M (2015) A column evaluation of Appalachian coal mine spoils' temporal leaching behavior. Environ Pollut 204:39–47

Osberg PH, Tull JF, Robinson P, Hon R, Butler JR (1989) The acadian orogen. In: Hatcher RD, Jr, Thomas WA, Viele GW (eds) The Appalachian-Ouachita orogen in the United States. The geology of North America. Geol Soc Am Boulder Colorado F2:179–232

Palmer CA, Oman CL, Park AJ, Luppens JA (2015) The U.S. geological survey coal quality (COALQUAL) database version 3.0: U.S. Geological survey data series 975, p 43
Rankin DW (1993) The volcanogenic Mount Rogers Formation and the overlying Konnarock Formation: two late Paleozoic units in southwest Virginia. US Geol Surv Bull. 2029, 26 p.
Read JF (1989a) Evolution of Cambro-Ordovician passive margin, U. S. Appalachians. The geology of North America. Geol Soc Am Boulder Colorado F2:42–57
Read JF (1989b) Controls on the evolution of the Cambro-Ordovician passive margin, U.S. Appalachians. In: Crevello PD, Read JF, Wilson JL, Sarg JF (eds) Controls on carbonate platform and basin development: SEPM special publication 44:146–165
Read JF, Eriksson KA (2016) Paleozoic sedimentary successions of the Valley and ridge and plateau of Virginia. Geology of Virginia. Virginia Museum of Natural History. In: Bailey CM, Sherwood WC, Eaton LS, Powars DS (eds) The Geology of Virginia. Virginia Museum Nat Hist Spec. Pub. 18: 17–54.
Reed JS, Spotila JA, Eriksson KA, Bodnar RJ (2005) Burial and exhumation history of Pennsylvanian strata, central Appalachian basin: an integrated study. Basin Res 17:259–268
Rice CL, Survey KG (1984) Sandstone units of the Lee Formation and related strata in eastern Kentucky. US G.P.O, Washington DC iv, p 53
Rice CL, Schwietering JF (1988) Fluvial deposition in the central Appalachians during the early Pennsylvanian> US Geol Surv Bull 1839-B: 1–9.
Ruppert LF, Trippi MH, Slucher ER (2014) Correlation Chart of Pennsylvanian Rocks in Alabama, Tennessee, Kentucky, Virginia, West Virginia, Ohio, Maryland, and Pennsylvania Showing Approximate Position of Coal Beds, Coal Zones, and Key Stratigraphic Units. Chapter D-2. In: Ruppert LF, Ryder RT (eds.) Coal and Petroleum Resources in the Appalachian Basin: Distribution, Geologic Framework, and Geochemical Character. US Geological Survey Professional Paper 1708.
Schneider JW, Lucas SJ, Barrick JE (2013) The Early Permian age of the Dunkard Group, Appalachian Basin, USA, based on spiloblattinid insect biostratigraphy. Int J Coal Geol 119:8–92
Schweinfurth, S.P., (2009) An introduction to coal quality. In: Pierce BS, Dennen KO (eds.) The National Coal Resource Assessment Overview: U.S. Geological Survey Professional Paper 1625–F, Chapter C, 16 p
Scotese C (2003) A continental drift flipbook. J Geol 112:729–741
Secor DT, Snoke AW, Dallmeyer RD (1986) Character of the Alleghanian Orogeny in the southern Appalachians. Geol Soc Am Bull 97:1345–1353
Simpson EL, Eriksson KA (1990) Early Cambrian progradational and transgressive sedimentation patterns in Virginia: an example of the early history of a passive margin: J Sed Pet 60:84–100
Skousen JG, Daniels WL, Zipper CE (2021) Soils on Appalachian coal-mined lands. Chapter 4 in: Zipper CE, Skousen JG (eds) Appalachia's Coal-Mined Landscapes. Springer
Skousen J, Renton J, Brown H, Evans P, Leavitt B, Brady K, Cohen L, Ziemkiewicz P (1997) Neutralization potential of overburden samples containing siderite. J Environ Qual 36(3):673–681
Skousen JG, Zipper CE (2021) Coal mining and reclamation in Appalachia. Chapter 3 in: Zipper CE, Skousen JG (eds) Appalachia's Coal-Mined Landscapes. Springer
Sobek AS, Skousen JG, Fisher SE, Jr. (2000) Chemical and physical properties of overburdens and minesoils. In: R.I. Barnhisel et al. (ed.) Reclamation of drastically disturbed lands. Agronomy 41. American Society of Agronomy, Madison, WI. p. 77–104
Tabor NJ, Montanez IP (2002) Shifts in late Paleozoic atmospheric circulation over western Pangea: insights from pedogenic mineral $\delta^{18}O$ compositions. Geology 30:1127–1130
Trippi MH, Ruppert LF, Attanasi ED, Milici RC, Freeman PA (2014) Appalachian basin bituminous coal; Sulfur content and potential sulfur dioxide emissions of coal mined for electrical power generation. In: Ruppert LF, Ryder RT (eds) Coal and petroleum resources in the Appalachian basin; Distribution, geologic framework, and geochemical character: U.S. Geological Survey Professional Paper 1708, 68 p.
USDA-NRCS, (2014) Keys to Soil Taxonomy, 12th edn. USDA-Natural Resources Conservation Service, Washington, DC

Van Der Voo R (1983) A plate tectonics model for the Paleozoic assembly of Pangea based on paleomagnetic data. In: Hatcher RD, Jr, Williams H, Zietz I (eds) Contributions to the Tectonics and Geophysics of Mountain Chains. Geol Soc Am Mem 158: 19–24.

Wehr F, Glover L III (1985) Stratigraphy and tectonics of the Virginia-North Carolina Blue Ridge: evolution of a late Proterozoic-early Paleozoic hinge zone. Geol Soc Am Bull 96:285–295

Wizevich MC (1991) Sedimentology and Regional Implications of Fluvial Quartzose Sandstones of the Lee Formation, Central Appalachian Basin. Ph.D. Dissertation, Virginia Tech, Blacksburg, Virginia, p 237

Wizevich MC (1992) Sedimentology of Pennsylvanian quartzose sandstones of the Lee Formation, central Appalachian basin: fluvial interpretation based on lateral profile analysis. Sediment Geol 78:1–47

Woodward NB (ed) (1985) Valley and Ridge Thrust Belt: balanced structural sections Pennsylvania to Alabama. Studies in Geology #12, University of Tennessee, Knoxville.

Wortman GL, Samson SD, Hibbard JP (2000) Precise U-Pb zircon constraints on the earliest magmatic history of the Carolina terrane. J Geol 108:321–338

Wyrick GW, Borchers JW (1981) Hydrologic effects of stress-relief fracturing in an Appalachian valley. US Geol Surv Water Supply Paper 2177.

Coal Mining and Reclamation in Appalachia

Jeff Skousen and Carl E. Zipper

Abstract Coal is produced in Appalachia via both surface and underground mining. Mining methods have changed and advanced over nearly two centuries, improving worker safety and environmental performance. Reclamation was not practiced in the early days of mining, but laws to regulate mining were enacted and performance standards have been developed since the 1960s. Today, mining operations are heavily regulated and the mining and reclamation processes are fully integrated. Mine planning and design, as well as reclamation procedures to establish a suitable post-mining land use, are now required. Reclamation includes landscape reconstruction, mine-site revegetation, and associated activities intended to avoid or mitigate environmental impacts. Today's coal miners manage mine spoils to limit water contamination by acid mine drainage or other dissolved constituents. Reclamation of coal surface mines is intended to prepare the land for post-mining land use, usually hayland-pasture, forest, or wildlife habitat. Coal mining in Appalachia has had extensive impacts on the landscape and those impacts continue today despite federal and state reclamation regulatory requirements, but at far-reduced levels relative to early mining.

Keywords Environmental disturbance · Mine spoil · Revegetation · Valley fill

1 Introduction

Coal mining has contributed to the Appalachian coalfield economy for more than a century (Zipper et al. 2021). The coal mining industry provides good-paying jobs, often in places with otherwise limited economic opportunities. Coal mining also supports mining communities in other ways. Mining firms support mine-service and supply businesses; while employees of those businesses and the miners themselves spend money locally, supporting retail businesses, medical services, and other forms

J. Skousen (✉)
West Virginia University, Morgantown, WV, USA
e-mail: jskousen@wvu.edu

C. E. Zipper
Virginia Tech, Blacksburg, VA, USA

C. E. Zipper and J. Skousen (eds.), *Appalachia's Coal-Mined Landscapes*,
https://doi.org/10.1007/978-3-030-57780-3_3

of economic activity. Some mining firms have also been major contributors to civic and charitable causes in the communities where they operate.

Coal mining, however, also influences coalfield communities in other ways. Despite significant worker health and safety advances, coal mining remains a hazardous occupation (Zipper et al. 2021; Gohlke 2021). Coal mining affects the environment both within and beyond the mine site's borders, even when the best reclamation practices are used.

Here, we provide a brief overview of coal mining, mine reclamation methods, and related activities in Appalachia.

2 Coal Mining

In its essence, coal mining is a simple process. The coal lies beneath the ground, embedded among layers of sedimentary rocks. Surface miners remove the overlying or surrounding rocks and then remove the coal. Underground miners excavate the coal while leaving the overlying rock in place to the extent that is possible.

Throughout most of Appalachia's bituminous coalfield, coal seams are oriented within a few degrees of horizontal, but the region's anthracite seams usually occur at steeper angles. Appalachian coal seams can be as much as 3 m or more in thickness. The thicker seams were initially targeted for mining, and most have largely been mined out during the past century. Today, most mining operations are accessing coal seams 1 or 2 m thick. Miners can mine even thinner coals on large surface mines when extracting multiple coal seams, and in underground mines by extracting associated rock along with the coal.

2.1 Underground Mining

Today's underground mining is far different from methods used in the early 1800s when US underground mining began. The coal industry has developed advanced technologies for excavating coal from underground.

Miners initiate a new mine by clearing soil and rock materials from the intended mine entry. Where flat-lying coal seams are within steep terrain, mine entries often occur on hillsides where coal seams outcrop. Where coal seams are not accessible from the surface directly, mine entries may be established as slope mines or mines with vertical shafts (Fig. 1).

As the excavations advance, both mining and human safety require that rock overlying the mine cavity remain in place. Early miners inserted wood posts and pillars to hold up the mine "roof" of overlying rock and, in some cases, to support horizontal timbers or crossbars to support rock over wider passages. These practices changed with the introduction of roof-bolting in the 1940s (Mark 2002). Roof-bolting secures the rock immediately above the cavity, which is often less-stable than strata

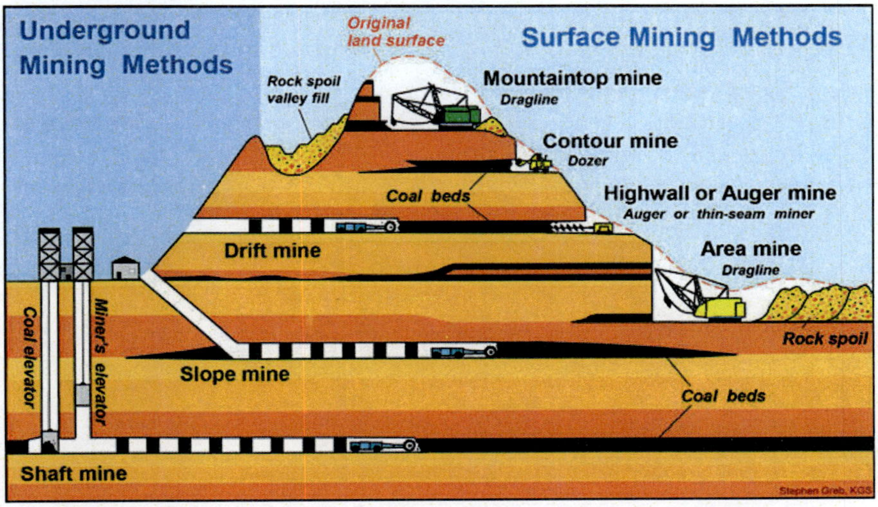

Fig. 1 Generalized schematic representing coal mining methods employed in Appalachia (*Source* Courtesy of Stephen Greb. Copyright 2020 Kentucky Geological Survey)

further above, by securing long steel rods in holes that have been drilled a meter or more into the overlying rock. These rods hold steel bearing plates against the roof that support the thin rock layers that often occur directly above the mine cavity.

Methods for advancing the mine also consider overburden stability. Room and pillar mines advance by excavating parallel linear channels into the coal seam and then connect those channels while leaving coal "pillars" in place to support the overburden (Fig. 2 top). This method removes coal inefficiently by leaving more than 50% of the coal as pillars. In time, miners developed a technique called pillar-retreat mining which proceeds initially by advancing room-and-pillar excavations into the coal seam, but then "retreats" back toward the mine entry by partially or fully extracting the coal pillars. Although the inevitable result is overburden collapse and potential subsidence at the land surface, the miners' intent is to abandon mined-out areas before that occurs.

Workers advance the underground mine by breaking coal from its rock-like form into smaller fragments that can be transported out of the mine. Originally, this was done using steel spikes and hammers. In the late 1800s, miners began using explosives to fracture the coal (Mainero and Verakis 2010). Miners would "undercut" the coal, meaning to excavate a thin horizontal channel beneath the coal-seam segment intended for fragmentation. Then, they would drill holes into the coal face, insert explosive charges, and explode the charges to fragment the coal. Originally, the undercutting and drilling operations were performed with hand tools. Beginning in the 1870s, mechanization introduced air-powered and electric tools to help miners with these operations (Garges 2003).

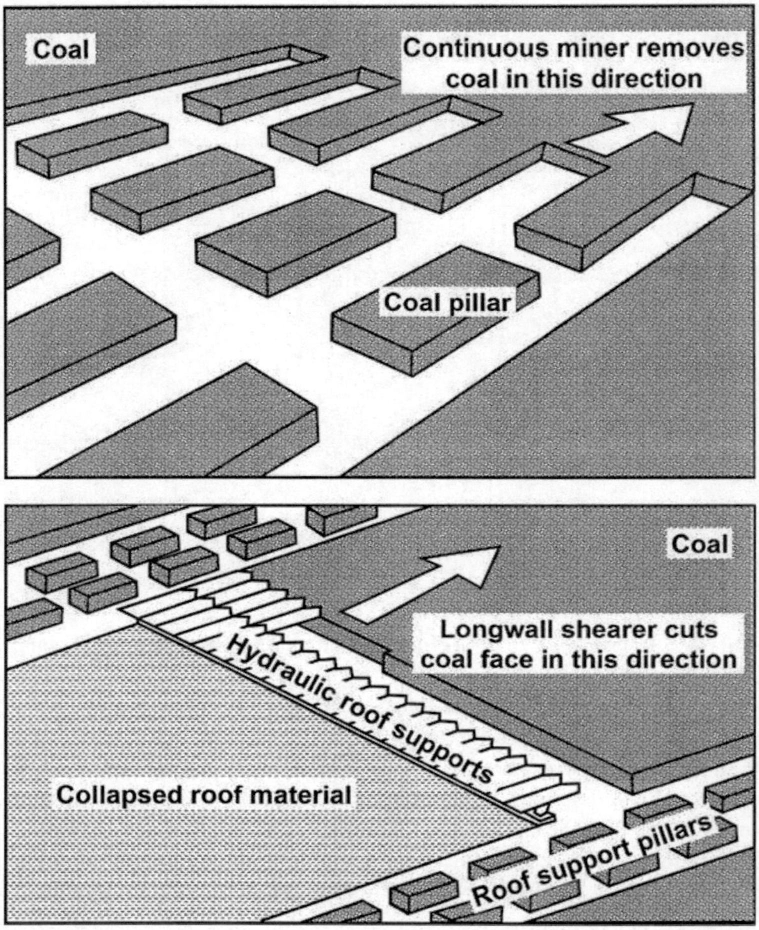

Fig. 2 The two primary underground mining methods used in Appalachia: Room-and-pillar mining (above) and longwall mining (below). Both figures show the coal mining operation as it occurs beneath overlying rock materials, which are not shown. "Pillar Retreat" occurs as a variant of room-and-pillar mining, when the direction of mining is reversed and pillars are partially or fully removed. (Figure by Kathryn Haering)

As the coal is broken, it must be removed from the mine. Early miners loaded coal manually into wheeled carts that were pulled by animals, often dogs or mules, out of the mine. An early innovation was to install iron rails to reduce the coal wagons' rolling resistance. Electric engines were introduced for coal hauling beginning in the 1880s. Beginning in the 1910s, mechanized scoopers began replacing human labor for loading the coal, and belt conveyors were introduced to transport the coal from some mines. Gasoline- and diesel-powered equipment has also been used for coal transport. The 1930s saw the introduction of battery-powered shuttle cars for coal movement within the mine (Garges 2003).

Two equipment innovations greatly enhanced underground coal production. The first was the continuous miner, a machine with a carbide-steel-tipped cutter drum that shears off the coal as small fragments, causing them to fall onto a conveyor that carries it to the machine's back end where it can be accessed for loading and transport. Continuous miners were introduced in 1948 and their use was accelerated by the 1969 Coal Mine Health and Safety Act to replace traditional mining methods that used explosives (Garges 2003). More recently, remote continuous miners have come into use. These machines can be operated by a person standing away from the machine itself, thus reducing operator hazards. Continuous miners can extract coal from seams thinner than the vertical space required for mineworkers by cutting into mine rock immediately above or below the coal seam; that rock is then separated from the coal by coal processing.

Longwall mining was introduced in the USA in 1951 (Garges 2003). A longwall miner employs a shearer that moves back and forth along the face of a coal seam, typically extending over hundreds of meters (Fig. 2 lower). The loosened coal falls onto a conveyor belt for automated transport out of the mine. Longwall mining machines include self-advancing hydraulic roof supports that place a steel barrier over the workspace at the coal face, protecting workers and equipment from falling rock. The longwall mine machine removes coal while advancing from the mine interior toward the entry and allows for "controlled subsidence," the relatively rapid collapse of the overburden into the excavated cavity. Longwall systems greatly increase coal production relative to the number of workers, but their use is restricted to large blocks of relatively thick coal, generally 1.5 m or greater. By the mid-2000s, longwall panels averaged nearly 300 m wide by 3,000 m long (Weisdack and Kvitkovich 2005).

An essential worker-safety concern underground is the need for adequate ventilation. The exchange of interior mine air with fresh air from outside replenishes oxygen, removes gases emitted by the exposed coal including methane and carbon monoxide, and removes airborne particles produced by drilling and fracturing of rocks and coal. Exposure to airborne particles is a worker safety hazard and excessive exposures cause the debilitating health condition known as "black lung" (Almberg et al. 2018). As well as providing breathable air, ventilation reduces the risks of mine explosions that can be caused if a spark or flame ignites airborne methane and combustible coal particles. Early mines relied on natural drafts for ventilation, often placing several mine openings to enable air movement. In the eighteenth and nineteenth centuries, some mines installed a furnace that would cause hot air to rise through a vent to the outside, thus drawing outside air into the mine through an alternate shaft. Some early mines used steam engines to power fans, which were replaced by electric fans in the twentieth century (Reed and Taylor 2007). The mechanization of underground mining, including the use of roof-bolting drills and continuous miners, increased airborne-particle production within the mines and, thus, the need for effective dust control and ventilation (Reed and Taylor 2007). Today, that equipment often includes mechanisms intended to limit airborne dust, such as a water-mist spray apparatus installed on rotating drums of continuous and longwall mining machines. Today's mines are required to provide workers with specified volumes of fresh air, which is accomplished by drawing air through the tunnels and shafts with large fans and by using in-mine air-flow barriers to direct the fresh-drawn air to work areas.

Other essential underground mine functions include subsidence and water control. The term "subsidence" means a lowering or caving of geologic strata and land surface when caused by the excavation of underground support. Underground mines are engineered to prevent subsidence from occurring until after the miners have left the area. When mines extend beneath towns, roadways, streams, and other important surface features, large pillars of intact coal are typically left for support. Subsidence-related surface depressions and cracks can damage buildings and other infrastructure, and sometimes entire blocks of land will subside. Subsidence effects at the land surface are prevalent and obvious where underground mines remove coal from less than 30 m below the land surface elevation. If the underground mine is deeper, subsidence is less obvious at the surface and, if evident, usually occurs as cracks or small openings. At still greater depths depending on the types and thickness of overlying rocks and mine excavation, subsidence effects are often even less evident at the surface (Fig. 3), although underground rock structures and groundwater flows are affected.

Underground mining affects groundwater. Mines located below the water table will become inundated unless waters are drained or pumped from the mine workings. Even above-water-table mines can suffer from water problems, as waters infiltrating the earth above may drain into the mine cavity. Water supplies to homes can be affected when mining interferes with aquifers accessed by household wells (Zipper et al. 1997), while subsidence can harm piping that carries waters from public water supplies to homes if present. Above-drainage underground mines can affect surface waters by collecting groundwaters that drain into the mine cavities and then out of those cavities through discrete discharges to the surface. Depending on the composition and chemistry of the rocks inside the mine, the water emanating from the mine

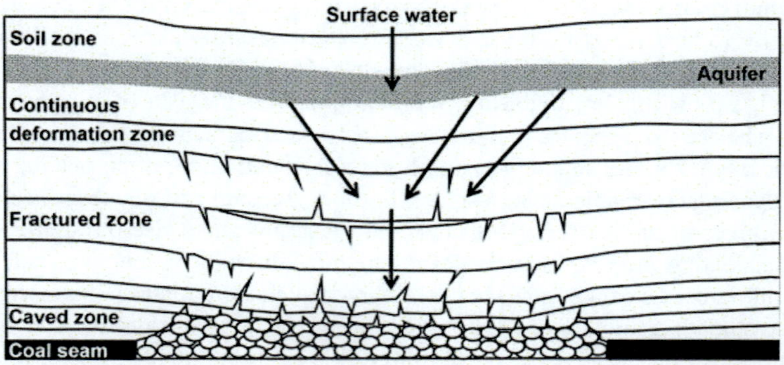

Fig. 3 Effects of subsidence caused by underground coal excavation on overlying geologic strata and groundwater movement (Figure by Kathryn Haering, partially based on Peng and Chiang 1984)

can carry pollutants to surface waters (Mack et al. 2010). Unlike the early miners, modern miners are required by law to control the quality of water being discharged to the outside.

2.2 *Surface Mining*

Surface miners access coal by removing the overburden, which is the soil and rock overlying the coal seam and when excavated is known as mine "spoil." Early surface mines used equipment such as bulldozers to remove geologic material overlying coal seams that outcropped on slopes. After World War II, surface-mine production increased due to technological improvements (Allsman 1968), including the use of rotary drilling and ammonium nitrate fuel-oil explosives for fracturing overburden rock. Dynamite, the predominant blasting method prior to the 1940s, was more dangerous and expensive. Another advance was the shift to larger and more sophisticated excavators, and to haulage trucks with increasing size, mobility, and power. For example, Caterpillar introduced a 100-ton haul truck in 1974, a 150-ton truck in 1984, a 240-ton truck in 1991, and a 400-ton capacity truck in 1998 (Orlemann 2000). Other forms of mining equipment used in Appalachia such as wheel-loaders and dozers also increased in size over this period. As a result, mining firms were able to increase the scale of their operations, and surface mine disturbances of land surface were extended over larger areas.

The "stripping ratio" refers to the ratio of overburden moved versus the amount of coal extracted; as the stripping ratio increases, so does the mining cost per ton of coal. Hence, miners seek to manage stripping ratios based on economic considerations such as equipment operation costs and coal sale prices. The technical advances of surface mining during recent decades enabled increased stripping ratios and, hence, increased surface-mine production and scale. These changes also allowed surface miners to conduct additional mining in areas that had been mined previously with smaller equipment and lower stripping ratios, a process known as "remining."

Contour mining is a common form of Appalachian surface mining that extracts coal from outcrops along the sides of steep hills (Fig. 1), often following the coal seam around the hillside. The overburden above the coal seam is removed and then the coal is removed. Early contour miners simply pushed overburden over the side of the hill downslope of the coal seam, thereby creating a near-vertical highwall above the mine "bench," and a steep outslope of material that had been cast down the slope; this form of mining was known as "shoot and shove" (Fig. 4). Shoot-and-shove miners relied on a rock drill to prepare holes for explosives, and dozers and other equipment to push or dump the loose and fractured earth material over the outslope. Highwalls produced by contour mining at that time were relatively small, usually less than 30 m. The passage of reclamation laws during the 1970s, including the federal Surface Mining Control and Reclamation Act of 1977 (SMCRA), essentially eliminated shoot-and-shove contour mining. Today's contour miners typically backfill highwalls with overburden and restore the mined landscape's approximate

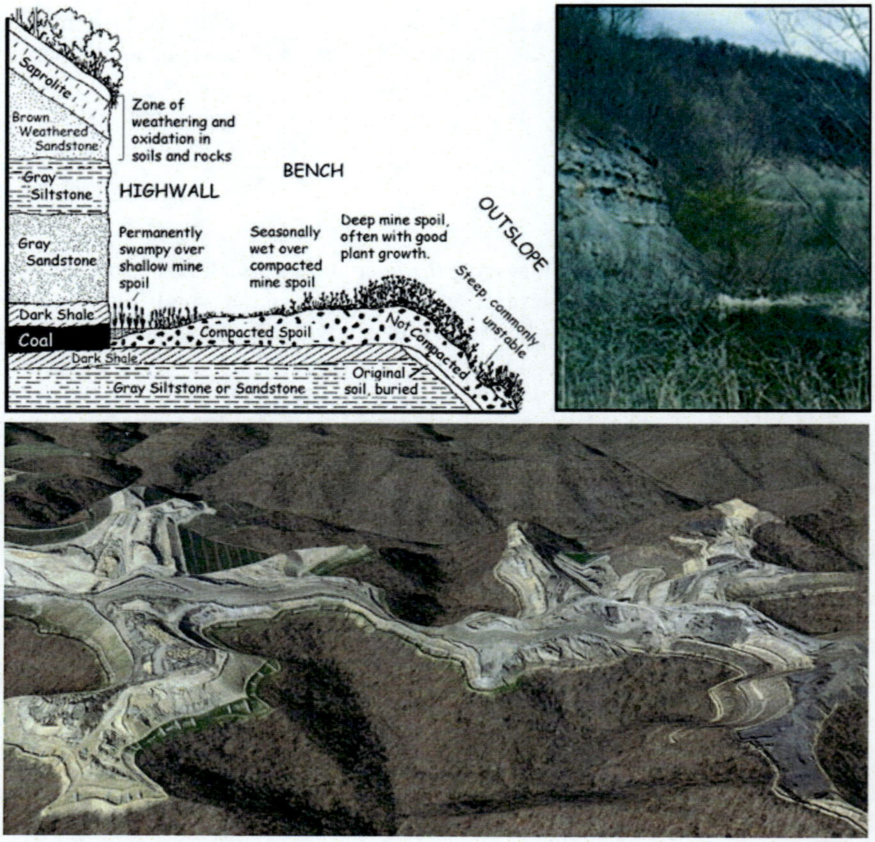

Fig. 4 (Above left) Diagrammatic representation of pre-SMCRA "shoot and shove" contour mining (from Daniels and Zipper 2018; by Kathryn Haering). (Above right) Photo of "shoot and shove" highwall (photo: Jeff Skousen). (Below) A mountaintop removal mining operation in process; horizontal dimension of the area shown is approximately 3.7 km (Image from Google Earth)

original contour (AOC; Fig. 5). Following SMCRA's passage, expanding sizes and capacities of equipment enabled remining of pre-SMCRA mine sites, often by taking additional cuts from the hillsides and covering the pre-SMCRA highwalls with spoils generated by the new mining.

Coal is extracted from less-mountainous areas of Appalachia using area-mining techniques, or in some cases as open-pit mining. These mines access coal by excavating overburden from the land surface and stacking the mine spoil on an adjacent area. The mining operation proceeds in this manner across the coal-bearing landscape, often drilling and fracturing the rock with explosives, and then using large track-mounted shovels known as draglines to move the spoils. Prior to SMCRA, this type of mining commonly produced landscapes comprised of parallel ridges of spoil

Fig. 5 (Top left) A contour mine reclaimed to approximate original contour with grasses and legumes in West Virginia. (Top right) A mountaintop mine reclaimed to approximate original contour with herbaceous and woody vegetation in West Virginia. (Photos by Jeff Skousen)

left by the dragline. Modern area mines, however, replace the spoil in the mined-out pits, and then regrade and revegetate the land surface.

A variant of area mining, known as mountaintop mining (sometimes called mountaintop removal mining), removes an entire ridge or summit of overburden to access underlying coals, often extracting multiple seams while removing mountaintops to depths of 100 meters or more (Fig. 1). This mining method has been practiced since the late 1950s (Plass 1967) but underwent significant expansion during the 1990s for reasons that included the expanded scale of mining equipment and coal processing facilities (Copeland 2015). Historically, mountaintop mines pushed or otherwise placed much of the mine spoils on downslope areas below the lowest coal seam, like shoot-and-shove contour mines. Under SMCRA, overburden can be replaced to rebuild the mountain and achieve approximate original contour or the area may be reclaimed to an alternative landform if the post-mining land use meets the applicable SMCRA requirements. Large areas mined using these methods occur today in central Appalachia and some occur as extensive plateau-like landforms extending over multiple-km^2 areas of coal extraction and adjacent valley fills constructed with spoils removed from the mountaintops (US EPA 2011b; Fig. 4).

Fracturing of intact geologic material causes volumetric expansion. Hence, Appalachian mining operations often produce spoil volumes in excess of what is needed to reconstruct the original contours. Such materials are described as "excess spoil" and must be disposed of by the mining operation. In Appalachia, much of this excess has been placed in valley areas downslope of the actual mining in structures known as valley fills (US EPA 2011b) (Fig. 6). Valley fills are constructed with internal drains that are intended to ensure that environmental waters entering the fill structure can drain through the fill material, with the intent of avoiding saturated-spoil conditions that could lead to geotechnical instability. As such, valley fills exert a disproportionate influence on surface water quality, relative to both unmined areas and to mined areas more generally (Cormier et al. 2013a and b).

Fig. 6 (Left) A valley fill while under construction at a southwestern Virginia surface mine. (Right) A completed valley fill in southern West Virginia. (Images from Google Earth)

3 Coal Processing

Raw coal from the mine typically requires processing to separate rock and similar materials from the marketable coal that is sold to customers. The processed coal must meet purchase contract characteristics such as caloric heat values along with maximum ash and sulfur contents. Such processing is generally required for underground-mine coal and processing is often required for surface-mined coals from seams that incorporate streaks (partings) of non-marketable materials. Surface mines operating in thick seams of higher quality coals are often able to separate coal from associated rock prior to loading. The vast majority of Appalachian coals, however, require processing; hundreds of coal processing operations are located throughout Appalachia (Fig. 7 left).

In early times, coal was processed using physical means, including manual removal of non-coal materials identified by visual inspection and "hand picking." This activity typically occurred near the mine opening. The non-coal material, often called "coal waste," "refuse," "gob," or "slate", was often discarded by leaving it in unmanaged piles on the landscape. The longer the mine operated and the more coal it produced, the further those waste piles extended from the mine opening. In some cases, large piles of gob were left on steep slopes and in an unstable condition. Today, many such piles of discarded materials remain scattered through Appalachian landscapes; while others have been stabilized or removed through the Abandoned Mine Land reclamation program (see below).

Modern coal mines use more sophisticated coal-processing methods. The extraction process from underground mines produces a mix of fractured rock and coal fragments that must be separated prior to shipping the marketable coal. One common processing method, called flotation, relies on material-density differences. The long-chain hydrocarbons that predominate within coal's molecular structure have lower densities than most rock-forming minerals. The raw material produced by the mine

Fig. 7 A coal processing facility adjacent to a rail line (Photo by C. Zipper), and (right) a coal refuse disposal area occupying > 100 ha in southern West Virginia, with some refuse in the lower center of the photo having been covered with soil and revegetated. The dark-colored area in the photo's upper center is a coal slurry impoundment (Image from Google Earth)

is placed into a liquid bath of water mixed with liquid hydrocarbons and other chemical additives such that coal particles remain buoyant while rock fragments sink to the bottom. Physical means such as screening are then used to separate coal fragments from the remaining liquid, which is called coal slurry. Such separations cannot be accomplished with 100% efficiencies, however, as both the waste-rock materials and the slurries produced by coal processing typically retain some coal. The wastes are disposed of by placing them in large landfill-like structures located near the processing plant (Daniels and Stewart 2000). Slurries may be processed further, such that their liquids can be recycled through the processing plant, or they may be disposed of (Aken et al. 2015). Typical methods of disposal include placement in large open-air ponds, often designed as impoundments within the refuse-disposal facilities (Fig. 7) or underground injection into old mine voids (Quaranta et al. 2014). Coal refuse disposal areas vary in size, with some such facilities occupying areas extending over hundreds of ha. Such facilities are subject to regulation by statutes including SMCRA and the Clean Water Act.

4 Reclamation

Reclamation includes landscape reconstruction, mine-site revegetation, and environmental impact mitigation. In a narrow sense, reclamation is not essential to mining since it occurs following mineral extraction; hence, it is possible to extract and sell coal while performing no reclamation. Without reclamation, however, coal mining can cause severe negative effects on environmental resources both on and beyond the mine site. Off-site effects are exacerbated by water and wind, especially in mountainous terrain such as in much of Appalachia. Hence, reclamation is generally seen as essential to societal welfare and, in the modern-day, is required by law to be conducted as mining progresses.

History demonstrates that government regulation is necessary to ensure the reclamation of mine sites. Such is especially the case when reclamation costs exceed the reclaimed land's economic value or when the coal mining firm does not own the land being mined; both are common occurrences in Appalachia. If in the absence of mandated reclamation, some mining firms were to perform reclamation voluntarily while others did not, the reclaiming firms would suffer financial disadvantages in the marketplace.

4.1 Prior to Federal Law

Early in Appalachia's coal mining history, miners often left lands with no effort towards reclamation (e.g. Plass 1967). Hence, as surface coal mining was expanding, the states began passing reclamation laws beginning with West Virginia in 1939, followed by Pennsylvania, Ohio, and Kentucky in the 1940s and 1950s (Imhoff et al. 1976). These laws required miners to register with the state and to apply for and obtain mining permits. Reclamation requirements varied among states but typically required covering excavated areas with soil or overburden, and grading and revegetating the land surface (Plass 2000). In some states, miners were required to post performance bonds to ensure reclamation.

By the mid-1960s, approximately 3600 km^2 had been surface-mined for coal in the seven Appalachian states extending from Tennessee north to Pennsylvania and Ohio; but > 60% of those mined areas were described as having experienced reclamation that was not "adequate" to mitigate negative effects (USDI 1969). Surface mining for coal was expanding rapidly during that era. By the late-1970's, the land affected by surface mining in those same seven states was estimated at 9300 km^2, most of which had been mined for coal (Johnson and Paone 1981).

Most of the surface mining prior to federal law in central Appalachia was shoot-and-shove contour mining (USDI 1969). Some of the mine sites produced during that era became naturally revegetated over time. Plant communities on sites with no human-induced revegetation, favorable soils, and stable landforms developed productive and diverse plant communities with multiple native species (Skousen et al. 2006). Similarly, where native trees were planted on the older mine sites with favorable soils, they often survived and grew well (Gorman et al. 2001; Skousen et al. 1990; Rodrigue and Burger 2004). Such results occurred when mine spoils from the upper 10–20 m of overburden were left on the land surface. Such spoil materials had been weathered by environmental processes, and thus had become more favorable to native plants and forest trees relative to unweathered spoils originating from greater depths (Zipper et al. 2013). Also, spoils from early surface mining were often left in an uncompacted condition more favorable to native vegetation growth than the compacted mine soils constructed immediately following SMCRA (Angel et al. 2005). However, soil conditions on the older mine sites were not always favorable, especially when mining left acidic strata exposed at the surface. Hence, state-level reclamation laws generally mandated revegetation and miners generally seeded non-native plants that could colonize a wide variety of mine soil materials (Brenner et al. 1984; Wade et al. 1985).

Pre-SMCRA mining, in some cases, created environmental hazards. The mine spoil materials pushed over the outslope in some cases remained in an unstable condition, prone to landslide. Such conditions were exacerbated by the steep slopes of central Appalachia and in locations where the spoils were prone to saturation with environmental waters, such as upland valleys. By the mid-1960s, approximately 32,000 km of contour-mine highwall had been created, with more than 3,000 km of outslope characterized by "massive slides" (USDI 1969). Where spoils unfavorable to vegetation were exposed by mining, they often revegetated slowly, if at all (Skousen et al. 2021). A common condition hindering revegetation occurred when mining processes turned the land "inside out" by leaving spoils originating from close to the coal seam on the land surface. These materials were the last materials to be excavated prior to coal removal and were often elevated in acidic and coal-like minerals with poor revegetation potentials. This condition occurred as a consequence of mining when no reclamation was performed. The unstable land surfaces on steep outslopes below contour-mine benches also created revegetation difficulties.

Other major problems with pre-SMCRA mining concerned water. Results of poor revegetation included excessive erosion and water-borne movement of the exposed mineral material into streams, in some cases clogging those streams and causing local flooding (USDI 1969). Other forms of water pollution occurred when sulfur-bearing spoils were excavated and exposed to air and water, causing acid drainage and mobilizing acid-soluble trace metals (Kruse Daniels et al. 2021). A 1965 survey found that more than 60% of 318 sampled Appalachian coalfield streams were affected by mining-origin sulfate, acidity, and other pollutants (Biesecker and George 1966). A 1990 study found 10,000 km of streams in the Appalachian region to have been strongly affected by acid mine drainage, mostly from pre-SMCRA mining (Herlihy et al. 1990). Underground mining disrupted aquifers, often causing loss of water in household wells (NRC 1981).

By the1970s, all Appalachian states had established reclamation laws and required revegetation, but reclamation requirements varied among states. In retrospect, when considering the eventual federal statute (SMCRA), coal mining environmental controls during the era of state regulation were often considered as inadequate (e.g. Rochow, 1979, wrote of "… the disastrous consequences resulting from the historic failure of the states to regulate coal mining effectively"). Part of this effect can be attributed to the fact that mineable coal deposits transcend state boundaries, which caused individual states to avoid establishing strict regulatory programs that might incentivize mining firms to re-locate their economic activity into other states where reclamation was less costly.

4.2 Current Reclamation Laws and Practices

The Surface Mining Control and Reclamation Act (SMCRA) of 1977 established federal control over coal-mine reclamation with a federal-state oversight system. State agencies can implement SMCRA under the oversight of a federal agency, the USDI Office of Surface Mining Reclamation and Enforcement (OSM). States choosing to enforce SMCRA are said to have "primacy" and such states establish state-level laws and regulations that are consistent with federal standards but need not be identical and must be no less strict.

Under SMCRA, firms must apply for and obtain a federal permit prior to mining coal. The permit specifies practices to comply with performance standards that require miners to establish vegetation adequate to control erosion, to produce reclaimed lands that are stable geotechnically, and to reclaim surface mines to a suitable approved post-mining land use. Reclaimed lands must also provide "adequate drainage" and must lack exposure of acid-forming or other "toxic" geologic materials that would interfere with revegetation or impair water quality.

The SMCRA also requires miners to file with the regulatory agency a performance bond to ensure the agency's ability to execute mandated reclamation should the firm fail to do so. This may be achieved by placing financial assets under temporary control of the agency prior to land disturbance (Cheng and Skousen 2017). However, alternative forms of bonding are often employed. In some cases, the mining firm may provide financial assurances, such as a surety bond provided by an insurance firm; or, if the mining firm is judged by the agency to be financially sound, it may pledge of its own corporate assets in a practice known as "self bonding." In some states, firms are able to participate in a risk-based bonding approach called "bond pools" by paying partial bond amounts into a pool of resources that would support reclamation should any pool participant fail to execute reclamation obligations.

In addition to SMCRA, other legal requirements govern coal mining and reclamation in the US. All waters leaving the mine site must satisfy water-quality standards established under the Clean Water Act. Similarly, mining-induced stream disturbances such as those caused by valley fills must comply with Clean Water Act requirements for dredging and filling of streams. When protected species or their habitat are present within the permit or potentially affected waters, mining and reclamation must comply with the Endangered Species Act.

4.3 Reclaiming Surface Mines to Achieve Post-mining Land Use

Under SMCRA, surface miners must reclaim the land such that it can support post-mining land uses similar to that preceding mining, or higher or better use. The post-mining land use must also be "ecologically sound" and "compatible with the surrounding region." Most land mined for coal in central Appalachia supported forest or agriculture prior to mining; hence, common post-mining land uses under SMCRA are wildlife habitat, hay and pastureland, and forestland (US GAO 2009). In some cases, mined lands have been reclaimed to serve public uses, such as residential and industrial uses, parklands, lands used for hunting and fishing, and recreational areas such as parks and ball fields. Post-mining land uses that require surface development, such as industrial, commercial, and residential, are generally considered to be "higher and better" than extensive land uses such as agriculture, forestry, or wildlife habitat. Land uses that require management, such as hayland or pasture, are considered "higher and better" than forestry and wildlife land uses.

Hayland-Pasture: Hayland and pasture are common post-mining land uses, especially in areas with gently sloping terrain favorable to agriculture. Haylands may be established on near-flat or gently sloping lands that allow management and harvesting with agricultural equipment. Pastures can be established on lands that are steeper but not so severe as to prevent fertilization, liming, and other management such as invasive vegetation control. Pastureland must also enable livestock access to water. Once the post-mining landform has been determined, reclamation practices required for these uses start with the construction of a mine soil suitable for forage and hay production. Suitable materials for mine soil construction include salvaged topsoil or spoils that lack excessive coarse rock. Such materials may include unweathered spoils from alkaline geologic strata, since hay and forage grasses can be productive on slightly alkaline or circumneutral mine soils, or they may include moderately acidic materials such as weathered spoils that lack excessive acid-forming minerals (Daniels et al. 2016). Once placed, the mine soil materials are graded and then fertilized, limed, and revegetated, typically with a hydroseeder (Skousen and Zipper 2018).

Fertilization, liming, and overseeding are necessary activities for pasture management, especially if topsoil-substitute rock materials were used to construct the mine soils. When managed competently, many reclaimed mine areas can support hay crops and livestock forages (Ditsch and Collins 2000).

Forest: Forests are native vegetation throughout the Appalachian coalfield and are commonly displaced by mining (Wickham et al. 2007; Drummond and Loveland 2010). Reestablishing forest ecosystems after mining creates societal value by producing "ecosystem services" such as carbon storage, water quality protection, and water flow regulation (Zipper et al. 2011b). In recent years, regulatory agencies have encouraged a reclamation method known as the Forestry Reclamation Approach (FRA) which enables the reestablishment of forests after mining (Burger et al. 2005; Zipper et al. 2011b). Practices suitable for hayland-pasture reclamation are not adequate for restoring forests (Angel et al. 2005), and many mining firms have either fully or partially applied the FRA in reclamation.

Forest vegetation grows on a variety of unmined landforms, including those that are steeply sloping. Hence, forests can be restored after mining on almost any landform, but suitable mine soils are essential for effective reforestation. Mine soils comprised of weathered spoils and/or salvaged natural soils often have chemical properties suitable for forest re-establishment (Skousen et al. 2011; Zipper et al. 2013). In order to ensure favorable physical properties, such mine soils must be constructed with adequate depth (generally > 1.2 m) while ensuring minimal compaction by mining equipment; hence, they should be graded minimally, if at all. After soils are placed, they are seeded with tree-compatible herbaceous plants, species that are not fast-growing and competitive but will establish vegetative groundcover to control erosion (Burger et al. 2009; Franklin et al. 2012) and planted with seedlings of multiple native-tree species (Davis et al. 2010, 2012).

Mine soils that will aid planted tree survival and growth are essential to mine reforestation. Research demonstrates that unweathered alkaline spoils hinder growth of planted trees relative to growth rates achieved on mine sites reclaimed with weathered spoils and in natural forests (Zipper et al. 2013). As the trees grow, they establish canopy shading that can aid in restricting site occupation by sun-loving invasive plants (Zipper et al. 2019). Proper mine soils, with properties similar to those of native forests and with rough-graded and uneven surfaces will also aid germination and establishment by native plants that enter the area as seed. The fast-growing trees supported by suitable mine soils will also aid recruitment of native forest plants by producing mast (e.g., nuts and fruit) to attract seed-carrying wildlife and providing perches for seed-carrying birds. Reclamation procedures that aid native species establishment are essential to forest re-establishment since native forests support far more species than are typically seeded and planted by miners (Burger et al. 2005; Zipper et al. 2011b).

Wildlife Habitat: Mine sites are best prepared for wildlife habitat post-mining land uses using procedures similar to those used to re-establish forests but by planting trees and shrubs tailored for the intended post-mining habitat (Wood et al. 2013). A wildlife specialist may be engaged to design a habitat plan that incorporates plant species for specific wildlife-habitat purposes. Such plans may include blocks of vegetation intended for use by certain birds or mammals. In some cases, those plans include corridors of vegetation that enable access by target species to or through the mined area. Since woody vegetation is a primary component of wildlife habitat, FRA techniques are often used to reclaim mined areas for wildlife habitat. Wildlife habitat plans often incorporate water sources, such as impoundments placed in small valleys or, when possible, flowing streams.

Other Post-Mining Land Uses: The majority of Appalachian mined lands have been reclaimed to support hayland and pasture, unmanaged forest, and wildlife habitat as post-mining land uses, but other uses are possible. For example, research demonstrates that herbaceous crops, such as switchgrass (*Panicum* virgatum) and miscanthus (*Miscanthus x gigantea*), and fast-growing woody crops suitable for use as biofuels are possible on mine sites (Brinks et al. 2011; Caterino et al. 2017; Marra et al. 2013; Nobert et al. 2016; Scagline-Mellor et al. 2018). Due to a lack of biomass markets, such post-mining land uses are rarely if ever implemented by active miners.

There has been some activity in converting mine sites to developed post-mining land uses, such as commercial and industrial sites or recreational areas. Such activity can occur when the mining area is located suitably for such use and with favorable terrain such that building areas can be developed. When such post-mining land uses are planned with high-value buildings, reclamation procedures must ensure that stable building-support surfaces are created. This can be achieved by uniformly and deeply compacting the mine-spoil fill materials during placement, which are costly procedures (Zipper and Winter 2009). Other factors influencing a mine site's suitability for high-value development include access to public roads, water, utilities, and waste disposal.

A study of eastern Kentucky and West Virginia mine permits through 2008 found that < 10% of permits specified post-mining land uses other than wildlife habitat, forest, or hayland and pasture for any part of the disturbed area (US GAO 2009).

4.4 Other Reclamation

Underground Mine Openings: Underground mines affect land surfaces directly at mine openings, including those used for worker access, coal removal, and air exchange. A common method of accessing underground coal is to take a cut from the mountainside to create a flat bench, and to tunnel into the mountainside at that location. A working area is created in areas adjacent to the mine opening (Fig. 8). Such working areas and access roads are permitted under SMCRA.

The working area and access roads that will be taken out of use are reclaimed after mining is complete. Reclamation operations include equipment removal, closure, and covering of the mine opening with soil and/or rock material, and covering highwall-like areas with additional rocks and soils. Following these operations, the disturbed areas are prepared for revegetation by removing any remaining excess coal or similar materials, applying topsoil in some cases, loosening exposed soil areas that have become compacted, and revegetating with herbaceous species via hydroseeding.

Coal Refuse: SMCRA regulations require that coal-refuse piles be constructed in a manner that is geotechnically stable, similar to surface mines. Once refuse disposal has been completed within a given area, SMCRA requires that the area be revegetated. SMCRA regulations require that exposed refuse be covered with at least

Fig. 8 An underground mine opening (upper left) and supporting facilities, including water retention pond (lower left) and coal stockpile (right). (Image from Google Earth)

1.2 m of non-toxic plant-growth material, such as suitable mine spoil or topsoil, in most cases. However, facility operators can obtain a variance from this requirement if able to demonstrate that the refuse materials are able to support adequate vegetation with a reduced thickness of soil cover, or with no soil cover if adequate vegetation can establish and grow on the refuse materials themselves (Daniels et al. 2010).

Coal refuse varies widely in physical and chemical properties influencing revegetation potential, such as fragment-size distribution and acidity (Stewart and Daniels 1992). Most refuse materials, when fresh, have characteristics that hinder vegetation, including dark-colored surfaces that become hot when exposed to sunlight, high salinity, and low contents of plant-growth nutrients. Exposure to the ambient environment and consequent weathering, however, improves the refuse materials' suitability for plants (Daniels et al. 2010).

Coal refuse reclamation can be costly because of the need to obtain, transport, and apply soil-cover materials in quantities sufficient for revegetation of their large areas. A common strategy is to construct refuse fills from the bottom up and to revegetate the lower slopes as the disposal fill expands upwards. As an example, the bottom section of the refuse fill shown in Fig. 7 (right, lower center of photo) has been revegetated, even as the upper segment of the fill remains accessible and available for additional disposal.

5 Water Management

Land disturbances by coal mining influence surface waters draining from the mine site and, by extension, streams, and rivers receiving those waters. Rainwater infiltrates or runs over the mined-land surface. Groundwater from adjacent unmined areas may enter mining excavations and interact chemically with the mine spoils. Infiltrating rainwaters and groundwaters may drain into the subsurface voids created by underground mining and those waters must be discharged from the mine void while mining is occurring. Such discharges may continue after mining if the mine voids are higher in elevation than local streams. Mined-land structures such as valley fills are often constructed with drains intended to convey internal waters to discharge at the surface. After mining is completed, environmental waters continue to interact chemically with mining-disturbed materials. Hence, mining can influence both the quantity and the quality of waters entering surface-water streams both during and following mining.

In the USA, influences by coal mining on water resources are regulated under federal laws. The SMCRA requires that both surface mining and coal-refuse leachate "will not degrade" the water quality below standards established by state and federal law, including the Clean Water Act. All regulated mining is required by SMCRA to minimize "toxic" drainages and adverse effects on fish, wildlife, and environmental values. The SMCRA also requires regulatory agencies to "prevent material damage to the hydrologic balance outside the permit area" as they execute permitting responsibilities, and it requires both surface and underground miners to operate in a

Table 1 Technology-based effluent limitations applied to active coal mining operations and post-mining reclamation areas by Clean Water Act regulations (40 CFR 434).†

Pollutant, or discharge property	Maximum for any one day	Maximum average daily values—30 consecutive days
Iron, total		
Post 5/1984 sources	6.0 mg/L	3.0 mg/L
Pre 5/1984 sources	7.0 mg/L	3.5 mg/L
Manganese, total	4.0 mg/L	2.0 mg/L
Total suspended solids ‡	70.0 mg/L	35.0 mg/L
Settleable solids ‡	0.5 ml/L	n/a
pH	§	§

† Not all standards are applied under all conditions; alternate effluent limitations may be established and/or applied under conditions defined by Clean Water Act
‡ Applies only to post-mining areas, "not to be exceeded."
§ Within the range of 6.0–9.0 at all times

manner that will "minimize the disturbances to the prevailing hydrologic balance" both on and off the mine site. Whereas current law prohibits underground mines from causing subsidence that will influence above-ground streams, pre-1970s mining in some cases had such effects. Similarly, numerous surface-water streams have been impacted by surface mining, as miners may excavate, redirect while mining, and then reconstruct stream channels. Also, surface miners may cover stream beds with mine spoils while constructing valley fills.

Water Quality Protection: All waters leaving active mine sites pass through sedimentation ponds that are intended to limit the suspended solids in discharge waters. Ponds are sized in light of anticipated pumping rates for underground mines, and for areas of disturbed land contributing to each discharge point for surface mines. Waters discharged from pond outflows at permit boundaries are regulated as point-sources. At a minimum, such waters must achieve the effluent-limitation standards described by Table 1, although more stringent conditions may be applied when necessary to assure Clean Water Act compliance by the receiving stream.

Acidity and Related Pollutants: Primary water-quality concerns for coal mining operations involve acidity. The SMCRA, for example, includes multiple clauses that require control or limitation of "toxic" water pollutants. Such phrasing refers to acid waters and to acid-soluble metals that may be mobilized by such waters. The primary method used by modern surface mines to control acidity is overburden management. By law, miners are required to characterize the geologic materials prior to mining. This is achieved by conducting test borings through the geologic materials to be disturbed and testing each stratigraphic layer for acid-production and acid-neutralization potential. The standard method of testing such materials, acid-base accounting (Skousen 2017), characterizes each rock stratum for potential to produce acidity via oxidation of reduced-sulfur minerals and potential to produce alkalinity through mineral reactions such as dissolution of carbonates. Surface miners manage spoils with the intent of ensuring that acid production within any given structure or area is offset by alkalinity sufficient to neutralize that acidity. Should the mine produce acidic drainage despite such precautions, water-treatment procedures may be employed to neutralize acidity and retain acid-soluble metals. Such treatment may consist of neutralization with an alkaline chemical reagent, i.e., "active" treatment, or by "passive" treatment utilizing geochemical, biological, and/or gravitational processes such as constructed wetlands (Kruse Daniels et al. 2021).

Under today's regulations, acid-drainage treatment is considered a temporary measure to be employed during active disturbance only, with the presumption that mine reclamation and closure will cause acid-production to cease. This can occur if miners identify acid-producing strata prior to mining, and reclaim the disturbed areas in a manner that isolates those strata from hydrologic processes and/or neutralizes whatever acidity they produce. When pre-mine testing determines potential overburden materials contain reduced-sulfur minerals but lack offsetting alkalinity, today's regulations prevent mining from occurring unless those materials can be isolated to prevent acid production. If the quality of waters draining from a reclaimed mine site fails to satisfy regulatory standards, the mining operator must continue water treatment. Miners are released from mine-permit obligations only when waters leaving the site satisfy regulatory standards without treatment. In past years, however, much mining occurred in net-acidic strata without being subjected to today's regulatory controls, and acid-mine-drainage waters are common in some parts of Appalachia (Herlihy et al. 1990).

Non-Acid Water Contaminants: The primary non-acid contaminants of concern are elevated mining-induced salinity, often called total dissolved solids or TDS, and measured by proxy as water conductivity, and the trace-element selenium (Se; Clark et al. 2021).

Aquatic biota in streams with elevated TDS originating from mining often occur in an altered condition, relative to low-TDS streams in non-mining areas, with reduced richness and diversity observed with salinity levels as low as 300–500 $\mu S\ cm^{-1}$ (Pond et al. 2008; US EPA 2011a; Cormier et al. 2013a; Timpano et al. 2018). Findings of reduced aquatic diversity in high-TDS streams are sufficiently common to suggest that elevated TDS is causing biotic impairments (Merovich et al. 2021). Pre-mine testing procedures are available for TDS (Daniels et al. 2016), indicating a potential for mining operators to manage TDS using techniques similar to those used for acidity, i.e. to identify high-TDS strata via pre-mining testing and to isolate those strata during reclamation. Such procedures are not employed commonly, perhaps because TDS in mine-water discharges is not currently subject to federal regulation. Experimental efforts to achieve such outcomes have achieved modest success, with discharge water conductivity below levels that would otherwise be expected but still high (> 1000 $\mu S\ cm^{-1}$) relative to levels associated with biological effects (Hester et al. 2019). Efforts to control TDS through mine-spoil handling and placement are limited by the prevalence in the Appalachian coalfield of unweathered coal-bearing strata prone to relatively high levels of TDS production (Daniels et al. 2016).

The water contaminant Se also occurs commonly in mining-influenced Appalachian waters, in some cases at levels of regulatory concern and with the potential to impair biota (Lindberg et al. 2011; Arnold et al. 2014; US EPA 2016; Whitmore et al. 2018). Like TDS, Se-release potentials vary widely among spoils (Clark et al. 2018) and suggest high-Se strata identification and isolation as a means for potential control. Selenium in mine-water discharges may also be controlled by water-treatment methods (Kapoor et al. 2007; Skousen et al. 2017).

Stream Reconstruction: Surface coal mining operations commonly mine through or place excess spoils in valleys occupied by streams. Because Appalachian mining often occurs in mountainous areas of high relief, streams disturbed or filled by mining are often low-order or headwater streams, and in some cases flow intermittently. The Clean Water Act allows such impacts if appropriate compensatory mitigation is employed to avoid loss of stream-ecosystem structure and function regionally. Hence, permits authorizing in-stream mining or filling usually require either reconstruction of the stream that is subject to disturbance or restoration of stream function at another location within the same ecological region. Off-site restoration often takes place on coal-mined areas with water channels that were reclaimed prior to the early 2000s and requirements for compensatory mitigation. Research demonstrates that such reconstructions do carry out natural-stream functions, but often at a reduced level relative to unmined streams (Northington et al. 2011; Petty et al. 2013; Krenz et al. 2016, 2018).

6 Air Management

Air-quality management for the ambient environment does not receive great emphasis on most mine sites, but airborne dust emissions from Appalachian mines have come to public interest recently because of potential public-health implications (Kurth et al. 2014; Nichols et al. 2015).

On surface mines, unpaved roads are commonly watered, oiled, or otherwise treated to reduce airborne particulate emissions by vehicular travel. Other sources of airborne dust from mine sites may include the drilling, blasting, and spoil-movement operations required for overburden excavation including coal loading, hauling, and dumping; and wind erosion from unvegetated areas and from coal stockpiles. The SMCRA requires mined areas to be reclaimed "as contemporaneously" as possible or practicable, a practice that can be expected to reduce the potential for wind-borne particle emissions; but we are aware of no studies that have documented the degree or timing of such effects. Similarly, we are aware of no studies that have documented volumes, characteristics, or fate of airborne particles discharged to the atmosphere by underground mine ventilation.

7 Abandoned Mines

Lands mined prior to SMCRA and left with inadequate reclamation are known as abandoned mine lands (AML). The SMCRA defines AML as lands and associated waters that were "adversely affected by past coal mining." The vast majority of AML are in Appalachia and they include a variety of resources and landscape conditions. The AML are classified for reclamation priority: Priority 1 and 2 AML features are hazardous to the public health and safety, whereas Priority 3 sites are those that create environmental hazards only.

The SMCRA established a tax on active coal mines to generate revenues for reclamation of AML sites. The OSM administers the funds, distributing resources to the states in response to authorizations by the federal budget. States achieve reclamation of AML sites by issuing contracts to private sector firms. Reclamation of AML is also achieved when mining firms obtain permits for new coal mining on previously mined areas with AML features (called remining). With modern mining equipment, it is sometimes possible to extract additional coal from such areas while eliminating AML features and hazards such as degraded water quality and unstable slopes.

Much progress has been made in reclaiming AML under SMCRA. Thousands of hazardous pre-SMCRA abandoned mine features have been eliminated as a result of the AML program (Table 2). Common forms of hazards eliminated due to AML fund expenditures include openings to abandoned underground mines, unstable and combustion-prone coal-refuse piles, and coal processing facilities abandoned by now-bankrupt companies as well as abandoned surface-mine features such as unstable highwalls and outslopes. Among the significant effects of the AML program has been the provision of public water to numerous residents whose home water supplies were damaged by pre-SMCRA underground mining. Despite the expenditure of nearly $3 billion for AML reclamation over more than four decades, numerous public health, public safety, and environmental problems remain as a result of pre-SMCRA coal mining in Appalachia (Table 2).

8 Outcomes of Coal Mining and Reclamation in Appalachia

Extensive coal mining has left its mark in Appalachia, despite reclamation efforts. Coal mining has disturbed extensive areas, both prior to and since SMCRA's passage in 1977. Surface coal mining disturbed approximately 9300 km^2 from 1930 to 1980 in the United States, with much of that disturbance in Appalachia (Johnson and Paone 1981; Plass 2000). Despite extensive reclamation of AML from that era enabled by SMCRA, numerous hazards to human safety and the environment, resulting from pre-SMCRA coal mining remain evident (Table 2).

Since SMCRA's passage, approximately 10,000 km^2 (data from US OSMRE 2012, 2018), about 6.5% of the > 150,000 km^2 Appalachian coalfield area, have been disturbed by surface coal mining, including remining of pre-SMCRA mined

Table 2 Abandoned mine land inventory totals for seven Appalachian states (Tennessee to Pennsylvania)

Problem type	Problems addressed	Costs to date ($ million)	Problems remaining	Costs remaining ($ million)
Hazardous and other problematic highwalls	920 km	487	1,518 km	872
Dangerous piles, embankments, slides	7,721 ha	754	6,376 ha	365
Other spoil and refuse problems	7,456 ha	203	10,978 ha	225
Fires, underground and surface	4,048 ha	198	1,584 ha	861
Underground mine openings	2,906	110	8,436	81
Subsidence problem areas	1,750 ha	268	1,826 ha	262
Clogged streams	2,495 km	63	8,659 km	58
Hazardous and polluted waters	10,718	105	4,025	101
Damaged water sources	9,852	524	5,209	2,974
Other problems	2,345	48	1,942	29
Total		2,759		5,829

Data are for AML features endangering public health and safety (Priority 1 and 2), and for AML-related environmental problem areas (Priority 3) in direct association with Priority 1 and 2 problems only as of December 2019.†

† *Data source* US OSM, https://amlis.osmre.gov/Summaries.aspx

areas. Most post-SMCRA lands have been reclaimed to support hayland-pasture, wildlife habitat, and forest post-mining land uses. Little study has been performed to determine how well SMCRA's goal of reclaiming land to post-mining uses with capabilities "equal to or better than" those preceding mining have been met. Some post-SMCRA lands are supporting viable agricultural enterprises (Ditsch and Collins 2000), but the frequency and extent of such areas are not known. It is known that many areas surface-mined for coal prior to the early 2000s were reclaimed to a nominal hayland-pasture use but were not put in use for that purpose (Torbert and Burger 2000). Landforms of areas surface-mined for coal are often altered by the creation of valley fills and by the reconstruction of contours varying from the original landscape (Ross et al. 2016). Coal refuse disposal fills are also common features in the Appalachian coalfield and one study identified > 200 refuse fills with slurry impoundments were active as of the year 2000 (Greenberg 2017), while another identified 1,821 spoil fills, and 270 refuse fills, occupying a combined area of 23,000 ha in West Virginia alone (Shank 2010).

The Appalachian region's natural vegetation is forest but few mined lands were reclaimed to support forest vegetation during SMCRA's early decades (Angel et al. 2005). A 2005 survey of 25 post-SMCRA reclaimed areas found only two with forest-like vegetation, none in managed use, and more than 50% of the areas sampled occupied by exotic invasive plants (Zipper et al. 2011a). Another study found that 13% of Virginia areas mined and reclaimed since the early 1980s were covered by the exotic invasive shrub Autumn olive (*Elaeagnus umbellata*) (Oliphant et al. 2017). Since the mid-2000s when the Forestry Reclamation Approach was described (Burger et al. 2005), reforestation has been widely applied by industry. Multiple experimental studies have shown growth and productivity of native forest trees on mine soils, with the level of productivity based on mine soil material, compaction, and competition from other invading species (Dallaire and Skousen, 2019; Zipper et al. 2013). Few if any studies, however, have been conducted to determine the extent to which mining firms are constructing mine soils for the forest as recommended, despite the more extensive availability of unweathered spoils on most modern mine sites, or are achieving forest-tree growth rates consistent with pre-mining forest productivity.

Water resources are also widely influenced by Appalachian mining. Acid drainage was formerly a major problem throughout the region (USDI 1969). Although modern mining practices and regulatory policies control acid mine drainage more effectively today than in earlier times, acid drainage from older mines continues to affect water resources in some areas, especially northern Appalachia (Skousen et al. 2019). Elevated TDS and selenium are also mining-related water quality problems, as multiple studies demonstrate influence by these mining-origin constituents in Appalachian waters. If Appalachian coal mining continues at current levels or declines, influence by these pollutants on regional water resources will likely dissipate over time (Daniels et al. 2016; Clark et al. 2018) but is likely to extend to multiple decades (Evans et al. 2014). Within central Appalachia, more than 1900 km of headwater streams, approximately 4% of regional first- and second-order headwater streams, were covered or otherwise lost to surface coal mining prior to 2002 (US EPA 2011b), while > 1300 km of streams were affected similarly prior to 2010 within West Virginia alone (Shank 2010).

Underground mining effects are also extensive but are far-less studied relative to surface mining. Underground mines have produced approximately three times the coal tonnage that has come from surface mines (Zipper 2020), suggesting that the extent of areas undermined far exceeds surface-mining disturbances. Subsidence induced by underground mining can affect surface features such as buildings, roads, and streams and has done so historically, although in recent years it appears that those effects have been well controlled and are in decline. Underground mines can also affect groundwater hydrology, water yields of household wells, and groundwater chemical quality (Hobba 1993; Zipper et al. 1997; Booth 2006; Winters and Capo 2004); while also affecting surface water quality and flows (Lambert et al. 2004; Mack et al. 2010; Cravotta et al. 2014).

References

Aken BV, Quaranta JD, Mack B, Yu H, Ducatman A, Ziemkiewicz P (2015) Environmental contaminants in coal slurry intended for underground injection in the state of West Virginia. J Environ Eng 141:05014004. https://doi.org/10.1061/(ASCE)EE.1943-7870.0000874

Allsman P (1968) Current and future status of surface mining. Chapter 1. In: Surface mining. Am Insti Min Metall Petro Eng, New York

Almberg KS, Halldin CN, Blackley DJ, Laney AS, Storey E et al (2018) Progressive massive fibrosis resurgence identified in U.S. coal miners filing for black lung benefits, 1970–2016. Annals Amer Thoracic Soc 15:1420–1426. https://doi.org/10.1513/AnnalsATS.201804-261OC

Angel P, Davis V, Burger J, Graves D, Zipper C (2005) The Appalachian regional reforestation initiative. US Office of Surface Mining, Appalachian Regional Reforestation Initiative, Forest Reclamation Advisory Number 1

Arnold MC, Lindberg TT, Liu YT, Porter KA, Hsu-Kim H et al (2014) Bioaccumulation and speciation of selenium in fish and insects collected from a mountaintop removal coal mining-impacted stream in West Virginia. Ecotoxicology 23:929–938. https://doi.org/10.1007/s10646-014-1236-4

Biesecker JE, George JR (1966) Stream quality in Appalachia as related to coal-mine drainage, 1965. US Geol Surv Circular 526

Booth CJ (2006) Groundwater as an environmental constraint of longwall coal mining. Environ Geol 49:796–803. https://doi.org/10.1007/s00254-006-0173-9

Brenner FJ, Werner M, Pike J (1984) Ecosystem development and natural succession in surface coal mine reclamation. Environ Geochem Health 6:10–22. https://doi.org/10.1007/BF02072661

Brinks J, Lhotka J, Barton C, Warner R, Agouridis C (2011) Effects of fertilization and irrigation on American sycamore and black locust planted on a reclaimed surface mine in Appalachia. Forest Ecol Manage 261:640–648. https://doi.org/10.1016/j.foreco.2010.11.018

Burger J, Graves D, Angel P, Davis V, Zipper C (2005) The forestry reclamation approach. US Office of Surface Mining, Appalachian Regional Reforestation Initiative Advisory 2

Burger J, Davis V, Franklin J, Zipper C, Skousen J et al. (2009) Tree-compatible groundcovers for reforestation and erosion control. US Office of Surface Mining, Appalachian Regional Reforestation Initiative Advisory 6

Caterino B, Schuler J, Grushecky S, Skousen J (2017) Surface mine to biomass farm: growing shrub willow (*Salix* spp.) in northeastern West Virginia–first year results. J Am Soc Mining Recl 6(1):1–14. https://doi.org/10.21000/jasmr17010001

Cheng L, Skousen J (2017) Comparison of international mine reclamation bonding systems with recommendations for China. Intl J Coal Sci Technol 4:67–79. https://doi.org/10.1007/s40789-017-0164-3

Clark EV, Zipper CE, Daniels WL, Keefe MJ (2018) Appalachian coal mine spoil elemental release patterns and depletion. Appl Geochem 98:109–120. https://doi.org/10.1016/j.apgeochem.2018.09.016

Clark EV, Zipper CE, Soucek DJ, Daniels WL (2021) Contaminants in Appalachian water resources generated by non-acid-forming coal-mining materials. Chapter 9. In: Zipper CE, Skousen JG (eds) Appalachia's coal-mined landscapes. Springer

Copeland C (2015) Mountaintop mining: background on current controversies. Congressional Research Service RS21421. Washington, DC

Cormier SM, Suter GW, Zheng L (2013a) Derivation of a benchmark for freshwater ionic strength. Environ Toxicol Chem 32:263–271. https://doi.org/10.1002/etc.2064

Cormier SM, Wilkes SP, Zheng L (2013b) Relationship of land use and elevated ionic strength in Appalachian watersheds. Environ Toxicol Chem 32:296–303. https://doi.org/10.1002/etc.2055

Cravotta CA, Goode DJ, Bartles MD, Risser DW, Galeone DG (2014) Surface water and groundwater interactions in an extensively mined watershed, upper Schuylkill River, Pennsylvania, USA. Hydrol Proc 28:3574–3601. https://doi.org/10.1002/hyp.9885

Dallaire K, Skousen J (2019) Early tree growth in reclaimed mine soils in Appalachia USA. Forests 10:549. https://doi.org/10.3390/f10070549

Daniels WL, Stewart BR (2000) Reclamation of Appalachian coal refuse disposal areas. In: Barnhisel RL et al (eds) Reclamation of drastically disturbed lands. Am Soc Agron, Madison, WI, pp 433–459

Daniels WL, Stewart BR, Zipper CE (2010) Reclamation of coal refuse disposal areas. Va Coop Extension Pub 460–131

Daniels WL, Zipper CE (2018) Creation and management of productive minesoils. Va Coop Extension pub 460–121

Daniels WL, Zipper CE, Orndorff CW, Skousen JG, Barton CD, McDonald L, Beck M (2016) Predicting total dissolved solids release from central Appalachian coal mine spoils. Environ Pollut 216:371–379. https://doi.org/10.1016/j.envpol.2016.05.044

Davis V, Franklin J, Zipper C, Angel P (2010) Planting hardwood tree seedlings on reclaimed mine land in Appalachia. US Office of Surface Mining, Appalachian Regional Reforestation Initiative advisory 7

Davis V, Burger J, Rathfon R, Zipper C, Miller C (2012) Selecting tree species for reforestation of Appalachian mined land. US Office of Surface Mining, Appalachian Regional Reforestation Initiative advisory 9

Ditsch D, Collins M (2000) Reclamation considerations for pasture and hay lands receiving 66 cm or more precipitation annually. In: Barnhisel RL et al (eds) Reclamation of drastically disturbed lands. Am Soc Agron, Madison, WI, pp 241–271

Drummond MA, Loveland TR (2010) Land-use pressure and a transition to forest-cover loss in the eastern United States. Bioscience 60:286–298. https://doi.org/10.1525/bio.2010.60.4.7

Evans DM, Zipper CE, Donovan PF, Daniels WL (2014) Long-term trends of specific conductance in waters discharged by coal-mine valley fills in central Appalachia, USA. J Amer Water Res Assoc 50:1449–1460. https://doi.org/10.1111/jawr.12198

Franklin J, Zipper CE, Burger JA, Skousen JG, Jacobs D (2012) Influence of planted ground cover on forest restoration of eastern US coal surface mines. New Forest 43:905–924. https://doi.org/10.1007/s11056-012-9342-8

Garges T (2003) Underground mining technology evolution. In: William N (ed) Poundstone lecture, Department of Mining Engineering, West Virginia University. 13 Nov 2003

Gohlke JM (2021) Human health in coalfield communities of Appalachia. Chapter 12. In: Zipper CE, Skousen JG (eds) Appalachia's coal-mined landscapes. Springer

Gorman J, Skousen J, Sencindiver J, Ziemkiewicz P (2001) Forest productivity and minesoil development under a white pine plantation versus natural vegetation after 30 years. p 103–111. In: Proc Am Soc Surf Min Reclam. https://doi.org/10.21000/jasmr01010103

Greenberg P (2017) Disproportionality and resource-based environmental inequality: an analysis of neighborhood proximity to coal impoundments in Appalachia. Rural Sociol 82:149–178. https://doi.org/10.1111/ruso.12119

Herlihy AT, Kaufman PR, Mitch ME, Brown DD (1990) Regional estimates of acid mine drainage impact on streams in the mid-Atlantic and southeastern United States. Water Air Soil Poll 50:91–107. https://doi.org/10.1007/BF00284786

Hester ET, Little KL, Buckwalter JD, Zipper CE, Burbey TJ (2019) Variability of subsurface structure and infiltration hydrology among surface coal mine valley fills. Sci Total Environ 651:2648–2661. https://doi.org/10.1016/j.scitotenv.2018.10.169

Hobba WA (1993) Effects of underground mining and mine collapse on the hydrology of selected basins in West Virginia. US Geol Surv Water Supply Paper 2384

Imhoff E, Friz T, LaFevers J (1976) A guide to state programs for the reclamation of surface mined areas. US Geol Surv Circular 731. Reston VA. p 33

Johnson W, Paone J (1981) Land utilization and reclamation in the mining industry, 1930-80. US Bureau of Mines Info Circular 8862, Pittsburgh, PA

Kapoor A, Tanjore S, Virarghavan T (2007) Removal of selenium from water and wastewater. Intl J Environ Studies 49:137–147. https://doi.org/10.1080/00207239508711016

Krenz RJ, Schoenholtz SH, Zipper CE (2016) Riparian subsidies and hierarchical effects of ecosystem structure on leaf breakdown in Appalachian coalfield constructed streams. Ecol Engineer 97:389–399. https://doi.org/10.1016/j.ecoleng.2016.10.012

Krenz RJ, Zipper CE, Schoenholtz SH (2018) Periphyton structure and function in constructed headwater streams of the Appalachian coalfield. Freshw Sci 37:780–794. https://doi.org/10.1086/700621

Kruse Daniels N, LaBar J, McDonald LM (2021) Acid mine drainage in Appalachia: sources, legacy, and treatment. Chapter 8. In: Zipper CE, Skousen JG (eds) Appalachia's coal-mined landscapes. Springer

Kurth LM, McCawley M, Hendryx M, Lusk S (2014) Atmospheric particulate matter size distribution and concentration in West Virginia coal mining and non-mining areas. J Expo Sci Environ Epidemiol 24:405–411. https://doi.org/10.1007/s10653-014-9669-5

Lambert DC, McDonough KM, Dzombak DA (2004) Long-term changes in quality of discharge water from abandoned underground coal mines in Uniontown Syncline, Fayette County, PA, USA. Water Res 38:277–288. https://doi.org/10.1016/j.watres.2003.09.017

Lindberg TT, Bernhardt ES, Bier R, Helton AM, Merola RB, Vengosh, Di Giulio RT (2011) Cumulative impacts of mountaintop mining on an Appalachian watershed. Proc Natl Acad Sci 108:20929–20934. https://doi.org/10.1073/pnas.1112381108

Mack B, McDonald L, Skousen J (2010) Acidity decay of above-drainage underground mines in West Virginia. J Environ Qual 39:1043–1050. https://doi.org/10.2134/jeq2009.0229

Mainero RJ, HC Verakis (2010) A century of Bureau of Mines/NIOSH explosives research. In: Proceedings, 2010 SME annual meeting, Phoenix, AZ. ISBN 9781617380822

Mark C (2002) The introduction of roof bolting to U.S. underground coal mines (1948–1960): a cautionary tale. In: Proc, 21st Int Conf Ground Control in Mining. Morgantown WV, p 150–160

Marra M, Keene T, Skousen J, Griggs T (2013) Switchgrass yield on reclaimed surface mines for bioenergy production. J. Env Qual 42:696–703. https://doi.org/10.2134/jeq2012.0453

Merovich GT, Hitt NP, Merriam ER, Jones JW (2021) Response of aquatic life to coal mining in Appalachia. Chapter 10. In: Zipper CE, Skousen JG (eds) Appalachia's coal-mined landscapes. Springer

National Research Council (NRC) (1981) Coal mining and ground-water resources in the United States. US Bureau of Mines, Washington, DC

Nichols CE, Shepherd DL, Knuckles TL, Thapa D, Stricker JC et al (2015) Cardiac and mitochondrial dysfunction following acute pulmonary exposure to mountaintop removal mining particulate matter. Am J Physiol Heart Circ Physiol 309:H2017–H2030. https://doi.org/10.1152/ajpheart.00353.2015

Nobert H, McGill D, Grushecky S, Skousen J, Schuler J (2016) Salix spp. as a biomass crop: investigating its potential on mined lands and the use of biochar as a soil amendment. J Am Soc Mining Recl 5(2):58–76. https://doi.org/10.21000/jasmr16020058

Northington RM, Benfield EF, Schoenholtz SH, Timpano AJ, Webster JR, Zipper CE (2011) An assessment of structural attributes and ecosystem function in restored Virginia coalfield streams. Hydrobiologia 671:51–63. https://doi.org/10.1007/s10750-011-0703-7

Oliphant AJ, Wynne RH, Zipper CE, Ford WM, Donovan PF, Li J (2017) Autumn olive (*Elaeagnus umbellata*) presence and proliferation on former surface coal mines in eastern USA. Biol Invasions 19:179–195. https://doi.org/10.1007/s10530-016-1271-6

Orlemann EC (2000) Caterpillar chronicle: the history of the greatest earthmovers. MotorBooks International, p 168, ISBN 0760336733

Peng S, Chiang HS (1984) Longwall mining. Wiley, New York, p 708

Petty JT, Gingerich G, Anderson JT, Ziemkiewicz PF (2013) Ecological function of constructed perennial stream channels on reclaimed surface coal mines. Hydrobiologia 720:39–53. https://doi.org/10.1007/s10750-013-1619-1

Plass W (1967) Land disturbances from strip-mining in eastern Kentucky: Hazard coal reserve district. US Forest Service NE-71

Plass W (2000) History of surface mining reclamation and associated legislation. In: Barnhisel RL et al (eds) Reclamation of drastically disturbed lands. American Soc Agronomy, Madison, WI, pp 1–20

Pond G, Passmore M, Borsuk F, Reynolds L, Rose C (2008) Downstream effects of mountaintop coal mining: comparing biological conditions using family- and genus-level macroinvertebrate bioassessment tools. J North Am Benthol Soc 27:717–737. https://doi.org/10.1899/08-015.1

Quaranta JD, Mack B, Van Aken B, Ducatman A, Ziemkiewicz P (2014) Practical application of dilution analysis for estimating groundwater quality effects due to coal slurry injection into underground mine voids. Mine Water Environ 33:353–361. https://doi.org/10.1007/s10230-014-0274-8

Reed WR, Taylor CD (2007) Factors affecting the development of mine face ventilation systems in the 20th century. p 1–9. In: 2007 SME annual meeting and exhibit, DENVER, CO, preprint 07–071. Littleton, CO: Society for Mining, Metallurgy, and Exploration, Inc

Rochow KWJ (1979) The far side of paradox: state regulation of the environmental effects of coal mining. WV Law Rev 81:559–593

Rodrigue JA, Burger JA (2004) Forest soil productivity of mined land in the midwestern and eastern coalfield regions. Soil Sci Soc Am J 68:833–844. https://doi.org/10.2136/sssaj2004.0833

Ross MRV, McGlynn BL, Bernhardt ES (2016) Deep impact: effects of mountaintop mining on surface topography, bedrock structure, and downstream waters. Environ Sci Technol 50:2064–2074. https://doi.org/10.1021/acs.est.5b04532

Scagline-Mellor S, Griggs T, Skousen J, Wolfrum E, Holaskova I (2018) Switchgrass and giant miscanthus biomass and theoretical ethanol production from reclaimed mine lands. BioEnergy Res 11:562–573. https://doi.org/10.1007/s12155-018-9915-2

Shank M (2010) Trends in mining fills and associated stream loss in West Virginia 1984–2009. WV Dep Environm Protection. https://tagis.dep.wv.gov/tagis/projects/Mining_Fill_trends_in_West_Virginia_1984-2009.pdf

Skousen JG (2017) A methodology for geologic testing for land disturbance: acid-base accounting for surface mines. Geoderma 308:302–311. https://doi.org/10.1016/j.geoderma.2017.07.038

Skousen J, Call C, Knight R (1990) Natural revegetation of an unreclaimed lignite surface mine in east-central Texas. Southwestern Naturalist 35:157–163. https://doi.org/10.2307/3672042

Skousen JG, Daniels WL, Zipper CE (2021) Soils on Appalachian coal-mined lands. Chapter 4 In: Zipper CE, Skousen JG (eds) Appalachia's coal-mined landscapes. Springer

Skousen J, Ziemkiewicz P, McDonald L (2019) Acid mine drainage formation, control and treatment: approaches and strategies. Extractive Industries and Society, DOI. https://doi.org/10.1016/j.exis.2018.09.008

Skousen J, Ziemkiewicz P, Venable C (2006) Tree recruitment and growth on 20-yr-old, unreclaimed surface mined lands in West Virginia. Intl J Mining, Recl Environ 20(2):142–154. https://doi.org/10.1080/17480930600589833

Skousen JG, Zipper CE (2018) Revegetation species and practices. Va Coop Extension Pub 460–122

Skousen J, Zipper C, Burger J, Barton C, Angel P (2011) Selecting materials for mine soil construction when establishing forests on Appalachian mine sites. U.S. Office of Surface Mining, Appalachian Regional Reforestation Initiative Advisory 8

Skousen J, Zipper C, Rose A, Ziemkiewicz P, Nairn R, McDonald L, Kleinmann R (2017) Review of passive systems for acid mine drainage treatment. Mine Water Environ 36:133–153. https://doi.org/10.1007/s10230-016-0417-1

Stewart BR, Daniels WL (1992) Physical and chemical properties of coal refuse from southwest Virginia. J Environ Qual 21:635–642. https://doi.org/10.2134/jeq1992.00472425002100040018x

Timpano AJ, Schoenholtz SH, Soucek DJ, Zipper CE (2018) Benthic macroinvertebrate community response to salinization in headwater streams in Appalachia USA over multiple years. Ecol Indicators 91:645–656. https://doi.org/10.1016/j.ecolind.2018.04.031

Torbert J, Burger J (2000) Forest land reclamation. In: Barnhisel RL et al (eds) Reclamation of drastically disturbed lands. American Soc Agronomy, Madison, WI, pp 371–398

United States Environmental Protection Agency (US EPA) (2016) Aquatic life ambient water quality criterion for selenium—Freshwater 2016. EPA 822-R-16-006. Washington, DC
US Department of Interior (USDI) (1969) Surface mining and our environment
US Government Accounting Office (US GAO) (2009) Surface coal mining characteristics of mining in mountainous areas of Kentucky and West Virginia. GAO-10.21
US Environmental Protection Agency (US EPA) (2011a) A field-based aquatic life benchmark for conductivity in central Appalachian streams. Office of Research and Development. EPA/600/R-10/023F
U.S. Environmental Protection Agency (US EPA) (2011b) The effects of mountaintop mines and valley fills on aquatic ecosystems of the central Appalachian coalfields. Office of Research and Development. EPA/600/R-09/138F
Wade GL, Thompson RL, Vogel WG (1985) Success of trees and shrubs in an 18-year-oid planting on mine spoil. US Forest Service, Research Paper NE-567
Weisdack GV, Kvitkovich JF (2005) Importance of longwall mining to the coal industry. Min Eng 57:21–26
Whitmore KM, Schoenholtz SH, Soucek DJ, Hopkins WA, Zipper CE (2018) Selenium dynamics in headwater streams of the central Appalachian coalfield. Environ Tox Chem 37:2714–2726. https://doi.org/10.1002/etc.4245
Wickham JD, Riitters KH, Wade TG, Coan M, Homer C (2007) The effect of Appalachian mountaintop mining on interior forest. Landscape Ecol 22:179–187. https://doi.org/10.1007/s10980-006-9040-z
Winters WR, Capo RC (2004) Ground water flow parameterization of an Appalachian coal mine complex. Groundwater 42:700–710. https://doi.org/10.1111/j.1745-6584.2004.tb02724.x
Wood P, Larkin J, Mizel J, Zipper CE, Angel P (2013) Reforestation to enhance Appalachian mined lands as habitat for terrestrial wildlife. US Office of Surface Mining, Appalachian Regional Reforestation Initiative Advisory 10
Zipper C, Winter S (2009) Stabilizing reclaimed mines to support buildings and development. Va Coop Extension Pub 460–130
Zipper CE, Balfour W, Roth R, Randolph J (1997) Domestic water supply impacts by underground coal mining in Virginia, USA. Environ Geol 29:84–93. https://doi.org/10.1007/s002540050107
Zipper CE, Burger JA, McGrath J, Rodrigue J, Holtzman G (2011a) Forest restoration potentials of coal mined lands in the eastern United States. J Environ Qual 40:1567–1577. https://doi.org/10.2134/jeq2011.0040
Zipper C, Burger J, Skousen J, Angel PN, Barton CD et al (2011b) Restoring forests and associated ecosystem services on Appalachian coal surface mines. Environ Manage 47:751–765. https://doi.org/10.1007/s00267-011-9670-z
Zipper CE, Burger JA, Barton CD, Skousen JG (2013) Rebuilding soils on mined land for native forests in Appalachia, USA. Soil Sci Soc Am J 77:337–349. https://doi.org/10.2136/sssaj2012.0335
Zipper CE, Angel P, Adams MB, Sanderson T, Sena K, et al. (2019) The forestry reclamation approach aids in managing invasive exotic plants on active mine sites. U.S. Office of Surface Mining, Appalachian Regional Reforestation Initiative Advisory 17
Zipper CE, Adams MB, Skousen JG (2021) The Appalachian coalfield in historical context. Chapter 1. In: Zipper CE, Skousen JG (eds) Appalachia's coal-mined landscapes. Springer
Zipper CE (2020) Appalachian coal data. VTechData, University Libraries, Virginia Tech.https://doi.org/10.7294/bda6-7r15

Soils on Appalachian Coal-Mined Lands

Jeff Skousen, W. Lee Daniels, and Carl E. Zipper

Abstract Mine soils are anthropogenic soils formed in mine spoils. Mine soils on older sites were not purposefully placed and slowly changed with soil-forming processes to form plant growth media. Some early mine soils were made of materials such as large rocks, coal remnants, carbonaceous shales, and acid sandstones which are poorly suited for productive vegetation. Later mine soils were constructed with spoil materials selected to serve as plant growth media or the original topsoil. Mine soils constructed with redistributed topsoil and weathered spoils generally have properties more similar to those of natural soils than do mine soils constructed from unweathered rock originating from deep below the pre-mining land surface. Primary influences on mine soil properties are the materials used for construction, the degree of compaction caused by mining equipment, and the time over which they have developed. Their suitability for managed land uses and productive and desirable plant communities can be assessed by observing the vegetation present and by evaluating physical and chemical properties. Mine soils' suitability for livestock agriculture, forest tree growth, and other common uses and associated hydrologic properties vary widely. Over extended time, most mine soils become more like natural soils but retain characteristics unique to their origin.

Keywords Mine spoil · Parent material · Soil genesis · Soil morphology · Weathering

1 Introduction

Soil occurs across the Earth's surface as a mixture of weathered soil-sized (<2 mm) mineral grains, rock fragments (>2 mm) and organic materials with pore spaces that allow for gas exchange and water-holding capacity. In most environments, soils are

J. Skousen (✉)
West Virginia University, Morgantown, WV, USA
e-mail: jskousen@wvu.edu

W. L. Daniels · C. E. Zipper
Virginia Tech, Blacksburg, VA, USA

C. E. Zipper and J. Skousen (eds.), *Appalachia's Coal-Mined Landscapes*,
https://doi.org/10.1007/978-3-030-57780-3_4

rich biological media that host animals, microbes, and plant roots. Soils are critical components of natural ecosystems and support human society via their production of plant materials for food, fiber, and livestock. From an ecosystem functions perspective, soils provide water absorption and release to flowing streams, water filtration to remove pollutants, sequestration of carbon, nutrient cycling and a wide range of habitat values. Thus, soils play a foundational role in natural ecosystems that are essential to human welfare.

Soils develop over time on both natural landscapes and lands mined for coal. On Appalachia's natural landscapes, soils have had thousands of years to evolve. Complex layering and properties that develop in response to the parent material, climate, topography, and vegetation differentiate soils from raw geologic materials. In contrast, mine soils are much younger and have not developed the complex properties of mature native soils.

Mine soils evolve from the materials left or placed on the surface by mining operations. On modern mines, these materials are subjected to grading, liming and fertilization, herbaceous vegetation seeding, and in some areas, planting of trees and shrubs. As time passes, environmental processes (e.g., rainfall infiltration and geochemical weathering), biological processes (e.g., plant growth and litter decay), and ecological processes (e.g., species invasions and plant community changes) cause soils on mine sites to develop greater complexity, and in most cases to become more like natural soils.

Here we describe mine soils that occur on Appalachian mined landscapes, processes that occur within them, and procedures for assessing their potential to support human welfare and ecosystem functions.

2 Construction

2.1 Early Mining

Surface mines before 1960 in Appalachia rarely performed deliberate soil reconstruction (Skousen and Zipper 2021). Rather, the mining area was first cleared of timber, and then miners used explosives and bulldozers to move soil and overburden (geologic strata overlying coal seams) away from the mineable coal. This process created mine spoils, i.e., surficial deposits of loose rock and mineral fragments created by blasting. In sloped areas, spoils were pushed over the side of the hill, a mining method called "shoot-and-shove." Topsoil that existed at the surface before mining was generally the first material pushed downslope and covered by additional spoils. The resulting "outslopes" were typically left unvegetated, and therefore subject to erosion (Fig. 1).

During these early years, little or no effort was made to manage or limit environmental effects by controlling spoil placement or native soil salvage. Hence, the final landscape developed in response to initial spoil placement, as well as spoil settling

Fig. 1 A landscape created by "shoot-and-shove" contour mining in steep terrain and left without reclamation (left) and mine soils on the same mine site (right) (Photos by Carl E. Zipper)

and consolidation, movement of unstable materials downslope, and processes caused by running water such as sheet and gully erosion.

If materials left on the surface were favorable to plant growth, plants from nearby lands, often forests, quickly colonized and established. As soil and vegetation matured, natural ecological succession processes produced a plant community with species from adjacent areas (Wade et al. 1985; Skousen et al. 1994; Ashby 2006). If surface materials contained toxic or acid-producing materials, were heavily compacted, or were otherwise unfavorable to plants, native forest plant colonization was inhibited and a different set of plant species, including exotic non-native species if present on nearby areas, tended to invade and occupy the mine site (Gorman et al. 2001; Johnson and Skousen 1995; Zeleznik and Skousen 1996). Mine soils on flatter benches were often heavily compacted as they were convenient locations for equipment operations (Daniels and Amos 1985); while outslope steepness and instability also caused hazardous conditions on some sites.

Mine soils on such mine sites are highly variable. Since spoils were often placed on the surface randomly, older mine soils may be composed of any portion of the excavated overburden and may or may not be well vegetated. If a plant community developed and grew well, natural processes have caused organic matter to be incorporated into the mine soil. Over time, these mine soils usually developed thin, weakly developed surface horizons (Ciolkosz et al. 1985; Daniels and Amos 1985). Early mine soils generally have a coarse texture and rock fragments of various sizes at the surface and at depth.

These older mine soils are used in a variety of ways, and some have experienced weathering such that rock fragments have broken down into silt- and sand-sized particles with time. Mine soils with favorable properties can support agricultural grasses and legumes and may be suitable for use as livestock grazing or for hay production.

Mine soils derived from softer shales with alkaline chemistry and containing few rock fragments may have been plowed and planted with field crops; while moderately acidic and uncompacted mine soils may have developed productive tree stands (Rodrigue and Burger 2004). Often, however, plant productivity on these early mine soils has been impaired by factors such as coarse texture, compaction, poor water-holding-capacity, low fertility, and in some cases high acidity. Lands with such soils are generally unmanaged and not in productive uses.

2.2 *After SMCRA*

A major change in mine soil construction practices occurred following 1977 when the Surface Mining Control and Reclamation Act (SMCRA) became law. The SMCRA required miners to return the landscape to "approximate original contour," unless a variance was obtained, and to develop an appropriate post-mining land use. Rock materials were often so pulverized by blasting that mine spoils were comprised of a significant amount of fine earth fraction. This led to the idea of blending spoils into constructed mine soils appropriate for various land uses.

The SMCRA's performance standards [Sects. 515(b)(5) and (6)] require topsoil salvage prior to mining and its replacement on the land surface during reclamation unless a variance is obtained. If the topsoil (A + E horizon) is less than 15 cm thick and a topsoil substitute (defined below) has not been approved, federal regulations (30 CFR 816.22) require the operator to remove the topsoil along with unconsolidated material immediately below and place the mixture as topsoil.

The SMCRA and its regulations allow selected and fractured overburden to be used as substitute or to supplement for topsoil if the operator demonstrates that material to be equally or more suitable for post-mining vegetation. The term "topsoil substitute" describes rock spoils used to construct mine soils in this manner.

2.3 *Construction Materials*

Three types of material are generally available for use in mine soil construction:

- Soils: These contain weathered mineral fragments and organic matter, and are the most favorable material for most agricultural, forestry, and other plant-dependent post-mining land uses. Best practices when restoring natural ecosystems are to move soil materials directly from the excavation to the reclamation area, thus maintaining viability of seeds, plant propagules, and microorganisms; a practice known as "direct haul." Common practices, however, were to place excavated soils in stockpiles for temporary storage and later re-spreading. Although regulations require salvage of the native soil O, A and E horizons, typical practice when salvaging soils in steep terrain was/is to excavate all surface material (to depths as much as 30 cm) that could be moved easily with a dozer.

- Weathered spoils (also known as pre-weathered): These materials are found directly below subsoil horizons and generally extend from 10 to 20 m in depth from the surface (Fig. 2). Weathered rock strata have been affected visibly by environmental processes due to rainwater infiltration from the surface and are often brownish or yellowish in color due to iron oxidation (Eriksson and Daniels 2021). Generally, they break down into soil-like materials with fewer coarse fragments and have less capacity to generate dissolved salts than deeper, unweathered rocks (Clark et al. 2021). However, weathered spoil lacks the organic matter, plant nutrients, and living organisms present in salvaged soil materials.
- Unweathered spoils: These materials occur below the zone of environmental weathering, often comprising most of the overburden material disturbed on modern mines, and vary widely in physical and chemical properties. Some unweathered rock strata break down to soil-like materials relatively quickly if exposed to air and water (are said to "slake"), whereas others are more durable and thus create rockier soils. They also vary widely in chemical properties, ranging from alkaline to highly acidic (Eriksson and Daniels 2021). These materials are usually gray to black in color due to their lack of iron oxidation.

Mine soils are constructed from these materials using various methods. Prior to SMCRA, any material from the overburden column may have ended up on the surface (uncontrolled placement). Following SMCRA, native soils were salvaged and placed at the surface or topsoil substitutes were identified, selected, and used in a similar manner, a method known as controlled placement (Fig. 3). Mine soil construction using salvaged soils was common in northern Appalachia where landowners often

Fig. 2 An exposed highwall on an Appalachian mine site, with weathered rocks in the highwall's upper half being darker in color than the unweathered rocks below (Photo by Carl E. Zipper)

Fig. 3 A dozer spreading salvaged soil materials over mine spoil (left); salvaged soil storage piles (foreground, right) and a mine site backfill where salvaged soil has been partially applied (background, right) (Photos by Jeff Skousen)

desired the reclaimed land for cropping and haying, and where native soils were often thicker and the topography less steep than in central Appalachia. Some mines employed controlled placement by identifying specific geologic strata as topsoil substitutes based on properties such as low acidity and the tendency to decompose into soil-like fine fragments. These substitutes were commonly selected for topsoil placement in central Appalachia and were spread over the reclaimed land surface in varying depths, typically in the range of 0.25–1 m. On some mines in central Appalachia's steep landscapes, large units of unweathered geologic strata were permitted as topsoil substitutes, and this material was used for reconstruction of land contours and served as the plant growth media.

The original reconstruction material has a major influence on subsequent mine soil properties and suitability for specific land uses. Mine soils constructed with salvaged soils are often presumed to have land use capabilities like those soils present prior to mining if compaction is avoided. Few if any peer-reviewed studies, however, have been conducted to test that presumption in the Appalachian region. The properties and productivity of mine soils constructed from rock spoils vary widely and in response to the construction materials, both shortly following construction and for many years following (Table 1).

Topsoil substitutes constructed from unweathered rock are commonly circumneutral to alkaline (e.g., pH 7.0–8.0) when first emplaced, while weathered non-pyritic rocks are often moderately acidic (Roberts et al. 1988b; Torbert et al. 1990; Emerson et al. 2009; Miller et al. 2012; Sena et al. 2015). Although some Appalachian mine spoils are pyritic and, hence, highly acidic ($pH<5.0$), these materials are typically avoided when intentionally constructing mine soils. Grasses and legumes grow well in alkaline spoils, even without the lime applications that are often required to maintain productive pastures on Appalachia's somewhat acidic natural soils (Roberts et al.

Table 1 Selected surface properties (0–5 cm) of mine soils constructed from partially weathered sandstone, unweathered siltstone, and a natural forest soil from the same region measured using conventional soil analysis techniques

Source material	Partially weathered sandstone		Unweathered siltstone		Natural forest soil
Mine soil age (years)	1–2	26	1–2	26	
Coarse fragments (%)	65	53	79	77	15
<2 mm: Sand (%)	69	62	59	31	55
<2 mm: Silt (%)	22	25	30	52	20
<2 mm: Clay (%)	9	13	11	17	25
Soil pH	5.5	5.6	7.5	6.5	5.0
Electrical conductivity (μS cm^{-1})	0.15	0.16	0.29	0.25	0.01
Organic matter, Walkley Black (%)[a]	1.4	7.7	3.3	11.7	7.5
Soil N (%)	0.4	1.5	0.8	1.9	2.3
Bicarbonate-extractable P (mg kg^{-1})	78	20	74	15	2.35
Acid-extractable P (mg kg^{-1})[a]	28	15	46	38	15
Extractable Fe oxides (%)	0.35	1.29	0.17	1.18	2.35

After an initial fertilization, the soils were sampled one year following construction for all properties except soil N (sampled two years following construction) and 26 years later. Data from Nash (2012), Nash et al. (2016), and Jeff Skousen for natural soil

[a]Common soil analysis techniques that give incorrect or misleading results when applied to mine soils constructed from rock spoils

1988c). Hence, such materials are often used as topsoil substitutes when mine soils are constructed to support hay lands and pastures (Ditsch and Collins 2000), while the moderately acidic mine soils created with weathered spoils are more favorable for the native forest trees (Zipper et al. 2013). Some soil properties that differentiate mine soils from native soils, such as high coarse fragment contents and alkaline pH, are often mitigated over time by soil weathering (Nash et al. 2016), but sometimes they persist for decades (e.g., Wilson-Kokes et al. 2013a).

2.4 Mine Soil Compaction

The degree of compaction applied to the mine soil at construction will have a fundamental influence on its properties. Soil compaction is necessary for certain post-mining land uses such as airport runways, paved roadways, parking areas, and buildings with rigid structures require stable support. Most mines, however, establish post-mining land uses requiring plant growth and plant community development.

Plants extend roots into the soil to obtain water and nutrients essential to their survival and growth. Excessive compaction causes high soil bulk density which slows and limits the development of functional rooting systems by physically impeding root extension. These effects are more severe for woody plants such as forest trees that require root access to deeper soils than for grasses with their shallower rooting systems. Dense soils also hinder plant growth by restricting movement of water and air, both within deeper soil layers and between subsurface and the surface. Unless reclamation operations were undertaken with the intent of avoiding it (Sweigard et al. 2007b), excessive soil compaction can be a problem because of the size of equipment that is typically used to redistribute bulk spoils and final mine soil materials on operating mines.

Unless mitigated by mechanical means (Sweigard et al. 2007a; Burger et al. 2013), equipment-induced soil compaction, especially in the subsurface, can influence vegetation over long terms (Daniels and Amos 1985; Haering et al. 2004). Effects of compaction by mining equipment applied solely to the surface may not be as persistent due to mitigating effects such freeze-induced expansion and subsequent thawing (Guebert and Gardner 2001), especially in rocky mine soils (DeLong et al. 2012).

3 Morphology and Genesis

Mine soils differ in properties and morphology from natural soils. Much of Appalachia is covered by forests and forest soils with A horizons less than 20 cm in thickness underlain by B and C horizons of varying thicknesses (Fig. 4). On rocky

Fig. 4 Examples of natural soil profiles in Appalachia: a pasture soil with a 15-cm-thick A horizon underlain by a 70-cm B horizon (left); and a forest soil with a 20-cm-thick O-A horizon sequence underlain by bedrock (right) (Photos by Jeff Skousen)

ridges, natural forest soils may be thinner with very little soil overlying bedrock, but native soils can be quite thick on plateau-like uplands, on lower side-slopes, and in concave drainage ways where local slope debris and water-transported materials accumulate.

Appalachian mine soils constructed from rock spoils are strongly influenced by their original rock type. Rock fragments are typically high (≥50%), while pH and available cations vary widely by original rock type. In general, mine soils constructed from weathered strata are often lower in rock fragment content, lower in pH, and higher in silt + clay content than mine soils from unweathered non-pyritic spoils originating from deeper strata (Haering et al. 1993). Bridging voids (open air-filled spaces between larger spoil fragments; Fig. 5) and highly compacted subsoil zones are commonly described in younger mine soils. Strongly contrasting layers are typically present where controlled placement was used to construct the mine soil surface. Such layering is particularly notable when either salvaged topsoil or pre-weathered topsoil substitute materials were placed over unweathered materials (Fig. 6).

Unless constructed as separate layers, mine soils lack horizons or distinct layers initially. As organic matter accumulates, an ^A horizon (with the "^" symbol designating human-transported materials) typically forms and then extends downward into the underlying and relatively unaltered spoils (^C horizon). Where construction materials are relatively uncompacted, rooting can easily extend to 50 cm or more within several years (Haering et al. 2004). Horizonation below the A, however, is typically muted relative to natural soils (Fig. 7).

When mine soils constructed from rock spoils are well vegetated, weak ^A horizons may develop in times as short as one to two years, and usually by five years, due to combined effects of particle aggregation and plant-derived organic matter additions (Daniels and Amos 1985; Haering et al. 1993). Well-developed ^A horizons

Fig. 5 Examples of mine soil profiles constructed from rock spoils via uncontrolled placement. Left example is composed mostly of fine-textured material and with low pH and poorly suited for plant growth. Right example is composed of large durable rock fragments with large void spaces and limited soil-sized material, which is poorly suited for vegetation (Photos by Jeff Skousen)

Fig. 6 Profiles of mine soils constructed with controlled placement of "topsoil" salvaged as mixtures of the original A horizon with subsoil materials having been placed at <0.5-m depths over rock spoils (left and center) and placed at greater depth (right) (Photos by Jeff Skousen)

Fig. 7 Mine soils in place for 15 or more years with herbaceous vegetation (Photos by Jeff Skousen)

with organic matter contents comparable to those of native soils have been reported after 26 years (Table 1; Nash et al. 2016). Transitional A–C horizons designated as ^AC have been reported eight to 13 years following mine soil construction (Ciolkosz et al. 1985; Haering et al. 1993), while cambic-like ^B horizons have been reported for mine soils 20 years of age and older (Ciolkosz et al. 1985; Haering et al. 2004). In certain scenarios, particularly where the original spoils were derived from finer-textured rocks (e.g., shales and mudrocks), or where the spoils were pre-weathered before placement, weakly developed subsoils (^Bw horizons) can develop within decades (Ciolkosz et al. 1985) along with weak but observable blocky soil structure (Haering et al. 1993). Although strong B horizons have not been reported in mine soils, the illuvial processes responsible for strong B horizon development in natural soils do occur as geochemical weathering of mine spoil minerals causes release of soluble constituents (Clark et al. 2018), rock fragments disintegrate to form smaller particles (Nash et al. 2016), and downward particle redistribution occurs (Roberts et al. 1988b).

4 Mine Soil Properties and Weathering Influences

Mine soil construction materials exert primary influence on mine soil properties, especially during the early stages of soil formation. Other influences include time since construction, the degree of compaction applied during and after construction, and vegetation. To evaluate mine soils for suitability to support managed land uses (Zipper et al. 2021), their physical and chemical properties can be analyzed using laboratory methods and assessed in the field.

A first step in mine soil evaluation is a site visit. If a mine soil has little vegetation growing after one or two growing seasons, it likely contains problematic minerals (either highly alkaline or acidic) or unusually high rock contents or was heavily compacted. Important properties can be determined by conducting field assessments and obtaining soil samples from each area with a different soil type (such as areas with differing surface soil colors or drainage patterns) and differing vegetation; or a grid sampling pattern can be used (Burger et al. 2013). Areas with sparse or discolored vegetation should be sampled separately. Procedures for obtaining soil samples described in soil testing laboratory publications can be applied to mine soils. Where a relatively thin layer of material has been placed on the surface as a plant growth medium, subsoils should be sampled in addition to surface soils. Soil samples can be sent to a soil testing laboratory for analysis.

4.1 Rock Fragment Content

The amount and size of rock fragments in mine soils can be a use-limiting property. Particles >2 mm (larger than sand-sized particles) are known as rock fragments or coarse fragments which, when greater than 2 or 3 cm in size, can be problematic for agricultural uses. Rock fragments in mine soils can range from close to 0 to >80% (Figs. 5, 6, 7 and 8). High rock contents can limit a mine soils' capability to hold and supply plants with water and nutrients, and can limit plant rooting (Fig. 6). Forestry and wildlife habitat land uses rely on woody plants that are more tolerant of higher rock fragment contents than many crop species. A southwestern Virginia study found the combination of rock fragment content and mine soil effective depth above heavily compacted subsoils, an indicator of the soil volume available for plant rooting, to influence growth rates of eastern white pine (*Pinus strobus*) (Torbert et al. 1988). However, mine soil-rock fragment contents alone, when varying over a range from 4 to 82%, were found to have a negative influence on the growth rates of forest tree stands with multiple species (Rodrigue and Burger 2004). In mine soils

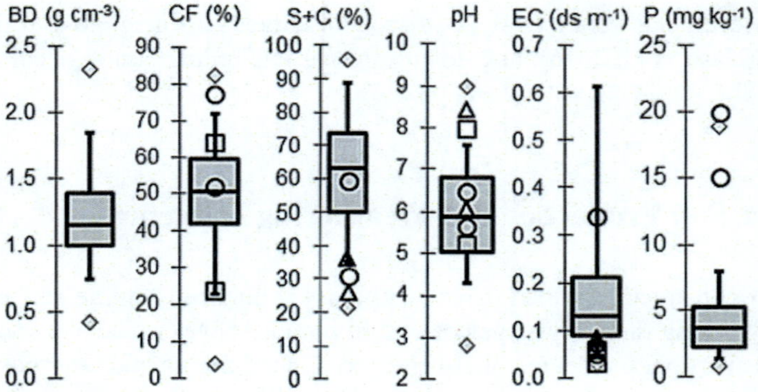

Fig. 8 Aged mine soil properties (BD—bulk density; CF—coarse fragment content; S + C—silt + clay fraction of <2 mm particles; soil pH; EC—electrical conductivity; and extractable P) and variability as indicated by various studies. Box plots represent 5th, 25th, 50th, 75th, and 95th percentiles of measured values from 225 mine soil samples on 25 mine sites ranging from 6 to 25 years of age in 4 states (Zipper et al. 2011). Symbols represent maximum and minimum measured values from that study, and from 28-, 8-, and 8-year-old mine soils constructed from rock spoils from Nash et al. (2016), Sena et al (2015), and Wilson-Kokes et al. (2013a), respectively. EC values are for 2:1 soil:water mixtures, and soil P values were determined by bicarbonate extraction. The maximum EC value from Zipper et al. (2011) is off scale

constructed from non-durable rocks, rock fragment content can decline over time due to continuing weathering, slaking and disintegration (Nash et al. 2016; Table 1). However, such processes may not have significant influences even over near-decadal time scales on mine soils constructed from durable rock (Wilson-Kokes et al. 2013a).

Rock fragment contents can be assessed visually at the surface but more accurately by excavation. If a soil sample is obtained, rocks can be separated by sieving with a 2 mm (approximately 1/10 inch) screen. For lands intended for use as non-tillage cropping or pasture, a rock fragment content of <30% of total soil mass is desirable. Trees can tolerate higher levels of rock fragments if soils are loose and rooting depth is adequate (Torbert et al. 1988; Rodrigue and Burger 2004). Agricultural cropping systems requiring tillage are not feasible on most Appalachian mine soils because of high rock fragment contents.

4.2 *Texture*

Soil texture refers to the relative contents of sand, silt and clay particles in the <2 mm size fraction. Based on particle size distributions, soils are classified into one of 12 primary soil texture classes, such as loam, silty clay loam, or sandy loam. Mine soils composed of mostly sand-sized particles and rock fragments often do not hold and supply enough water and nutrients for plants to grow well. Clayey mine soils are often dense, cloddy, and sticky when wet, thereby restricting water penetration, water movement, and root development. The clay-sized fraction of mine soils may contain minerals not typically found in clays of natural soils due to the effects of rock blasting (Howard 2017). The best soils for plant growth are those with loamy textures which have moderate amounts of sand, silt, and clay. Some of the negative effects of extreme textures can be mitigated by adding organic matter, and such additions will aid in the maintenance of porosity in fine-textured soils and improve water and nutrient relations in coarse-textured soils. In mine soils, soil texture can change over time, becoming finer as a result of continuing slaking and disintegration of the geologic materials (Table 1; Nash et al. 2016).

A skilled soil scientist can make approximate assessments of soil texture in the field: sandy soils have a gritty feel while clayey soils, when moistened, feel sticky, slimy or smooth. Silty soils have a smooth floury feel. Alternatively, soil samples can be sent to a laboratory for textural analysis.

4.3 Bulk Density and Compaction

Bulk density refers to the mass of soil within a defined soil volume, and typically applies only to the soil-sized fraction. The densities of most natural soils vary between 1.1 and 1.5 g cm^{-3}, while densities of dry unbroken mine rock are often in the range of 2.4–2.6 g cm^{-3}. Compacted soils have less pore space and are denser than natural soils, and some mine soils can have density values >2 g cm^{-3} (Fig. 8). Water and air flow and movement are often severely restricted in compacted soils. Extension by tree roots can be limited by soil bulk densities greater than 1.6–1.7 g cm^{-3} (Pritchett and Fisher 1987; Gale et al. 1991). Mine soils with high bulk densities can be loosened by deep ripping and tillage (Sweigard et al. 2007a), but maintenance of reduced-density conditions induced by such treatments often requires establishment of plant rooting within a few years to offset subsequent reconsolidation and settling (Haigh et al. 2015). Remedial ripping is generally beneficial when re-establishing forest trees to mine soils that were previously revegetated with grasses and legumes (Burger et al. 2013; Zipper et al. 2021).

Compaction by mining equipment is a primary cause of elevated mine soil densities in Appalachia. Elevated mine soil densities can also occur due to gravity- and water-induced physical settlement and fine particle redistribution within the profile (Miller et al. 2012). Disintegration and slaking of finer-grained spoils (shales and mudrocks) can fill or bridge voids, and the physical migration of fines downward during major wetting events can contribute to mine soils' "self-compaction" behavior (Haigh and Sansom 1999). We observed elevated subsurface soil densities in southwestern Virginia mine soils (Roberts et al. 1988b; Nash et al. 2016) more than 30 years following their construction using low-compaction techniques, and the higher soil densities corresponded to soil settlement equivalent to about 15–20% of their original depth. This finding has significant implications for certain long-term land uses and may explain why compacted layers have been observed to "re-form" on even steeply sloping post-mining backfills.

Density of surface soils can be assessed in the field, as highly compacted and dense mine soils will resist insertion of hand-held digging tools. Field investigations can also employ soil density probes (penetrometers) or can be conducted by excavating soil and rock from a known volume, and then determining the soil-sized fraction's mass and volume (DeLong et al. 2012).

4.4 Color

Soil colors can reflect drainage, weathering, and development status, but must be interpreted with care for mine soils. Bright red and yellow colors signify highly weathered, oxidized, and favorable moisture conditions in natural soils, but rarely occur in mine soils constructed from rock spoils. In contrast, dull, gray colors in native soils often signify unoxidized conditions which are found in areas that are flooded or wet during most of the year. However, gray mine soils may have inherited the color from their origin as unweathered geologic materials and may not have drainage restrictions. Black or dark brown colors in the surface of natural soils can indicate high organic matter content, but black colors in mine soils may also indicate the presence of carbonaceous rock that is not favorable for plants and, when pyritic, can produce acidic conditions hostile to plants. Black materials on the surface can create hot and dry rooting environments unfavorable to plant growth. In such situations, mulching the land surface with organic materials will help to mitigate those conditions and promote plant growth.

4.5 Soil pH

Soil pH is a measure of the active soil acidity and an important indicator of soil chemical status. Soil pH is measured by mixing soil with water, usually as a 1:1 weight ratio, allowing the suspension to equilibrate for a defined period (often 15 min), and then using a pH probe to measure pH of the mixture. Natural soils used for agriculture often have pHs in the 5.5–7.5 range that is well suited for most crops. If topsoil was placed during reclamation, mine soil pH will generally be in the 5.0–6.5 range that is typical for natural soils in Appalachia. Soil pH's in the upper end of this range are ideal for grasses and legumes, while more acidic pH's within this range can be tolerated or managed by liming for hay and pasture production. If rock spoils were used for mine soil construction, the pH could vary from <4.0 to >8.0 depending on material chemistry. Rock spoils composed of alkaline-producing shales and sandstones may initially have pH levels 7.5–8.0, but on the land surface these will often decline gradually to the pH 6.0–7.0 range with continued weathering (Table 1). However, such decline does not always occur as some mine soils have maintained pH's of >7.9 for nearly a decade (Wilson-Kokes et al. 2013a; Sena et al. 2015). Mine soils constructed from weathered spoils or native topsoil are often moderately acidic (pH

4.5–6.5), both initially and over protracted periods (Table 1), and well suited for growth of native trees (Zipper et al. 2013). If construction materials contain pyritic shales or coal fragments, soil pH's may be as low as 3.0, and thus far too acidic for vegetation establishment and survival.

4.6 *Soluble Salts and Electrical Conductivity*

Many mine soils are constructed from rock spoils containing geologic minerals that weather to release soluble constituents (Clark et al. 2018). The electrical conductivity (EC) of a soil-water mixture is an indicator of soluble elements' combined concentration in soil solution. Primary soluble constituents released from mine spoils' geochemical weathering include the plant macronutrients Ca, Mg, K, S, and HCO_3^-, as well as Na and Cl. Trace elements, including some that are micronutrients such as Fe, Cu, Mn, Ni, and Zn, can also become soluble. Soluble salt levels are often elevated in young Appalachian mine soils constructed from unweathered rock spoils and typically decline over time with continued soil weathering (Sena et al. 2015). However, depending on the nature of the spoil material, such declines may require many years (Table 1).

While some soluble salts in soil solution are necessary for good plant nutrition, high concentrations can be toxic to plants or restrict nutrient and water uptake. If soluble salt levels at the surface remain significantly above what is typical for natural soils, a more detailed soil investigation is advised. Native Appalachian forest trees have been found to be especially sensitive to elevated soluble salts (Torbert et al. 1988; Andrews et al. 1998; Rodrigue and Burger 2004). When EC levels rise to greater than 2 dS m^{-1} (mmho cm^{-1}), the soluble salts are highly deleterious to most plants. Most soil testing laboratories will report measurements of EC or "soluble salts"; a soluble salt concentration of 640 ppm (mg L^{-1}) is typically equivalent to approximately 1.0 mmho cm^{-1}.

4.7 *Organic Matter and Soil Structure*

Organic matter derived from growing plants and soil animals, known as pedogenic, is essential to soil quality. Organic matter: (1) stores and releases essential plant nutrients, (2) aids in the development of soil structure for water and air movement, (3) enables soil hydrologic function including water infiltration, storage, and release back to plant roots, (4) aids in the development of soil as habitat for microbial life and burrowing animals, and (5) maintains porosity in fine-textured soils.

Mine soils constructed from salvaged soils are endowed with organic matter from that source, but mine soils constructed from rock spoils are initially devoid of pedogenic organic matter. Although it accumulates over time in developing mine soils

(Roberts et al. 1988a), full-profile mine soil organic matter typically remains lower than in native soils even after decades (Amichev et al. 2008; Acton et al. 2011).

Mine soil organic matter data derived from conventional soil measurement techniques can be misleading (Maharaj et al. 2007). For example, application of the Walkley–Black chemical oxidation procedure to the mine soils in Table 1 gave results much higher than expected for pedogenic organic matter alone, as these results also reflect inorganic carbon, fossilized organic carbon, and other easily oxidizable mineral forms usually not present in Appalachia's natural soils. When constructed from rock spoils, mine soils can be presumed to be essentially devoid of organic matter prior to revegetation and are very low in pedogenic organic matter during their first few years. Older mine soils can be assessed for approximate soil organic matter manually. The presence of a loose, friable, and dark-colored surface layer with extensive plant rooting will indicate organic matter accumulation.

Mine soils constructed from rock spoils occur as collections of individual particles when initially placed but typically develop soil structure with time as particles aggregate. Formation of structural aggregates also creates pores, both within and between the aggregates, that enable movement of water and air to and from the surface and within the soil mass. When lacking aggregation, the geologic particles of a mine soil can form monolithic masses with limited water and air movement, a condition known as massive structure. Organic matter accumulation and plant rooting enable development of structural aggregates. When supporting productive plant growth, mine soils can develop weak structure in the ^A horizon over times as short as a few years, but structural development of subsurface materials requires more time (Haering et al. 1993). Soil compaction or properties non-conducive to plant growth and rooting can inhibit structural development, resulting in retention of massive structures that limit water and air movement over extended times (Haering et al. 2004).

Additions of plant- and animal-derived organic matter to mine soils derived from rock spoils can have long-term positive effects on soil quality (Roberts et al. 1988c; Wilson-Kokes et al. 2013b), but such additions rarely occur. Additions of nutrient-rich organic matter such as biosolids to mine soils can increase fertility, water-holding capacity, and other properties essential to plant productivity, but should be incorporated into the soil for maximum effectiveness and for limiting potential negative water quality effects (Haering et al. 2000).

4.8 Fertility

Soil fertility refers the quantity and types of nutrients available for plant uptake. Levels of bioavailable soil nutrients are dependent on factors such as rock fragment content, the types of minerals present, soil texture, pH, and organic matter content. Mine soils constructed from salvaged topsoil will commonly have organic matter to supply essential plant nutrients, particularly N and P. Mine soils made from rock spoils will typically have high levels of the macronutrients Ca, Mg, K, and S, which are released as they experience geochemical weathering; however, the bioavailability of micronutrients may be adequate or lacking, depending on mine soil pH and mineralogy.

Mine soils, especially when constructed from rock spoils, are generally limited in plant-available N and P unless fertilized (Howard et al. 1988; Li and Daniels 1994). Even when fertilized during revegetation, such soils can become nutrient-limited due to loss of N by leaching and fixation of P to non-bioavailable forms (Daniels and Zipper 2018). Establishing N-fixing legumes on mine soils can aid replenishment of soil N. While research indicates that bioavailable soil N, primarily as organic forms, can accumulate over time in mine soils (Table 1; Zipper et al. 2011), bioavailable soil P accumulation via similar processes has not been documented. Although some P is present in common Appalachian mine spoil minerals, P availability in mine soils constructed from rock spoils is typically low due to factors that include fixation by secondary iron-oxide minerals that form via geochemical weathering (Howard et al. 1988; Table 1). Liming and fertilization will generally rectify poor soil fertility, as will the addition of nutrient-rich organic matter amendments (Roberts et al. 1988a). Surface mining regulations encourage the addition of lime, mulch, and fertilizer when needed to establish vegetation, but after several years, those fertility levels will decline and further additions are generally needed to sustain agricultural production. In forests and other non-intensively managed plant communities, establishment of functional nutrient cycles with soil organic matter accumulations will aid continued plant growth and community development (Daniels and Zipper 2018).

To evaluate the fertility, mine soil samples can be obtained and sent to qualified laboratories. Soil laboratory results, however, should be evaluated by soil specialists knowledgeable of mine soils because standard soil tests can give misleading results when applied to mine soils. Analysis for mine soil P using acidic extractants suitable for natural soils, for example, can cause dissolution of alkaline P-containing minerals and produce high estimates of supposedly plant-available soil P (Table 1), although such P would likely not become bioavailable in nature even over multiple years. Bicarbonate extraction of P, in contrast, does not dissolve those alkaline materials to release those P forms and is a more reliable predictor. Similarly, acidic extractants can yield misleadingly high estimates for soil Mg, Ca, and K. However, such results are not so consequential because those nutrients are typically abundant as bioavailable forms in Appalachian mine soils. Generally, when preparing mine soils for crops, fertilizers with N and P should be applied. Most blended fertilizers also apply K at the same time. When preparing land for high-value crops, a fertilizer application with micronutrients is advised. When mine lands are being prepared for reforestation, application of fertilizer with high P content is advised, but N applications are typically limited to relatively low levels so as to avoid stimulating herbaceous vegetative competition with young planted trees (Burger et al. 2013).

4.9 Variability

Mine soil properties are often more variable across landscapes than natural soils (Fig. 8). Where multiple topsoil substitutes have been used, surface properties may change abruptly across spatial gradients in response to changes of topsoil substitute sourcing. Mine soil construction materials may not change, however, in response to landscape conditions such as slope, landscape position, and aspect that cause predictable variation in natural soils. Even when surface materials remain consistent over a given area, subsoil materials may exhibit spatial variability due to variable material sourcing not evident at the surface. Soil property differences between surface soils and subsoils are also common, especially when topsoils or topsoil substitutes have been applied at thicknesses less than the full depth of the soil profile (Fig. 6). Even when mine soils are constructed from a uniform material, property differences with depth will occur as a result of surface accumulation of organic matter and depth-dependent differential weathering (Nash et al. 2016).

5 Hydrology

The extreme soil and geologic disturbances caused by coal surface mining also alter landscape hydrology. Appalachian mine sites often produce greater peak surface water stormflows and greater baseflows than do corresponding natural landscapes primarily due to greater water availability from decreased evapotranspiration and greater runoff caused by forest removal (Evans et al. 2015). Mine soil hydrology is fundamental to those effects which vary widely among mine sites.

Mine soils vary in their capacity to enable rainfall infiltration into the subsurface and to transmit that water to greater depths for eventual emergence as baseflows in receiving streams (Nippgen et al. 2017). Infiltration limitations, including those caused by soil compaction, can cause enhanced runoff and stormflows (Negley and Eshleman 2006; Simmons et al. 2008). Rainfall infiltration of fresh mine soils constructed from rock spoils can also be hindered by "soil crusting," a physical redistribution of mineral particles in response to rainfall impact (Daniels and Amos 1985), but this effect is usually mitigated with time by development of vegetation and consequent incorporation of organic matter into the soil surface.

Soil compaction can limit mine soils' capacities to transmit water. Such conditions at the surface can be mitigated over time by freeze-thaw and related surface processes (Guebert and Gardner 2001), but subsurface compaction, even within the rooting zone, can be more persistent (Daniels and Amos 1985; Haering et al. 2004). Surface infiltration can also be inhibited by subsurface drainage restrictions that cause the surface to become saturated during heavy rainfall. Such conditions have been observed on sloping mine sites with thin soil covers placed over subsurface mine spoils where the interface between the two materials inhibits further downward water flow (Guebert and Gardner 2001). Such conditions on hillslopes also give rise to enhanced discharge at the base due to subsurface interflows moving through the mine soil above the drainage restriction (Guebert and Gardner 2001). Excessive compaction of the subsurface prior to application of a topsoil or topsoil substitute layer may induce such conditions. The existence of subsurface drainage restrictions, if continuous across the mining area and below the rooting zone, would inhibit mine site hydrologic restoration by limiting water movement into the bulk mine spoil fill below (Evans et al. 2015). Such conditions have been observed on relatively flat land surfaces where mine soils with massive subsoil structures cause water-drainage restrictions, as indicated by development of wetland vegetation and hydric-soil features despite surface elevations well above local drainages (Haering et al. 2004). However, such conditions are not known to extend across large areas. Field investigations have revealed mine soils and landscapes constructed from rock spoils with highly heterogenous infiltration conditions, including occasional high-capacity channels from the surface through the mine soil that rapidly move saturated flows following major rain events into the underlying spoil fill (Clark and Zipper 2016; Greer et al. 2017).

Environmental processes including plant community development can cause rainwater infiltration capacities on mine soils to increase with time (Guebert and Gardner 2001; Shukla et al. 2004). Such effects have been documented on rock spoils in many parts of the world, although not to our knowledge in Appalachia (Evans et al. 2015). On mine soils lacking subsurface compaction, vegetation can influence hydrologic development. Forest trees, for example, have been found to extend rooting deeper into the subsoil than grasses, enabling subsurface macropore flow paths to extend deeper into the subsoil (Clark and Zipper 2016).

6 Classification

The earliest soil classification system for Appalachian mine soils was proposed by Sencindiver (1977) who developed a detailed proposal based upon the structure of USA Soil Taxonomy (Soil Survey Staff 1977). In this system, all mine soils occurred within the Entisol soil order (e.g., young A–C soils without B horizons). Sencindiver proposed that all mine soils be placed into the Spolents suborder and also developed a relatively detailed hierarchy of more detailed great group and subgroup taxons. This classification system was relatively specific to the Appalachian region, however, and a number of other classification proposals emerged in the 1980s and 1990s that also included urban soils and other anthropogenically disturbed materials (Sencindiver and Ammons 2000). Thus, while the Sencindiver criteria were commonly employed locally for research and descriptive purposes, Appalachian mine soils continued to be classified under existing Soil Taxonomy, all as Typic Udorthents (e.g., young soils in a humid region without B horizons).

By the early 2000s, however, the importance of recognizing human-altered and human-transported (HAHT) materials motivated the need for the improved taxonomic classification of mine soils in the USA (Galbraith 2018). Recent extensive revisions to Soil Taxonomy (Soil Survey Staff 2014) include a range of specific subgroups for HAHT soils including two directly applicable to Appalachian mine soils. Mine soils with laterally continuous highly compacted materials (e.g., densic contact) within 1 m of the surface now fall within the Anthrodensic subgroup of Udorthents, whereas soils formed in 50 cm or more in human-transported materials fall within the Anthroportic subgroup. Those soils that have developed cambic (^Bw) subsoil horizons fall into a different soil order (Inceptisols) and are classified as Anthrodensic/Anthroportic Dystrudepts (moderate to low pH) or Eutrudepts (high pH).

The US Natural Resource Conservation Service (NRCS) Cooperative Soil Survey classifies mine soils at a more-detailed level called the soil series, which is used to name local mapping units (Haering et al. 2005). Twelve mine soil series are currently recognized in the region, which vary primarily according to rock type and by rock fragment ratios, but which also incorporate texture, mineralogy, surface charge, pH and temperature classes (Table 2). Several of these series are mapped extensively in the Appalachian coal-mined landscape. For example, the Fiveblock series is dominantly mapped on moderate to high pH gray sandstones, whereas Sewell is common on brown highly acidic sandstone spoils. Kaymine is mapped over extensive finer-textured siltstone and mudrock strata while the majority of coal refuse (gob) piles are mapped as Itmann, unless they are covered in very thick (>1 m) spoils of different origin. It is interesting to note that although mine soils with cambic ^Bw horizons have been extensively described in the region and associated scientific literature (Ciolkosz et al. 1985; Haering et al. 1993, 2005), no Appalachian soil series have yet been proposed within the Inceptisol soil order. It is also quite clear from detailed field mapping and study (Haering et al. 2005) that somewhat poorly and poorly drained mine soils commonly occur across these landscapes, particularly on broad areas with gently convex surface landforms over compacted subsoils. While one such wetter series concept (Looneycreek) has been proposed and mapped in several counties in SW Virginia, it has never been officially correlated by USDA-NRCS.

7 Summary

Mine soils form in materials left or placed on the land surface by mining operations. Their properties vary widely and are strongly influenced by the materials from which they were constructed. When constructed with salvaged soil or geologic materials with properties suitable for plant growth, and without excessive compaction and when not limited by steepness, mine soils can be used for livestock agriculture, forest re-establishment, or other beneficial purposes. Mine soil properties and profiles change with time due to environmental processes such as geochemical weathering and the

Table 2 Mine soil series approved for use in classification and management of mine soils

Series	Taxonomic class	Competing series	Description and comments
Barkcamp	Loamy-skeletal, siliceous, semiactive, acid, mesic Typic Udorthents	Enoch	Siliceous, ultra-acidic (pH < 3.5); <18% clay in particle size control section; coarse fragments dominated by sandstone; well drained
Bethesda	Loamy-skeletal, mixed, active, acid, mesic Typic Udorthents	Cedarcreek	Acid; >18% clay in particle size control section; well drained
Cedarcreek	Loamy-skeletal, mixed, active, acid, mesic Typic Udorthents	Bethesda	Acid; silt loam, loam and sandy loam textures in particle size control section; well drained. Coarser textured than Bethesda. Mapped on sandstone and mudstone derived spoils
Enoch	Loamy-skeletal, siliceous, semiactive, acid, mesic Typic Udorthents	Barkcamp	Siliceous, ultra-acid (pH < 3.5); >18% clay in particle size control section; have acid shale as dominant coarse fragments; well drained. Mapped primarily on mine spoils and not on refuse piles
Fairpoint	Loamy-skeletal, mixed, active, nonacid, mesic Typic Udorthents	Kaymine and Fiveblock	Nonacid; >18% clay, with clay loam and silty clay loam textures in particle size control section; well drained. Mapped over a wide range of glaciated versus non-glaciated landscapes
Fiveblock	Loamy-skeletal, mixed, semiactive, nonacid, mesic Typic Udorthents	Fairpoint	Nonacid; <18% clay and allow sandy loam and loamy sand textures in particle size control section; >65% or more gray, neutral sandstone; somewhat excessively drained. Dominant mine soil mapped on "gray sandstones"

(continued)

Table 2 (continued)

Series	Taxonomic class	Competing series	Description and comments
Itmann	Loamy-skeletal, mixed, semiactive, acid, mesic Typic Udorthents	Sewell	Extremely Acidic (pH < 4.0) unless limed; <18% clay in particle size control section; derived from waste carbolithic rock from coal mining; >50% carbolithic rock fragments (black shales or coal refuse), somewhat excessively drained
Janelew	Loamy-skeletal, mixed, active, calcareous, mesic Typic Udorthents	Morristown	Neutral to moderately alkaline; formed in spoil derived from regolith with >65% mudstone with carbonates; well drained
Kaymine	Loamy-skeletal, mixed, active, nonacid, mesic Typic Udorthents	Fairpoint	Moderately acid to mildly alkaline; >18% clay with loam or silt loam textures in particle size control section; well-drained. Dominantly siltstone and mudrock derived. Not as fine textured in control section as Fairpoint
Morristown	Loamy-skeletal, mixed, active, calcareous, mesic Typic Udorthents	Janelew	Calcareous; formed in spoils derived from limestone, calcareous shale, sandstone, and siltstone with no one rock type making up more than 65% of the total; well-drained
Myra	Loamy-skeletal, mixed, superactive, calcareous, mesic Typic Udorthents	None directly; Morristown is similar	Superactive CEC class; calcareous; deep and well drained. Derived from high pH shale, siltstone and sandstones
Sewell	Loamy-skeletal, mixed, semiactive, acid, mesic Typic Udorthents	Itmann	Acid; <18% clay and > 65% or more brown sandstone rock fragments in particle size control section; somewhat excessively drained. Sewell is mapped on reclaimed lands while Itmann is found primarily on former refuse piles

Several series are common and widespread in the Appalachian Region, and productivity can be projected based on their properties influence of growing plants. Over extended time, most mine soils develop in ways that make them more like natural soils, but while also retaining characteristics unique to their origin. Mine soils are essential to ecosystem development and hydrologic processes occurring on Appalachia's coal-mined landscapes.

References

Acton PM, Fox JF, Campbell JE, Jones AL, Rowe H, Martin D, Bryson S (2011) Role of soil health in maintaining environmental sustainability of surface coal mining. Environ Sci Technol 45:10265–10272

Amichev BY, Burger JA, Rodrigue JA (2008) Carbon sequestration by forests and soils on mined land in the Midwestern and Appalachian coalfields of the U.S. Forest Ecol Manage 256:1949–1959

Andrews JA, Johnson JE, Torbert JL, Burger JA, Kelting DL (1998) Minesoil properties associated with early height growth of eastern white pine. J Environ Qual 27:192–198

Ashby WC (2006) Reflections of a botanist on reclamation to trees. Reclam Matters 3(Fall):5–10

Burger JA, Zipper CE, Angel PN, Hall N, Skousen J et al (2013) Establishing native trees on legacy surface mines. US Office of Surface Mining, Appalachian Regional Reforestation Initiative 11

Ciolkosz EJ, Cronce RC, Cunningham RL, Petersen GW (1985) Characteristics, genesis, and classification of Pennsylvania minesoils. Soil Sci 139:232–238

Clark EV, Daniels WL, Zipper CE, Eriksson K (2018) Mineralogical influences on water quality from weathering of surface coal mine spoils. Appl Geochem 91:97–106

Clark EV, Zipper CE (2016) Vegetation influences near-surface hydrological characteristics on a surface coal mine in eastern USA. CATENA 139:241–249

Clark EV, Zipper CE, Soucek DJ, Daniels WL (2021) Contaminants in Appalachian water resources generated by non-acid-forming coal-mining materials. In: Zipper CE, Skousen JG (eds) Chapter 9 in: Appalachia's coal-mined landscapes. Springer

Daniels WL, Amos DF (1985) Generating productive topsoil substitutes from hard rock overburden in the southern Appalachians. Environ Geochem Health 7:8–15. https://doi.org/10.1007/bf01875045

Daniels WL, Zipper CE (2018) Creation and management of productive minesoils. Virginia Cooperative Extension, 460–121

DeLong C, Skousen J, Pena-Yewtukhiw E (2012) Bulk density of rocky mine soils in forestry reclamation. Soil Sci Am J 76:1810–1815

Ditsch DC, Collins M (2000) Reclamation considerations for pasture and hay lands receiving sixty-six or more centimeters of precipitation annually. In: Barnhisel RI et al. (eds) Chapter 10, Reclamation of drastically disturbed lands. Am Soc Agronomy, Madison Wisconsin

Emerson P, Skousen J, Ziemkiewicz P (2009) Survival and growth of hardwoods in brown vs. gray sandstone on a surface mine in West Virginia. J Environ Qual 38:1821–1829

Eriksson KA, Daniels WL (2021) Environmental implications of regional geology and coal mining in the Appalachians. Chapter 2 in: Zipper CE, Skousen JG (eds) Appalachia's coal-mined landscapes. Springer

Evans DM, Zipper CE, Hester ET, Schoenholtz SH (2015) Hydrologic effects of surface coal mining in Appalachia (U.S.). J Am Water Resourc Assoc 51:1436–1452

Galbraith J (2018) Human-altered and human-transported (HAHT) soils in the U.S. soil classification system. Soil Sci Plant Nutr 64:190–199

Gale MR, Grigal DF, Harding RB (1991) Soil productivity index: predictions of site quality for white spruce plantations. Soil Sci Soc Am J 55:1701–1708

Gorman J, Skousen J, Sencindiver J, Ziemkiewicz P (2001) Forest productivity and minesoil development under a white pine plantation versus natural vegetation after 30 years. In: Proceedings of American society of mining and reclamation, pp 103–111

Greer BM, Burbey TJ, Zipper CE, Hester ET (2017) Electrical resistivity imaging of hydrologic flow through surface coal mine valley fills with comparison to other landforms. Hydrol Proc 31:2244–2260

Guebert MD, Gardner TW (2001) Macropore flow on a reclaimed surface mine: infiltration and hillslope hydrology. Geomorphology 39:3–4

Haering KC, Daniels WL, Feagley SE (2000) Reclaiming mined lands with biosolids, manures, and papermill sludges. In: Barnhisel RI et al. (eds) Chapter 10, Reclamation of drastically disturbed lands. Am Soc Agronomy, Madison Wisconsin

Haering KC, Daniels WL, Galbraith JM (2004) Appalachian mine soil morphology and properties: effects of weathering and mining method. Soil Sci Soc Am J 68:1315–1325

Haering KC, Daniels WL, Galbraith JM (2005) Mapping and classification of southwest Virginia mine soils. Soil Soil Sci Soc Am J 69:463–472

Haering KC, Daniels WL, Roberts JA (1993) Changes in mine soil properties resulting from overburden weathering. J Environ Qual 22:194–200

Haigh M, Reed H, Flege A, D'Aucourt M et al (2015) Effect of planting method on the growth of *Alnus glutinosa* and *Quercus petraea* in compacted opencast coal-mine spoils, south Wales. Land Degrad Dev 26:227–236

Haigh MJ, Sansom B (1999) Soil compaction, runoff and erosion on reclaimed coal-lands (UK). Int J Min Reclam Environ 13:135–146

Howard JL (2017) Anthropogenic soils. Springer, Progress in Soil Science

Howard JL, Amos DF, Daniels WL (1988) Phosphorous and potassium relationships in southwestern Virginia coal-mine spoils. J Environ Qual 17:695–700

Johnson C, Skousen J (1995) Minesoil properties of 15 abandoned mine land sites in West Virginia. J Environ Qual 24:635–643

Li RS, Daniels WL (1994) Nitrogen accumulation and form over time in young mine soils. J Environ Qual 23:166–172

Maharaj S, Barton CD, Karathanasis AD, Rowe HD, Rimmer SM (2007) Distinguishing "new" from "old" organic carbon in reclaimed coal mine sites using thermogravimetry: I Method Development. Soil Sci 172:292–301

Miller J, Barton C, Agouridis C, Fogel A, Dowdy T, Angel P (2012) Evaluating soil genesis and reforestation success on a surface coal mine in Appalachia. Soil Sci Soc Am J 76:950–960

Nash WL (2012) Long-term effects of rock type, weathering and amendments on Southwest Virginia Mine Soils. M.S. Thesis. Virginia Tech, Blacksburg

Nash WL, Daniels WL, Haering KC, Burger JA, Zipper CE (2016) Long-term effects of rock type on Appalachian coal mine soil properties. J Environ Qual 45:1597–1606

Negley TL, Eshlemen KD (2006) Comparison of stormflow responses of surface-mined and forested watersheds in the Appalachian Mountains, U.S.A. Hydrol Proc 20:3467–3483

Nippgen F, Ross MRV, Bernhardt ES, McGlynn BL (2017) Creating a more perennial problem? Mountaintop removal coal mining enhances and sustains saline baseflows of Appalachian watersheds. Environ Sci Technol 51:8324–8334

Pritchett WL, Fisher RF (1987) Properties and management of forest soils, 2nd edn. Wiley, New York

Roberts JA, Daniels WL, Bell JC, Burger JA (1988a) Early stages of mine soil genesis as affected by topsoiling and organic amendments. Soil Sci Soc Am J 52:730–738

Roberts JA, Daniels WL, Bell JC, Burger JA (1988b) Early stages of mine soil genesis in a SW Virginia mine spoil lithosequence. Soil Sci Soc Am J 52(716):723

Roberts JA, Daniels WL, Martens DC (1988) Tall fescue production and nutrient status on southwest Virginia mine soils. J Environ Qual 17:55–62

Rodrigue JA, Burger JA (2004) Forest soil productivity of mined land in the Midwestern and eastern coalfield regions. Soil Sci Soc Am J 68:833–844

Sena K, Barton C, Hall S, Angel P, Agouridis C, Warner R (2015) Influence of spoil type on afforestation success and natural vegetative recolonization on a surface coal mine in Appalachia, United States. Restor Ecol 23:131–138

Sencindiver JC (1977) Classification and genesis of minesoils. Ph.D. Diss. West Virginia Univ, Morgantown
Sencindiver JC, Ammons JT (2000) Minesoil genesis and classification. In: Barnhisel RI et al. (eds) Chapter 23, Reclamation of drastically disturbed lands. American Society of Agronomy, Madison, Wisconsin
Shukla MK, Lal R, Underwood J, Ebinger M (2004) Physical and hydrological characteristics of reclaimed minesoils in southeastern Ohio. Soil Sci Soc Am J 68:1352–1359
Simmons J, Currie W, Eshleman K, Kuers K, Monteleone S et al (2008) Forest to reclaimed land use change leads to altered ecosystem structure and function. Ecol Appl 18:104–118
Skousen J, Johnson C, Garbutt K (1994) Natural revegetation of 15 abandoned mine land sites in West Virginia. J Environ Qual 23:1224–1230
Skousen JG, Zipper CE (2021) Coal mining and reclamation in Appalachia. Chapter 3 in: Zipper CE, Skousen JG (eds) Appalachia's coal-mined landscapes. Springer
Soil Survey Staff (1977) Soil taxonomy: a basic system of soil classification for making and interpreting soil surveys. US Soil Conservation Service. Agricultural Handbook 436
Soil Survey Staff (2014) Keys to soil taxonomy, 12th edn. US Natural Resource Conservation Service
Sweigard R, Burger JA, Graves D, Zipper CE, Barton CD et al. (2007a) Loosening compacted soils on mine sites. US Office of Surface Mining, Appalachian Regional Reforestation Initiative 4
Sweigard R, Burger J, Zipper C, Skousen J, Barton C, Angel P (2007b) Low compaction grading to enhance reforestation success on coal surface mines. US Office of Surface Mining, Appalachian Regional Reforestation Initiative 3
Torbert JL, Burger JA, Daniels WL (1990) Pine growth variation associated with overburden rock type on a reclaimed surface mine in Virginia. J Environ Qual 19:88–92
Torbert JL, Tuladhar AR, Burger JA, Bell JC (1988) Minesoil property effects on the height of ten-year-old white pine. J Environ Qual 17:189–192
Wade GL, Thompson RL, Vogel WG (1985) Success of trees and shrubs in an 18-year-old planting on mine spoil. US Forest Service Res Paper NE-567
Wilson-Kokes L, Emerson P, DeLong C, Thomas C, Skousen J (2013a) Hardwood tree growth after eight years on brown and gray mine soils in West Virginia. J Environ Qual 42:1353–1362
Wilson-Kokes L, Emerson P, DeLong C, Thomas C, O'Dell K, Skousen J (2013b) Hardwood tree growth on amended mine soils in West Virginia. J Environ Qual 42:1363–1371
Zeleznik J, Skousen J (1996) Survival of three tree species on old reclaimed surface mines in Ohio. J Environ Qual 25:1429–1435
Zipper CE, Agouridis CT, Barton CD, Skousen JG (2021) Conversion options for mining-affected lands and waters in Appalachia. Chapter 7 in: Zipper CE, Skousen JG (eds) Appalachia's coal-mined landscapes. Springer
Zipper CE, Burger JA, Barton CD, Skousen JG (2013) Rebuilding soils on mined land for native forests in Appalachia. Soil Sci Soc Am J 77:337–349
Zipper CE, Burger JA, McGrath JM, Rodrigue JA, Holtzman GI (2011) Forest restoration potentials of coal mined lands in the eastern United States. J Environ Qual 40:1567–1577

Plant Communities on Appalachian Mined Lands

Kenton Sena, Jennifer A. Franklin, Rebecca M. Swab, and Sarah L. Hall

Abstract Since the mid-2000s, forest re-establishment has been a common post-mining reclamation goal for Appalachian mined lands, but mine reclamation and subsequent ecological processes are often not successful at establishing post-mining communities similar to native Appalachian forests. Pre-SMCRA, mined land was sometimes left barren for natural regeneration, and sometimes revegetated with varying levels of success. Much pre-SMCRA mined land is now revegetated with trees. In places, such plant communities have achieved above-ground biomass similar to forests in unmined areas, but they often attain low species diversity and are covered by non-native invasive species. Post-SMCRA, mined lands were typically seeded with herbaceous vegetation and in some cases also planted with shrubs or trees. Because of issues such as poor soil conditions, compaction, species selection, and vegetative competition, many such sites are in "arrested succession," and do not develop into forests on a decadal timeframe without intervention. Methods to reforest mined lands, developed in the mid-2000s, can be applied to establish functional forests on mined lands, but these approaches are not universally applied in Appalachia. Further management and restoration efforts for mined lands can increase biological productivity, improve habitat for wildlife and native plants, and increase provision of ecosystem services.

Keywords Surface mine · Mine reclamation · Reforestation · Restoration · Invasive species

K. Sena (✉)
University of Kentucky, Lexington, KY, USA
e-mail: kenton.sena@uky.edu

J. A. Franklin
University of Tennessee, Knoxville, TN, USA

R. M. Swab
Restoration Ecologist, Cumberland, OH, USA

S. L. Hall
Berea College, Berea, KY, USA

C. E. Zipper and J. Skousen (eds.), *Appalachia's Coal-Mined Landscapes*,
https://doi.org/10.1007/978-3-030-57780-3_5

1 Introduction

Surface mining has caused dramatic land cover changes throughout the Appalachian region (Wickham et al. 2007; Drummond and Loveland 2010), converting rich tracts of some of the world's most biodiverse temperate forests into less diverse and less valuable post-mining ecosystems. Approaches to and outcomes of surface mine reclamation have changed over time, especially as directed by policy. There are generally two eras, the 'tree law' era of reclamation, and the 'grass law' era—the divide formed by The Surface Mining Control and Reclamation Act of 1977 (SMCRA) which established federal regulatory controls over coal mining. Some pre-SMCRA mine sites were characterized by severe problems such as unstable soils prone to landslides and erosion (Plass 1966, 1967), acid mine drainage, and areas on which no vegetation grew, while others with soils well-suited for plant growth reverted or were planted to mixed forest and grass systems composed of species from the surrounding plant communities (Skousen et al. 1994). After SMCRA, surface mine reclamation often produced novel ecosystems dominated by non-native grasses, legumes, and shrubs (Simmons et al. 2008; Zipper et al. 2011a). Since the mid-2000s, reclamation techniques more favorable for reforestation of post-SMCRA land have been developed (Zipper et al. 2011b). Initial post-mining vegetative communities are structured by site conditions, especially soil quality and reclamation planting mixes. Over time, these communities are modified by soil and vegetation development processes such as accumulation of soil organic matter, mineral weathering, development of soil structure, colonization of plants from nearby lands, and competitive and facilitative interspecific interactions, as well as anthropogenic processes such as land management activities.

2 Appalachian Forests

The Appalachian region is home to one of the world's most biodiverse non-tropical ecosystems. The diverse forests of the Appalachian coalfield today are a result of many factors, including glacial advances and retreats during the Pleistocene and their concurrent climatic changes and species migrations. Around 10,000 years ago a marked warming trend began in eastern North America (Yahner 2000). The resulting humid climate, moderate temperatures, and varied physiography helped to shape the eastern deciduous forest, giving rise to diverse plant communities with more than 100 tree species (Yahner 2000). The mixed mesophytic forest region covers much of the Appalachian coalfield and is recognized for its high biodiversity (Braun 1950). As an example of the profound diversity of plant life found within this region, Ricketts et al. (1999) recognize 2487 vascular plant species in their Appalachian Mixed Mesophytic Forest ecoregion, which overlaps with USEPA (2020) level III Ecoregions 68, 69, and 70 (Fig. 1).

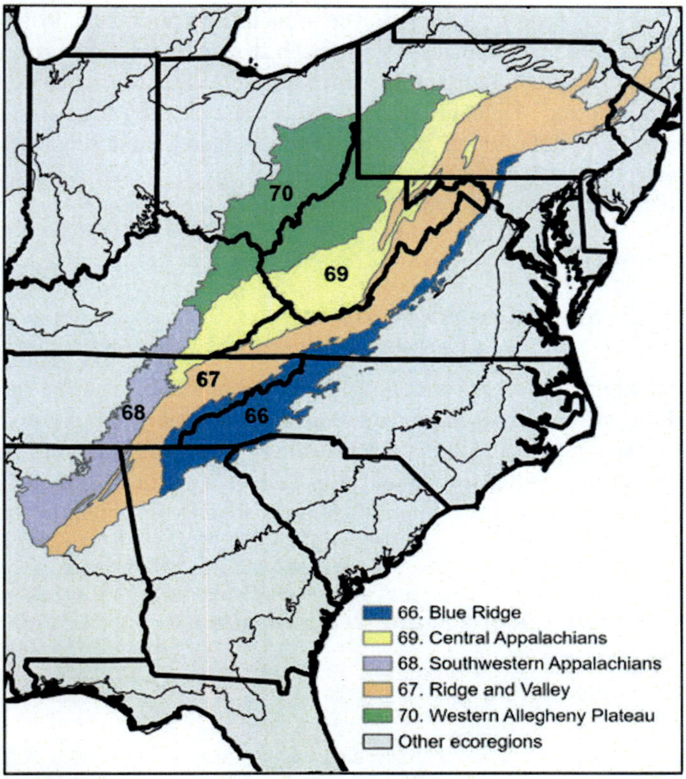

Fig. 1 Level III Ecoregions for the eastern U.S. (USEPA 2020) that overlap with the Appalachian coalfield

Within Braun's (1950) Appalachian Plateau province (which incorporates the Appalachian coalfield), a number of forest associations are common. Major contrasts in forest tree communities across landscapes are driven by factors such as topography, aspect, and latitude. North- and east-facing slopes are generally moist and characterized by species such as tulip poplar (*Liriodendron tulipifera)*, sugar maple (*Acer saccharum)*, northern red oak (*Quercus rubra)*, and black walnut (*Juglans nigra*), while south- and west-facing slopes tend to be drier, with species such as chestnut oak (*Quercus montana)* and hickories (*Carya* spp.). Similarly, drier ridgetops contain species tolerant of more xeric conditions (e.g., *Pinus* spp.) than those found in mesic valleys (e.g., *Magnolia* spp.). The region's northern latitudes and higher elevations are characterized by species that are adapted to colder winters, such as maples (*Acer* spp.) and eastern hemlock (*Tsuga canadensis*), while at southern latitudes conifers adapted to warmer climates are more common (e.g., shortleaf pine, *Pinus echinata*, and Virginia pine, *Pinus virginiana*). The predominant tree species in the region vary with these factors but are largely of the oak-hickory forest type (with American chestnut, *Castanea dentata*, having played a major role prior to its loss). The presence (or absence) of many species has also been shaped by prior disturbance history, with humans playing an important role since Native Americans appeared more than 12,000 years ago (e.g., Delcourt et al. 1999). Forest clearing for fuel, agriculture, tanning, charcoal, shipbuilding, and other purposes increased in scope and intensity as the human population spread. However, the thick seams of coal eventually became a driving factor for a different type of disturbance on the landscape.

3 Plant Communities After Mining

Surface coal mining disturbs pre-mining plant communities by clearing native forests and disturbing and displacing native soil. Post-mining plant communities often differ greatly from the forests that dominate Appalachia's unmined landscapes. Mine reclamation methods have varied throughout Appalachian's mining history (Skousen and Zipper 2021), as do the plant communities established by reclamation activities.

3.1 Pre-SMCRA Mining and Reclamation

3.1.1 Early Reclamation

Large areas of Appalachia were affected by coal mining prior to SMCRA (Skousen and Zipper 2021). In Kentucky, for instance, approximately 109,000 hectares of land was surface mined for coal before SMCRA (Thompson et al. 1986). In typical pre-law mining, topsoil was pushed aside with bulldozers and buried under overburden (Skousen and Zipper 2021; Skousen et al. 2021). Raw rock material from just above the coal was left exposed on the land surface to become plant growth media. In non-mountainous areas, the piles of mixed fine material and rock were often left in long, parallel ridges (spoil banks), with little vegetation (Bramble and Ashley 1955). On steep terrains, overburden was simply pushed downslope. Because overburden was typically placed as it was removed, mine-soil chemistry could vary dramatically across a site. Soils on research sites in Kentucky were found to vary in pH from 2.6 to 8.5, and from 18 ppm to less than 1 ppm of available phosphorus (Plass 1975; Thompson et al. 1986). Moisture content of post-mined soils was also highly variable, and often unfavorable for plant growth (Carter and Ungar 2002). In addition, soil and surface temperatures could be extremely high, potentially exceeding 60 °C in open sunlight, hot enough to cause heat injury to seedlings (Bramble and Ashley 1955; Bell and Ungar 1981). Pre-SMCRA spoil banks were often characterized by instability and erosion risk, especially in steep terrain. In northern Appalachia, extreme acidity in mine soils and water was common, with acid soil conditions less common in the southern part of the region (Bramble and Ashley 1955; Bell and Ungar 1981; Carter and Ungar 2002).

Such severe conditions were unfavorable for plant establishment from seeds, thus constraining natural revegetation and often resulting in unstable slopes that remained poorly vegetated for 30 years or more (Bramble and Ashley 1955). Soils below 4.0 pH were typically barren, but with time and weathering of exposed soils, soil conditions improved and plant growth and community development increased soil organic matter, and infrequent vegetated 'islands' developed (Croxton 1928; Bell and Ungar 1981).

3.1.2 Reclamation into Reforestation

In some areas, trees were planted or established unaided. However, the communities were less diverse than forests prior to mining, with different composition and lower density. For instance, plant communities in West Virginia varied based on the prevailing mine soil acidity and rock fragment content (Skousen et al. 1994; Johnson and Skousen 1995), and a Pennsylvania study found the pioneer community to be a tree-shrub mixture similar to that found on rock slides (Bramble and Ashley 1955). In areas where the spoils with favorable chemistry were placed without grading, native species colonized readily on the deep, loose mine soils. In some cases, tree height and diameter growth rates were greater than on the shallow natural soils of the surrounding area, but species composition still differed. Regeneration was dominated by faster-growing hardwoods (e.g., red maple, *Acer rubrum*; tulip poplar) rather than the oaks (*Quercus* spp.) typical of the surrounding forests (Frouz and Franklin 2014). Other areas, however, including the relatively flat benches created by excavation of coal seams where mining equipment operated, were left with highly compacted soil conditions that inhibited plant-root growth and vegetation development (Daniels and Amos 1981, 1985). One study found only 36.2% of milacre (4.05 m^2) quadrats to have a seedling—inadequate for successful reforestation within 35 years (Bramble and Ashley 1955).

3.1.3 Regulation and Practice

While the pre-SMCRA era has been called the 'tree law era' by some, laws and regulatory programs established by states during the decades prior to SMCRA differed greatly in their reclamation requirements (Imhoff et al. 1976). For example, in Kentucky prior to 1970, many areas were revegetated either naturally or by a combination of reclamation plantings and natural plant succession (Thompson et al. 1986). Reclamation efforts in Ohio were initially voluntary until a 1952 law required trees to be planted. After the 1952 law, Ohio Power, one of the larger mining companies in the state, developed a reclamation method with reforestation as the focus (Gary Kaster, Personal Communication 1969). Trees were planted on the mined spoils using a "three row" method utilizing early-, mid-, and late-successional species. The early successional species (black locust, *Robinia pseudoacacia;* and black alder, *Alnus glutinosa*) would establish quickly, the second row would develop into pulpwood (aspen, cottonwood (*Populus* spp.) and soft maples (*Acer* spp.)), and the third row would develop into the eventual climax forest of oaks, poplar, and ash (Gary Kaster, Personal Communication 1969). This method was used on thousands of acres in the largest strip-mined area in Ohio through the 1960s. In 1972, a new state reclamation law required the restoration of mined land to approximate original contours, replacement of soil on the land surface, a final vegetative cover of grass, and other requirements similar to those of SMCRA. In Virginia prior to SMCRA, miners established black locust on contour-mine outslopes along with three rows of eastern white pine (*Pinus strobus*) at the outer edge of mine benches, anticipating that the white pines would grow to provide visual cover for the highwall (Rodrigue 2001). In West Virginia prior to the enactment of SMCRA, conservation districts were often contracted by mining companies to reclaim mined areas (Gorman et al. 2001).

By the time SMCRA was passed in 1977, establishment of vegetative cover by methods that included tree planting had become relatively common (e.g., Thompson et al. 1986). Plant diversity was not a priority and planted areas were considerably lower in species diversity than in pre-mining forests (Holl and Cairns 1994). Typical reclamation goals on these sites included timber production, wildlife habitat, or aesthetics (Bramble and Ashley 1955). Reclamation performance was highly variable, but some pre-SMCRA sites developed healthy and productive forest tree communities (Fig. 2; Rodrigue et al. 2002).

In the best cases, successful tree plantings developed into forests with good potential for timber 20–55 years after planting (Rodrigue et al 2002). However, even successful plantings were frequently low-diversity and often dominated by fast-growing non-native conifer species (Rodrigue et al. 2002), which do not provide the same wildlife habitat benefits as the native hardwoods. Some plantings were of more diverse species mixtures focused on wildlife habitat, soil enhancement, or aesthetics, but many of these plantings included non-native species such as tree of heaven (*Ailanthus altissima*) or autumn olive (*Elaeagnus umbellata*) which are now considered invasive and no longer planted. On some of the harsher sites, reclamation efforts failed, and sites remained unvegetated—such as one highly acidic site where spoil banks were still bare 40 years after mining (Carter and Ungar 2002). Even on sites where above-ground timber volumes were similar to adjacent unmined sites, soil biology, microbial processes, and organic carbon contents were much slower to recover (Amichev et al. 2008; Frouz and Franklin 2014).

3.1.4 Issues Plaguing 'Tree Law' Reforestation Sites

While pre-SMCRA reforestation efforts were often successful at establishing tree canopy, other issues plagued these developing forest communities. Since pre-SMCRA mining left highwalls, poorly vegetated outslopes, and acid mine drainage, pre-SMCRA sites suffered from poor water quality and rough terrain (ODNR 2000). Additionally, invasive exotic shrubs and trees were commonly planted on reclamation sites (Plass 1975; Thompson et al. 1984; Wade et al 1985) as a means of dealing with harsh soil and site conditions, where many tree species native to Appalachia did

Fig. 2 A naturally established forest 48 years after mining on a pre-SMCRA site in Tennessee, dominated by mature early-successional trees such as yellow poplar. Spoil was loosely dumped, and the site was not graded or planted. (Photo by J. A. Franklin)

not establish and grow as well. For instance, a 1965 planting in Kentucky with 25 tree and 25 shrub species included only 10 native trees and six native shrubs (Plass 1975; Wade et al. 1985). As late as 1985, 20 years after planting, nearly all of the non-native species planted on this site, excepting only multiflora rose (*Rosa multiflora*), were still promoted by planting contractors for their "wide environmental tolerance," "role as soil builders," or ornamental value (Wade et al. 1985). This study also reported that autumn olive composed up to 75% of the vegetative cover on one site (Wade et al. 1985). As a result of these planting practices, the understories of many early-successional forests developing on pre-SMCRA mined lands have dense stands of non-native invasive plants. Furthermore, tall-growing non-native species such as tree of heaven and royal paulownia (*Paulownia tomentosa*) can become overstory components of developing forests. Overall, although some pre-SMCRA mined land forest communities may have achieved above-ground biomass similar to unmined areas (Amichev et al. 2008), most still differ in character from the region's native forests (Franklin and Rizza 2008), and may lack large trees, overstory species diversity, larger-sized downed woody debris, and standing snags; and may have understories lacking native species or dominated by invasive shrubs.

3.2 Early Post-SMCRA Sites

The passage of SMCRA ushered in a new era of surface mine reclamation focused on avoiding common pre-law mining problems by improving soil stability, reducing the risk of landslides, and minimizing impacts on water quality. Surface mines grew larger and placed compacted, alkaline mine spoils on the surface as mine soils (Haering et al. 2004). Mine sites were reclaimed using heavy soil compaction and competitive groundcover to reduce erosion risk (Angel et al. 2017; Simmons et al. 2008). This approach, while successful at remedying some of the most common and severe pre-SMCRA mining problems, rendered reclaimed sites unfavorable for the establishment and growth of native trees, effectively shifting post-mining vegetative communities from forests to ecosystems dominated by grasses, legumes, and shrubs. In general, vegetative communities developing on sites reclaimed under SMCRA that remain non-forested and unmanaged sometimes referred to as "legacy mines" (Burger et al. 2017b), are characterized by dense herbaceous cover and abundance of non-native species.

In the first few decades following SMCRA's passage, Appalachian surface mines were frequently vegetated with non-native species, including tall fescue (*Schedonorus arundinaceus*), sericea lespedeza (*Lespedeza cuneata*), and autumn olive (Fig. 3). These and other non-native species were known from pre-SMCRA revegetation practices (as noted above) to be good at rapidly establishing groundcover in unfavorable conditions (Yearsley and Samuel 1980; Auch et al. 2005). Due to soil compaction and highly competitive vegetation, post-SMCRA sites tend to remain unfavorable for native tree establishment for many years. A chronosequence study

Fig. 3 Surface mined land reclaimed under SMCRA, with characteristic plant community dominated by autumn olive (lighter-colored shrubs visible throughout the frame) and non-native grasses and legumes. (Photo by M. Barton)

of post-SMCRA reclaimed sites in Virginia found that mine sites had lower species richness and greater cover by herbaceous species than adjacent unmined forests (Holl and Cairns 1994; Holl 2002). Cavender et al. (2014) reported that post-SMCRA reclaimed sites 30–40 years after reclamation had less than 5% of cover consisting of native species, few of these trees. Other studies have found reclaimed sites to be dominated by tall fescue and sericea lespedeza (Sena et al. 2014; Clark and Zipper 2016), and frequently report abundant invasive shrubs such as autumn olive (Wood and Williams 2013; Oliphant et al. 2017). A 2005 survey across four states found plant communities resembling native forests to be significant contributors to vegetative cover on only 2 of 25 mine sites; while non-native plants comprised dominant vegetative cover on more than 50% of the surveyed area (Zipper et al. 2011a).

3.3 *The Forestry Reclamation Approach (FRA)*

Over time, reclamation practitioners became aware of the challenges facing surface mine reforestation under SMCRA, sparking collaborative research aimed at improving mine reforestation success. This effort was spearheaded by the Appalachian Regional Reforestation Initiative (ARRI), a group of researchers, practitioners, and regulators from across the region. Beginning in 2005, ARRI produced a series of advisories describing a method for re-establishing forest plant communities during reclamation, the Forestry Reclamation Approach (FRA) (republished in Adams 2017). Briefly, the FRA entails five steps: (1) select the best-available rooting medium, (2) minimize compaction, (3) minimize vegetative competition, (4) plant both early- and late-successional native tree species, and (5) employ best practices for tree planting.

The steps of the FRA have been tested at many experimental sites across the region in studies focused on factors influencing the growth and survival of native trees. Although native topsoil presents a ready source of native seeds and vegetative propagules that can aid reforestation (Hall et al. 2010), the shallow soils of Appalachian forests are often not used in reclamation. Thus, soils on reclaimed mines are often constructed from fractured overburden.

Overburden materials are classified as either weathered or unweathered. Weathered overburden extends from beneath the soil to depths generally in the range of 10–20 m prior to mining; these materials have been influenced by surface environmental processes such as oxidation and leaching to a greater extent than unweathered material located further below (Zipper et al. 2013). Researchers recommend the use of salvaged soil, either alone or mixed with weathered overburden, for soil construction on mine sites intended for reforestation where soil can be salvaged (Skousen et al. 2017). In studies conducted in multiple states, selective use of weathered sandstone rather than unweathered mine spoils improved planted tree growth (Torbert et al. 1990; Wilson-Kokes et al. 2013a, b; Sena et al. 2015, 2018), leading researchers to recommend weathered sandstone as the best available topsoil substitute when topsoil cannot be retained (Skousen et al. 2017). Additional studies reported that reducing soil compaction during reclamation improved planted tree growth (Cotton et al. 2012; Wilson-Kokes et al. 2013b), supporting recommendations to minimize compaction during soil placement. Compaction avoidance also aids erosion control by enabling greater rainfall infiltration.

The third step of the FRA—minimizing vegetative competition—is an important step of the process. On the one hand, seeding groundcover vegetation confers some benefits to planted trees through soil temperature reduction, organic matter accumulation, erosion control, increased water infiltration, and nitrogen fixation by legumes (Franklin et al. 2012). However, aggressive groundcover plants also compete with planted trees for vital resources: water, nutrients, and sunlight (Franklin et al. 2012). Overall, the FRA seeks to establish enough groundcover to control erosion but not so much that trees are hindered by competition for resources (Burger et al. 2017a). The FRA also calls for planting of both early-successional tree species and late-successional species that comprise the local mature forest canopy. Mine reforestation tree-planting mixes used for the FRA typically include multiple species from across the successional spectrum with the intent of producing a diverse forest community.

From 2004–2015, an estimated 95 million trees were planted on over 50,000 ha of surface-mined land in Appalachia (Barton et al. 2017). Experimental sites reclaimed in accordance with all FRA recommendations typically exhibit excellent growth and survival of planted trees. For example, tulip poplar growing in uncompacted weathered sandstone in Kentucky approached growth rates of tulip poplar regenerating from clearcuts in unmined forests (Cotton et al. 2012). Other experimental sites in Kentucky (Miller et al. 2012; Sena et al. 2015, 2018), West Virginia (Wilson-Kokes et al. 2013b), and Virginia (Fields-Johnson et al. 2012) exhibited good growth and survival by trees planted in accordance with FRA recommendations (Fig. 4). As planted trees grow on FRA-reclaimed sites, they can quickly establish canopy cover—critical for placing the site on a developmental trajectory toward native forest. In a Kentucky study, planted trees of multiple species (white oak, *Quercus alba*, northern red oak, tulip poplar, and green ash, *Fraxinus pennsylvanica*) exhibited excellent growth and survival on uncompacted brown sandstone and had established canopy closure in patches by the ninth growing season after planting (Sena et al. 2015). Although groundcover was not seeded as part of this experiment, brown sandstone plots were rapidly colonized by a variety of native and non-native species (70 species after three growing seasons), demonstrating that natural regeneration may quickly establish a diverse groundcover when soil and seed-dispersal conditions are favorable (Sena et al. 2015). Importantly, the naturally colonizing understory and groundcover species reported in this study included a number of invasive non-native species, such as sericea lespedeza, autumn olive, and royal paulownia. Sericea lespedeza was dominant in plot areas that had not yet achieved canopy closure and was suppressed as shade increased. These shaded areas were also characterized by a number of shade-tolerant native species, such as wild hydrangea (*Hydrangea arborescens*), demonstrating that natural regeneration is capable of restoring some native understory species when mine soil properties are favorable. However, invasive shade-intolerant species will likely remain, though

Fig. 4 (Foreground) Low-compaction experimental reforestation plots at the Starfire Mine in eastern Kentucky, approximately 20 years after planting. (Photo by M. Barton)

they tend to decline in abundance as canopy coverage increases. Forestry reclamation success may be improved by intentional management early in the reclamation process to prevent or limit the establishment of invasive plant species (Zipper et al. 2019).

Reclamation of mined land according to FRA recommendations may also present opportunities to aid in the recovery of native trees that have been historically extirpated (or nearly extirpated) by invasive pests and pathogens. For example, preliminary results indicate that Dutch-Elm-Disease resistant American elm (*Ulmus americana*) will be successful in mine reclamation plantings (Adams et al. 2015). Similarly, American chestnut, a formerly dominant canopy species throughout the eastern US that was brought to the brink of extinction by the introduction of two non-native pathogens (Anagnostakis et al. 2001), has exhibited vigorous growth rates on suitable experimental sites (Gilland and McCarthy 2012; Barton et al. 2015; French et al. 2017). If disease-resistant varieties of American elm and American chestnut can be developed successfully, surface-mine replanting efforts may facilitate reintroduction of these species to regional forests (Skousen et al. 2018).

While FRA practices have been applied during reclamation on some Appalachian coal surface mines since the mid-2000s, the effectiveness of routine applications by mining firms has not been well-studied, and the extent to which native tree cover is being re-established on Appalachian surface mine sites has not been determined. Placement of a suitable growth medium for trees, such as weathered mine spoils and/or salvaged soils, on the surface is an essential step in effective reforestation (Zipper et al. 2013). Availability of such materials on many modern mine sites, however, is often limited; and selective use of those materials as mine soils can require mining companies to bear extra costs. Hence, much FRA-like reforestation is actually conducted on unweathered spoils, which research suggests will result in reduced rates of growth and/or survival for the planted trees (Zipper et al. 2013). Observation indicates that slow-establishing mine reforestation sites are more vulnerable to invasions by non-native plant species which can suppress native trees and interfere with successful reforestation (Evans et al. 2013). Mine regulatory agencies commonly require densities of living trees to equal or exceed specific thresholds but do not assess growth rates or the presence of undesirable species such as invasive non-natives. Hence, the longer-term condition of mine sites reclaimed during the period of widespread FRA application is not yet known and represents a significant research need.

3.4 Other Post-SMCRA Reclamation

Many mined lands have been reclaimed using other methods and host plant communities that differ from those described above. Hayland and pasture are common post-mining land uses, especially in Appalachian regions that lack steep slopes (Ditsch and Collins 2000; GAO 2009). Such lands host plant communities dominated by forage grasses and legumes, but often include exotic invasives such as sericea lespedeza that are nearly ubiquitous in Appalachian mined areas. Furthermore, mine sites are suitable for establishing native warm-season grasses, such as switchgrass (*Panicum virgatum*), with potential for either forage or bioenergy application (Marra et al. 2013; Buckley and Franklin 2008).

A common non-forest post-mining land use is wildlife habitat (GAO 2009), with more than 10,000 hectares having been reclaimed to this post-mining land use in West Virginia alone (Kelley and Anderson 2009). Revegetation strategies for wildlife habitat lands vary throughout the region. In Virginia during the 1990s, for example, many such lands were revegetated with eastern white pine and black locust as trees and the exotics autumn olive and bicolor lespedeza (*Lespedeza bicolor*) as shrubs. Many such lands are densely vegetated today, albeit with species assemblages that pose little resemblance to native forests. In contrast, West Virginia practices for establishing wildlife habitat through the late 2000s included seeding reclaimed areas with a mixture of pasture-like grasses and legumes along with bicolor lespedeza (Kelley and Anderson 2009), in some cases also planting mast-producing but exotic invasive shrubs such as autumn olive. In 2017, the Ohio Department of Natural Resources began to offer landowners the option to use native prairie seed mixes in reclamation with the goal of increasing pollinator and other wildlife habitats (Swab et al. 2017). Current recommendations for establishing wildlife habitats include using the FRA to plant native trees and shrub species that have characteristics beneficial to specific wildlife species groups (Wood et al. 2017), but the plant-community outcomes of forestry reclamation practices intended to produce wildlife habitat have not been well-studied.

4 Mined-Land Vegetation Management

With the exception of lands used for haylands, pasture, or other agricultural products, Appalachian mined lands are rarely treated with management practices for the purpose of improving species composition, native plant cover, biological productivity, ecological function, or characteristics that would increase their societal value. As highly altered ecosystems, these lands present management challenges. Management goals for forested sites could include ecological endpoints such as restoration of a forest with characteristics similar to pre-mining forests, enhancement of timber production, or provisions for wildlife habitat.

4.1 Managing Forestry-Reclaimed Mine Sites

Active and adaptive management of reforested mine sites may be required to keep a recovery trajectory toward native forests. Development of a management plan with these objectives will require consideration of site characteristics in the context of the larger landscape. This could include an analysis of structural diversity, heterogeneity, and connectivity at the landscape scale (Lindenmayer et al. 2000). Targets for the restoration of biodiversity may be much lower on small sites, and the range of structural diversity as well as specific taxa may be less important in a landscape with high natural heterogeneity. Given limited resources for a monitoring program, the specific indicators used for assessing ecological structure and function must be carefully selected for the site and could include measures at different ecological scales (Vallauri et al. 2005).

Depending on the current state and age of the reforested site, management activities targeting restoration of native forest structure and function may include removing invasive species (Fig. 5), planting native understory species, creating wildlife habitat structures (Malone et al. 2019), and restoring historical fire regimes. Pre-SMCRA mined-land forests often established a tree canopy but had understory communities dominated by invasive shrubs and characterized by low abundance of native species. Preliminary results from one experimental restoration site suggest that invasive species removal can increase understory species richness and diversity (Table 1). This study also demonstrated additive benefits of following invasive species removal with native species (including shrubs, trees, and herbaceous species) planting (Malone et al. 2019). Non-native-species removal may be necessary for recovery of a diverse native plant community in the understory of developing mined-land forests, and planting native species afterwards may accelerate this recovery and ensure restoration of species of concern.

Fire is part of the natural disturbance regime of oak-dominated systems throughout the Appalachian region (Varner et al. 2016). However, over the past several decades, eastern deciduous forests have undergone ecological changes, including a decline in species richness and community shifts to greater dominance by fire-intolerant

Fig. 5 Understory of early-successional forest growing on reclaimed mined land in southeastern Ohio, dominated by the invasive shrub autumn olive. (Photo by R. M. Swab)

Table 1 Mean (± standard error) plant community metrics for a pre-SMCRA mine site in southeastern Ohio that was reforested using three reforestation treatments* during the second growing season after treatment**

	Reforestation treatment		
	Cleared, planted	Cleared, unplanted	Uncleared
Species richness	10.53 ± 1.01^a	11.15 ± 0.99^a	8.39 ± 1.52^b
Shannon–Wiener Diversity	1.84 ± 0.17^a	1.94 ± 0.13^a	1.56 ± 0.18^b
Understory canopy coverage in June	76.37 ± 12.22^a	76.25 ± 21.98^a	96.41 ± 2.55^b
Floral bloom richness in June	5.92 ± 2.78^a	3.50 ± 2.07^b	1.33 ± 0.89^c

*Some areas were cleared of invasive species and planted with woody canopy and understory species, herbaceous plugs, and herbaceous seeds. Some areas were cleared and left unplanted. Richness and diversity estimates represent total understory plant diversity, including native and non-native species

**Letters (abc) denote significant differences between treatments. All differences were significant to the $p < 0.001$ level. Data collected by Sarah Brown, summer 2019

species, especially maples (Alexander and Arthur 2010), due to the fire suppression policies of the mid-1900s (Nowacki and Abrams 2008). Prescribed fire can be used to promote the return of historically native fire-dependent communities. Thinning of dense young stands or other fuel reduction treatments may be needed along with fire to restore historic fuel characteristics and fire behavior (Brown et al. 2004). If target plant communities are fire-dependent (e.g., oaks, hickories, and pines), use of prescribed fire will enhance management outcomes.

Another important management goal may be to enhance wildlife habitat. While some of the management practices important for restoration of pre-mining forest conditions will also promote habitat quality, active and adaptive management will include specific activities to improve habitat for species of interest (Lituma et al. 2021). For example, creation of standing snags and brush piles may increase habitat for bats (underneath loose bark), solitary cavity-nesting pollinators (within dead stems), and small mammals and reptiles (underneath brush piles and among rocks). Artificial bat roosts are another alternative, such as bat boxes, which create roosting habitat in forests otherwise lacking the appropriate structure. Creation of vernal pools may provide amphibian breeding habitat. Approaches to increase wildlife habitat on reforested mine lands are detailed by Wood et al. (2017).

Finally, landowners may choose to manage reforested mined sites for timber production. As with management for wildlife habitat, some practices important for restoring pre-mining forest conditions will also enhance certain timber production outcomes. For example, use of prescribed fire will assist in the recovery of forests dominated by high-value oak species. If timber is a management objective, then management for timber stand improvement can begin after canopy closure. This involves removal of some trees to improve the growth of desired individuals (Casselman et al. 2007). Site-specific timber management plans can be developed in collaboration with professional foresters.

While these goals describe management of reforested mine sites, some landowners may wish to aid development of forest communities on legacy mine sites (Zipper et al. 2021). Successful reforestation on these sites requires mitigation of soil compaction and vegetative competition (Angel et al. 2011; Burger et al. 2017b). Generally, any soil-compaction mitigation treatment (e.g., disking, tilling, deep-ripping) prior to tree planting improves survival and/or growth of planted trees over time (McCarthy et al. 2008, 2010; Skousen et al. 2009; Burger and Evans 2010; Fields-Johnson et al. 2014). Methods for vegetation control include herbicide treatments (Fields-Johnson et al. 2014) and weed mats (Sena et al. 2014); but these approaches are inconsistently effective and may require significant cost and effort. Recently, some practitioners have reported successful preliminary control of competitive vegetation on legacy sites using a scraping method—a bulldozer is used to scrape the top layer of soil, including the invasive shrubs and grasses and their seedbank, to the perimeter of the reforestation site (Adams et al. 2019).

4.2 *Managing Reclaimed Sites as Grasslands*

Post-SMCRA legacy sites maintained as cool-season grasslands can be used for grazing (Fig. 6) and have been shown to provide habitat for grassland birds, some of which are threatened or declining species (Ingold 2002). However, most of the tall fescue planted on these sites likely harbors a fungal endophyte (*Neotyphodium*

Fig. 6 A post-SMCRA site in Tennessee reclaimed 12 years prior with the goal of providing wildlife habitat. Vegetation consists mainly of species planted during reclamation. Tall fescue and sericea lespedeza make up 87% of vegetative cover. (Photo by J. A. Franklin)

coenophialum) that is problematic for some grazing animals (Ball et al. 1993) and may inhibit natural succession (Rudgers et al. 2007). In addition, grassland birds may be threatened by incursion of invasive species such as autumn olive (Ingold and Dooley 2013). A satellite imagery study detected autumn olive on 12.6% of over 30,000 ha of reclaimed mined land in Virginia (Oliphant et al. 2017).

These sites can be improved by controlling non-native species and favoring native grasses and forbs to create prairie-like ecosystems for management as native grasslands. On reclaimed mine lands in Ohio, prairie species have been utilized in mine reclamation since at least the 1970s, effectively establishing native vegetation (Drake 1980; Rodgers and Anderson 1989; Schramm and Kalvin 1978; Swab et al. 2017) and even supporting endangered and threatened wildlife species (Lanoo et al. 2009). Prairies can provide native plant cover and pollinator habitat and have been shown to increase soil health (i.e., nutrient cycling, organic matter, soil microbial biomass) in abandoned agricultural lands (Allison et al. 2005; Matamala et al. 2008). Prairie species planted on reclaimed mine land did not show specific benefits for soil health after two years (Swab et al. 2017), although this is a short time frame for soil restoration and benefits may require longer time periods. Small prairie-like areas may have existed in Appalachia prior to European colonization; historical accounts suggest that such grassland systems were more common towards the southern end of the Appalachians. These consisted of several different community types, including eastern tallgrass prairie and oak savannah, all high in diversity (Noss 2012). However, active, and often intensive, management such as fire, mowing, or grazing, is required to maintain grasslands in predominantly native herbaceous vegetation that supports biodiversity. Therefore, prairie restoration may not provide the same benefits in non-managed sites.

5 Challenges Facing Vegetative Communities on Reclaimed Mine Sites

The trajectory of plant community development on reclaimed mine sites, while largely controlled by establishment conditions such as soil quality and species selected for planting, is also influenced by ongoing pressures that operate over various temporal and spatial scales. These pressures include invasive plant species (discussed throughout this chapter), climate change, non-native pests and pathogens, and herbivory. Forests in particular may be susceptible to threats, which may work individually or synergistically combine.

The primary climate trends influencing Appalachian forests are increasing temperatures (and temperature fluctuations) with accompanying shifts in plant phenology and changing precipitation patterns with drier fall weather and more frequent extreme rainfall events (Butler et al. 2015). The ranges of many forest species are predicted to either expand or contract in response to changing climate. Forest types that are predicted to be most adaptive, and with low to moderate vulnerability to climate change, include mixed oak-pine, dry and mesic oak forest, and mixed mesophytic forest (Butler et al. 2015), which are the forest types commonly targeted by FRA planting mixes in Appalachia.

As with relatively undisturbed forests, recovery of healthy plant communities on these sites may be challenged by non-native pests and pathogens, which have a long history of driving major ecological change in their introduced ranges. For example, American chestnut was functionally extirpated from eastern US forests by introduced pathogens in the early part of the twentieth century (Anagnostakis 2001), and native ash species (*Fraxinus* spp.) are currently being eliminated by the introduced emerald ash borer (*Agrilus planipennis*) (Larson and Vimmerstedt 1983; Zeleznik and Skousen 1996; Knight et al. 2010). Management of plant communities on mine sites must take these broad pressures into consideration.

Climate change, pests and pathogens, and herbivory may combine synergistically, potentially causing deforestation in the worst cases. Deer and elk browse has limited tree establishment in some areas and can strongly influence forest composition due to selective browsing (Brenner and Musaus 1977; Burney and Jacobs 2018; Franklin and Aldrovandi 2018). If browse limits seedling or sapling survival and growth rates while climate change and pests remove species, reclaimed forests, which tend to be low diversity and have dense stands of invasive species in the understory, may be converted to invasive shrublands. Active management focused on increasing diversity and controlling invasives may be necessary to maintain forests in the face of these threats.

6 Plant Community Development: Looking to the Future

Ecosystem succession is a natural process that causes changes in plant community composition over time. Reclaimed mine sites vary in substrate chemistry, reclamation planting composition, and proximity to sources of native or invasive vegetation to colonize; hence, successional trajectories vary as well. Where multiple species of native trees have been planted and soil conditions are favorable, succession can proceed rapidly and may produce a forest plant community on a trajectory toward those present on unmined areas. On sites with soils poorly suited to native trees or with dense and aggressive non-native vegetation, succession can be delayed or altered from natural-system processes (Zipper et al. 2011a; Cavender et al. 2014; Groninger et al. 2017).

Unless tree species characteristic of native forests are planted, the first plant communities to develop on mine sites are typically comprised of early successional species such as grasses, legumes, or woody species (fast-growing black locust and red maple), and will include non-native invasive plants. On large mine sites lacking successful reforestation, multiple decades or centuries may be required for regeneration of native species with heavy seeds that move slowly over landscapes, such as the oak and hickory species that dominate many native Appalachian forests. Succession on mine sites with soils favorable to native forest may follow a trajectory similar to what occurs on unmined sites. For example, Skousen et al. (2006) found multiple species of native trees from adjacent forests had colonized mine soils comprised of weathered spoils on pre-SMCRA mine outslopes. However, colonization of areas compacted by mining equipment was more limited. Also, the presence of non-native invasives which inhibit native-tree regeneration and persist within forest ecosystems can alter those successional processes (Evans et al. 2013). Poor regeneration in post-mining forests is likely due to a combination of invasive plant species, altered soils, and in some cases browsing by deer.

Some mine sites host non-forest plant communities dominated by non-native grasses and other invasives and are likely to remain in a state of "arrested succession" (Cavender et al. 2014) unless management interventions are successfully implemented. Dense non-native vegetation can greatly delay, suppress, or alter successional trajectories by limiting the establishment of native species and/or by altering the fire regime. When both soils and vegetation are highly altered (as on mine sites), the result can be a persistent degraded state or the development of novel ecosystems with plant communities comprised of species that do not occur together in undisturbed ecosystems (Cramer et al. 2008). Succession may occur in distinct stages initiated by extreme weather events (Bartha et al. 2003) or by other disturbances such as restoration activities. As an example of the latter, soil treatments to alleviate compaction on post-SMCRA mine sites in Tennessee resulted in the rapid replacement of agronomic grasses by native perennial forbs (Aldrovandi 2018; Franklin and Aldrovandi 2018).

Successional processes occur on mine sites but often differ from those observed in natural systems. Where mine-site characteristics are favorable for forest regeneration and appropriate management is applied, succession tends to proceed in a manner that produces ecosystems similar to native forest, although often with non-native understory components. Where mine sites are not reclaimed or managed to restore native plant communities or other post-mining land use, plant communities dominated by non-native invasive vegetation become established. Will such plant communities undergo succession to forest-like ecosystems dominated by native trees? Slowly, ecosystem functions will return. It is possible that plant communities on former mined lands will become indistinguishable from surrounding ecosystems but that outcome is not assured. The nature of and time required for recovery processes are unknown but can be influenced if interventions are applied by land managers.

References

Adams MB (ed) (2017) The Forestry Reclamation Approach: guide to successful reforestation of mined lands. US Forest Service, Gen Tech Report NRS-169

Adams MB, Angel P, Barton C, Slavicek J (2015) American elm in mine land reforestation. Reclamation Matters, Fall, pp 34–38

Adams MB, Sanderson T, Sena K, Barton C, Agouridis C, Angel P, Zipper C (2019) Managing invasive exotic plant species on legacy mine lands. US Office of Surface Mining, Appalachian Regional Reforestation Initiative 16

Aldrovandi MS (2018) Negative Interactions between Tree Seedlings and non-native grasses in a restoration setting: getting to the "Root" of ohe Problem, M.S. Theses, University of Tennessee

Alexander HD, Arthur MA (2010) Implications of a predicted shift from upland oaks to red maple on forest hydrology and nutrient availability. Can J Forest Res 40:716–726

Allison VJ, Miller RM, Jastrow JD et al (2005) Changes in soil microbial community structure in a tallgrass prairie chronosequence. Soil Sci Soc Am J 69:1412–1421

Amichev BY, Burger JA, Rodrigue JA (2008) Carbon sequestration by forests and soils on mined land in the Midwestern and Appalachian coalfields of the U.S. Forest Ecol Manag 256:1949–1959

Anagnostakis SL (2001) The effect of multiple importations of pests and pathogens on a native tree. Biol Invasions 152:225–233

Angel PN, Burger JA, Zipper CE, Eggerud S (2011) Reforesting unused surface mined lands by replanting with native trees. In: Proceedings of Forest and Conservation Nursery Associations. US Forest Service Proc RMRS-P-68, pp 10–15

Angel P, Davis V, Burger JA, Graves D, Zipper CE (2017) The Appalachian Regional Reforestation Initiative. In: Adams MB (ed) (2017) The Forestry Reclamation Approach: guide to successful reforestation of mined lands. US Forest Service, Gen Tech Report NRS-169

Auch T, Burger JA, Mitchem DO (2005) Hardwood stocking after five years on reclaimed mined land in central Appalachia: a preliminary analysis. In: Proceedings America society of mining and reclamation, Brekenridge CO, pp 24–39

Ball DM, Pedersen JF, Lacefield GD (1993) The tall-fescue endophyte: Evolution meets economic in the tale of the nation's most popular planted grass, which owes many of its qualities to a fungus toxic to livestock. Am Sci 81:370–380

Bartha S, Meiners SJ, Pickett ST, Cadenasso MT (2003) Plant colonization windows in a mesic old field succession. Appl Veg Sci 6:205–212

Barton C, Miller J, Sena K, Angel P, French M (2015) Evaluating the use of tree shelters for direct seeding of Castanea on a surface mine in Appalachia. Forests 6:3514–3527

Barton C, Zipper C, Burger J (2017) Preface. In: Adams MB (ed) (2017) The Forestry Reclamation Approach: Guide to Successful Reforestation of Mined Lands. US Forest Service, Gen Tech Report NRS-169.

Bell TJ, Ungar IA (1981) Factors affecting the establishment of natural vegetation on a coal strip mine spoil bank in southeastern Ohio. Am Midl Nat 105:19–31
Bramble WC, Ashley RH (1955) Natural revegetation of spoil banks in central Pennsylvania. Ecology 36:417–423
Braun EL (1950) Deciduous Forests of Eastern North America. Hafner Press, New York
Brenner FJ, Musaus MJ (1977) Selectivity of browse species by white-tailed deer on strip-mine lands in Mercer County, Pennsylvania. Proc Penn Acad Sci 51:105–108
Brown RT, Agee JK, Franklin JF (2004) Forest restoration and fire: principles in the context of place. Conserv Biol 18:903–912
Buckley DS, Franklin JA (2008) Early tree and ground cover establishment as affected by seeding and fertilization rates in Tennessee. In: Proc Am Soc Min Reclam, p. 180–191.
Burger J, Davis V, Franklin J, Zipper C, Skousen J, Barton C, Angel P (2017a) Tree-Compatible ground covers for reforestation and erosion control. In: Adams MB (ed) (2017) The Forestry Reclamation Approach: Guide to Successful Reforestation of Mined Lands. US Forest Service, Gen Tech Report NRS-169.
Burger JA, Evans DM (2010) Ripping compacted mine soils improved tree growth 18 years after planting. Proc Am Soc Min Reclam, Pittsburgh PA, pp 55–69
Burger J, Zipper CE, Angel P, Hall N, Skousen J, Barton C, Eggerud S (2017b) Establishing native trees on legacy surface mines. In: Adams MB (ed) (2017) The Forestry Reclamation Approach: Guide to Successful Reforestation of Mined Lands. US Forest Service, Gen Tech Report NRS-169.
Burney OT, Jacobs DF (2018) Species selection–A fundamental silvicultural tool to promote forest regeneration under high animal browsing pressure. Forest Ecol Manag 408:67–74
Butler PR, Iverson L, Thompson FR, Brandt L, Handler S, Janowiak M, Shannon PD, Swanston C, Karriker K, Bartig J, Connolly S (2015) Central Appalachians forest ecosystem vulnerability assessment and synthesis: a report from the Central Appalachians Climate Change Response Framework project. US Forest Service, Gen Tech Rep NRS-146..
Carter CT, Ungar IA (2002) Aboveground vegetation, seed bank and soil analysis of a 31-year-old forest restoration on coal mine spoil in southeastern Ohio. Am Midl Nat 147:44–59
Casselman CN, Fox TR, Burger JA (2007) Thinning response of a white pine stand on a reclaimed surface mine in southwestern Virginia. North J Appl for 24:9–13
Cavender N, Byrd S, Bechtoldt CL, Bauman JM (2014) Vegetation communities of a coal reclamation site in southeastern Ohio. Northeast Nat 21:31–47
Clark EV, Zipper CE (2016) Vegetation influences near-surface hydrological characteristics on a surface coal mine in eastern USA. CATENA 139:241–249
Cotton C, Barton C, Lhotka J, Angel PN, Graves D (2012) Evaluating reforestation success on a surface mine in eastern Kentucky. In: Proc Forest and Conservation Nursery Associations. US Forest Service Proc RMRS-P-68. pp. 16–23.
Cramer VA, Hobbs RJ, Standish RJ (2008) What's new about old fields? Land abandonment and ecosystem assembly. Trends Ecol Evol 23:104–112
Croxton WC (1928) Revegetation of Illinois coal stripped lands. Ecology 9:155–175
Daniels WL, Amos DF (1981) Mapping, characterization and genesis of mine soils on a reclamation research area in Wise County. Sympos Surface Min Hydrology, Sedimentology Reclam. Lexington, Kentucky, Virginia
Daniels WL, Amos DF (1985) Generating productive topsoil substitutes from hard rock overburden in the Southern Appalachians. Environ Geochem Health 1:8–15
Delcourt PA, Delcourt HR, Ison CR, Sharp WE, Henderson AG (1999) Forests, forest fires, & their makers: The story of cliff palace pond. Lexington, Kentucky Archeological Survey Education Series Number Four
Ditsch DC, Collins M (2000) Reclamation considerations for pasture and hay lands receiving sixty-six or more centimeters of precipitation annually. Chapter 10, in: Barnhisel RI, et al. (eds) Reclamation of Drastically Disturbed Lands. Am Soc Agronomy, Madison Wisconsin.
Drake LD (1980) Erosion control with prairie grasses in Iowa strip-mine reclamation. In: Proc 7th N Am Prairie Conf, Springfield MO, pp 4–6

Drummond MA, Loveland TR (2010) Land-use pressure and a transition to forest-cover loss in the eastern United States. Bioscience 60:286–298

Evans DM, Zipper CE, Burger JA, Strahm BD, Villamagna AM (2013) Reforestation practice for enhancement of ecosystem services on a compacted surface mine: path toward ecosystem recovery. Ecol Eng 51:16–23

Fields-Johnson CW, Burger JA, Evans DM, Zipper CE (2014) Ripping improves tree survival and growth on unused reclaimed mined lands. Environ Manage 53:1059–1065

Fields-Johnson CW, Zipper CE, Burger JA, Evans DM (2012) Forest restoration on steep slopes after coal surface mining in Appalachian USA: soil grading and seeding effects. Forest Ecol Manag 270:126–134

Franklin JA, Aldrovandi M (2018) Hardwood establishment on compacted mine tailings after subsoiling and native ground cover seeding. In: Kirschman, JE (comp) Proc 19th Biennial Southern Silvicultural Research Conf. US Forest Service Gen Tech Report SRS-234, p. 127–128.

Franklin JA, Rizza J (2008) Site productivity on reclaimed coal mine sites in eastern Tennessee. Proc Central Hardwood Conf, Lafayette, IN, Apr. 8-9, 2008

Franklin JA, Zipper CE, Burger JA, Skousen JG, Jacobs DG (2012) Influence of herbaceous ground cover on forest restoration of eastern US coal surface mines. New Forest 43:905–924

French M, Barton C, McCarthy B, Keiffer C, Skousen J, Zipper C, Angel P. 2017. Reestablishing American chestnut on mined lands in the Appalachian coalfields. In: Adams MB (ed) (2017) The Forestry Reclamation Approach: Guide to Successful Reforestation of Mined Lands. US Forest Service, Gen Tech Report NRS-169.

Frouz J, Franklin JA (2014) Vegetation and soil development in planted pine and naturally regenerated hardwood stands 48 years after mining. J Am Soc Min Reclam 3:21–40

Gilland KE, McCarthy BC (2012) Reintroduction of American chestnut (*Castanea dentata*) on reclaimed mine sites in Ohio: Microsite factors controlling establishment success. North J Appl for 29:197–205

Gorman J, Skousen J, Sencindiver J, Ziemkiewicz P (2001) Forest productivity and minesoil development under a white pine plantation versus natural vegetation after 30 years. Proc Am Soc Min Reclam. Albuquerque, NM, pp 103–111

Groninger J, Skousen J, Angel P, Barton C, Burger J, Zipper C (2017) Mine reclamation practices to enhance forest development through natural succession. In Adams MB (ed) (2017) The Forestry Reclamation Approach: Guide to Successful Reforestation of Mined Lands. US Forest Service, Gen Tech Report NRS-169.

Haering KC, Daniels WL, Galbraith JM (2004) Appalachian mine soil morphology and properties: Effects of weathering and mining method. Soil Sci Soc Am J 68:1315–1325

Hall SL, Barton CD, Baskin CC (2010) Topsoil seed bank of an Oak-Hickory forest in eastern Kentucky as a restoration tool on surface mines. Restor Ecol 18:834–842

Holl KD (2002) Long-term vegetation recovery on reclaimed coal surface mines in the eastern USA. J Appl Ecol 39:960–970

Holl KD, Cairns J (1994) Vegetational community development on reclaimed coal surface mines in Virginia. B Torrey Bot Club, 327–337.

Imhoff E, Friz T, LaFevers J (1976) A guide to state programs for the reclamation of surface mined areas. US Geological Survey Circular 731. Reston VA. 33 p.

Ingold DJ (2002) Use of a reclaimed stripmine by grassland nesting birds in east-central Ohio. Oh J Sci 102:56–62

Ingold DJ, Dooley JL (2013) Nesting success of grassland and shrub-nesting birds on the Wilds, an Ohio reclaimed surface mine. Ohio J Sci 111:37–41

Johnson C, Skousen J (1995) Minesoil properties of 15 abandoned mine land sites in West Virginia. J Environ Qual 24:635–643

Kaster G. Former Forestry Supervisor, American Electric Power, 1969–2006. Personal communication.

Kelley RW, Anderson RJ (2009) Mining reclamation for wildlife in West Virginia: A continuing challenge. Proc Am Soc Min Reclam. Billings, MT, pp 657–662

Knight KS, Herms DA, Cardina J, Long R, Rebbeck et al. (2010) Emerald ash borer aftermath forests: The dynamics of ash mortality and the responses of other plant species. In: Michler CH, Ginzel MD (eds) Proc Symposium on Ash in North America. US Forest Service, Gen. Tech. Rep. NRS-P-72.

Lannoo MJ, Kinney VC, Heemeyer JL et al (2009) Mine spoil prairies expand critical habitat for endangered and threatened amphibian and reptile species. Diversity 1:118–132

Larson MM, Vimmerstedt JP (1983) Evaluation of 30-year-old plantations on stripmined land in east central Ohio. Ohio Agric Res Development Center, Bulletin 1149.

Lindenmayer DB, Margules CR, Botkin DB (2000) Indicators of biodiversity for ecologically sustainable forest management. Conserv Biol 14:941–950

Lituma CM, Cox JJ, Spear S, Edwards JW, De La Cruz JL, Muller LI, Ford WM (2021) Terrestrial wildlife in the Appalachian post-mined landscape: status and opportunities. Chapter 6 in: Zipper CE, Skousen JG (eds) Appalachia's Coal-Mined Landscapes. Springer

Malone C, Swab RM, Nolan A (2019) Impact of invasive species removal on plant and Aranae diversity on reclaimed mine land. J Am Soc Min Reclam 8:48–65

Marra M, Keene T, Skousen J, Griggs T (2013) Switchgrass yield on reclaimed surface mines for bioenergy production. J Environ Qual 42:696–703

Matamala R, Jastrow JD, Miller RM, Garten CT (2008) Temporal changes in C and N stocks of restored prairie: implications for C sequestration strategies. Ecol Appl 18:1470–1488

McCarthy BC, Bauman JM, Keiffer CH (2008) Mine land reclamation strategies for the restoration of American chestnut. Ecol Rest 26:292–294

McCarthy BC, Gilland KE, Bauman JM, Keiffer CH (2010) Factors affecting performance of artificially regenerated American chestnut on reclaimed mine sites. Proc Am Soc Min Reclam, pp 582–597.

Miller J, Barton C, Agouridis C, Fogel A, Dowdy T, Angel P (2012) Evaluating soil genesis and reforestation success on a surface coal mine in Appalachia. Soil Sci Soc Am J 76:950–960

Noss RF (2012) Forgotten grasslands of the South: natural history and conservation. Island Press.

Nowacki GJ, Abrams MD (2008) The demise of fire and "mesophication" of forests in the eastern United States. Bioscience 58:123–138

Ohio Department of Natural Resources (ODNR) (2000) A report on the cumulative off-site impacts from a large area mine in southeast Ohio. Office of Surface Mining Oversight and Inspection.

Oliphant AJ, Wynne RH, Zipper CE, Ford WM, Donovan PF, Li J (2017) Autumn olive (Elaeagnus umbellata) presence and proliferation on former surface coal mines in Eastern USA. Biol Invasions 19:179–195

Plass WT (1966, 1967) Land disturbances from strip-mining in eastern Kentucky, I – V. US Forest Service Research Notes NE-52, NE-55, NE-68, NE-69, NE-71.

Plass WT (1975) An evaluation of trees and shrubs for planting surface-mine spoils. US Forest Service, Res Pap NE-317.

Ricketts TH, Dinerstein E, Olson DM, Loucks CJ, Eichbaum W et al (1999) Terrestrial Ecoregions of North America: A Conservation Assessment. Island Press, Washington, DC

Rodgers C, Anderson R (1989) Establishment of grasses on sewage sludge-amended strip mine spoils. In: Proc. N Am Prairie Conf , pp. 107–109.

Rodrigue JA (2001) Woody Species Diversity, Forest and Site Productivity, Stumpage Value, and Carbon Sequestration of Forests on Mined Lands Reclaimed prior to the Passage of the Surface Mining Control and Reclamation Act of 1977. M.S. Thesis, Virginia Tech, Blacksburg.

Rodrigue JA, Burger JA, Oderwald RG (2002) Forest productivity and commercial value of pre-law reclaimed mined land in the eastern United States. North J Appl for 19:106–114

Rudgers JA, Holah J, Orr SP, Clay K (2007) Forest succession suppressed by an introduced plant-fungal symbiosis. Ecology 88:18–25

Schramm P, Kalvin RL (1978) The use of prairie in strip mine reclamation. In: : Glenn-Lewin DC, Landers RQ (eds), Proc Fifth Midwest Prairie Conf, p 153

Sena K, Agouridis C, Miller J, Barton C (2018) Spoil type influences soil genesis and forest development on an Appalachian surface coal mine ten years after placement. Forests 9:780

Sena K, Angel H, Barton C (2014) Influence of tree shelters and weed mats on growth and survival of backcrossed chestnut seedlings on legacy minelands in eastern Kentucky. J Am Soc Min Reclam 3:41–63

Sena K, Barton C, Hall S, Angel P, Agouridis C, Warner R (2015) Influence of spoil type on afforestation success and natural vegetative recolonization on a surface coal mine in Appalachia, United States. Restor Ecol 23:131–138

Simmons J, Currie W, Eshleman KN, Kuers K, Monteleone S, Negley TL, Pohlad B, Thomas C (2008) Forest to reclaimed land use change leads to altered ecosystem structure and function. Ecol Appl 18:104–111

Skousen, J, Dallaire K, Scagline-Mellor S, Monteleone A, Wilson-Kokes L, et al. (2018) Plantation performance of chestnut hybrids and progenitors on reclaimed Appalachian surface mines. New Forests 49:599-611

Skousen JG, Daniels WL, Zipper CE (2021) Soils on Appalachian coal-mined lands. Chapter 4 in: Zipper CE, Skousen JG (eds) Appalachia's Coal-Mined Landscapes. Springer

Skousen J, Gorman J, Pena-Yewtukhiw E, King J, Stewart J, Emerson P, DeLong C (2009) Hardwood tree survival in heavy ground cover on reclaimed land in West Virginia: Mowing and ripping effects. J Environ Qual 38:1400-1409

Skousen J, Johnson C, Garbutt K (1994) Natural revegetation of 15 abandoned mine land sites in West Virginia. J Environ Qual 23:1224–1230

Skousen J, Ziemkiewicz P, Venable C (2006) Tree recruitment and growth on 20-year-old, unreclaimed surface mined lands in West Virginia. Int J Min Reclam Environ 20:142–154

Skousen JG, Zipper CE (2021) Coal mining and reclamation in Appalachia. Chapter 3 in: Zipper CE, Skousen JG (eds) Appalachia's Coal-Mined Landscapes. Springer.

Skousen J, Zipper C, Burger J, Barton C, Angel P (2017) Selecting materials for mine soil construction when establishing forests on Appalachian mined lands. In: Adams MB (ed) (2017) The Forestry Reclamation Approach: Guide to Successful Reforestation of Mined Lands. US Forest Service, Gen Tech Report NRS-169.

Swab RM, Lorenz N, Byrd S, Dick R (2017) Native vegetation in reclamation: Improving habitat and ecosystem function through using prairie species in mine land reclamation. Ecol Eng 108:525–536

Thompson RL, Vogel WG, Taylor DD (1984) Vegetation and flora of a coal surface-mined area in Laurel County, Kentucky. Castanea 49:111–126

Thompson RL, Vogel WG, Wade GL, Rafail BL (1986) Development of natural and planted vegetation on surface mines in southeastern Kentucky. Proc Am Soc Min Reclam, p. 145–154.

Torbert JL, Burger JA, Daniels WL (1990) Pine growth variation associated with overburden rock type on a reclaimed surface mine in Virginia. J Environ Qual 19:88–92

U.S. Environmental Protection Agency (US EPA) (2020) Level III and IV Ecoregions of the Continental United States. https://www.epa.gov/eco-research/level-iii-and-iv-ecoregions-continental-united-states

U.S. Government Accounting Office (GAO) (2009) Surface coal mining characteristics of mining in mountainous areas of Kentucky and West Virginia. GAO-10.21.

Vallauri D, Aronson J, Dudley N, Vallejo R (2005) Monitoring and evaluating forest restoration success. In: Mansourian S et al (eds) Forest Restoration in Landscapes. Springer, New York, pp 150–158

Varner JM, Arthur MA, Clark SL, Dey DC, Hart JL, Schweitzer CJ (2016) Fire in eastern North American oak ecosystems: Filling the gaps. Fire Ecol 12:1

Wade GL, Thompson RL, Vogel WG (1985) Success of trees and shrubs in an 18-year-old planting on mine spoil. US Forest Service, Res Pap NE-567.

Wickham JD, Riitters KH, Wade TG, Coan M, Homer C (2007) The effect of Appalachian mountaintop mining on interior forest. Landscape Ecol 22:179–187

Wilson-Kokes L, Emerson P, DeLong C, Thomas C, Skousen J (2013) Hardwood tree growth after eight years on brown and gray mine soils in West Virginia. J Environ Qual 42:1353–1362

Wilson-Kokes L, DeLong C, Thomas C, Emerson P, O'Dell K, Skousen J (2013a) Hardwood tree growth on amended mine soils in West Virginia. J Environ Qual 42:1363-1371

Wood PB, Williams JM (2013) Terrestrial salamander abundance on reclaimed mountaintop removal mines. Wildlife Soc B 37:815–823

Wood P, Larkin J, Mizel J, Zipper C, Angel P (2017) Reforestation to enhance Appalachian mined lands as habitat for terrestrial wildlife. In: Adams MB (ed) (2017) The Forestry Reclamation Approach: Guide to Successful Reforestation of Mined Lands. US Forest Service, Gen Tech Report NRS-169.

Yahner RH (2000) Eastern Deciduous Forest: Ecology and Wildlife Conservation. University of Minnesota Press, Minneapolis

Yearsley EF, Samuel DE (1980) Use of reclaimed surface mines by foxes in West Virginia. J Wildlife Manage 44:729–734

Zeleznik JD, Skousen JG (1996) Survival of three tree species on old reclaimed surface mines in Ohio. J Environ Qual 25:1429–1435

Zipper CE, Agouridis CT, Barton CD, Skousen JG (2021) Conversion options for mining-affected lands and waters in Appalachia. Chapter 7 in: Zipper CE, Skousen JG (eds) Appalachia's Coal-Mined Landscapes. Springer

Zipper C, Angel P, Adams MB, Sanderson T, Sena K et al (2019) The Forestry Reclamation Approach: An essential tool for controlling invasive exotic plants on active mine sites. US Office of Surface Mining, Appalachian Regional Reforestation Initiative, p 17

Zipper CE, Burger JA, Barton C, Skousen J (2013) Rebuilding soils on mined land for native forests in Appalachia, USA. Soil Sci Soc Am J 77:337–349

Zipper CE, Burger JA, McGrath JM, Rodrigue JA, Holtzman GI (2011) Forest restoration potentials of coal-mined lands in the eastern United States. J Environ Qual 40:1567–1577

Zipper CE, Burger JA, Skousen JG, Angel PN, Barton CD, Davis V, Franklin JA (2011) Restoring forests and associated ecosystem services on Appalachian coal surface mines. Environ Manage 47:751–765

Terrestrial Wildlife in the Post-mined Appalachian Landscape: Status and Opportunities

Christopher M. Lituma, John J. Cox, Stephen F. Spear, John W. Edwards, Jesse L. De La Cruz, Lisa I. Muller, and W. Mark Ford

Abstract Coal mining is an anthropogenic stressor that has impacted terrestrial and semi-aquatic wildlife in the Appalachian Plateau since European settlement. Creation of grassland and early-successional habitats resulting from mining in a forested landscape has resulted in novel, non-analog habitat conditions. Depending on the taxa, the extent of mining on the landscape, and reclamation practices, effects have ranged across a gradient of negative to positive. Forest-obligate species such as woodland salamanders and forest-interior birds or those that depend on aquatic systems in their life cycle have been most impacted. Others, such as grassland and early-successional bird species have responded favorably. Some bat species, as an unintended consequence, use legacy deep mines as winter hibernacula in a region with limited karst geology. Recolonization of impacted wildlife often depends on life strategies and species' vagility, but also on altered or arrested successional processes on the post-surface mine landscape. Many wildlife species will benefit from Forest Reclamation Approach practices going forward. In the future, managers will be faced with decisions about reforestation versus maintaining open habitats depending on the conservation need of species. Lastly, the post-mined landscape currently is the focal point for a regional effort to restore elk (*Cervus canadensis*) in the Appalachians.

C. M. Lituma · J. W. Edwards
West Virginia University, Morgantown, WV, USA

J. J. Cox
University of Kentucky, Lexington, KY, USA

S. F. Spear
The Wilds, Cumberland, OH, USA

J. L. De La Cruz
Virginia Tech, Blacksburg, VA, USA

L. I. Muller
University of Tennessee, Knoxville, TN, USA

W. M. Ford (✉)
U.S. Geological Survey, Blacksburg, VA, USA
e-mail: wmford@vt.edu

C. E. Zipper and J. Skousen (eds.), *Appalachia's Coal-Mined Landscapes*,
https://doi.org/10.1007/978-3-030-57780-3_6

Keywords Bats · Forest-obligate birds · Forest reclamation approach · Elk · Grassland birds · Salamanders

1 Setting the Context for Terrestrial Wildlife

The Appalachian region covers approximately 530,138 km^2 and extends more than 1,600 km from southwestern New York to northeastern Mississippi encompassing the Blue Ridge, Ridge and Valley, and Appalachian Plateau provinces, along with adjacent portions of the Interior Low Plateau provinces (Boettner et al. 2014). Although the region includes 13 states, the majority of coal recently was or is currently being produced from the 211,372 km^2 Appalachian Plateau portions of Alabama, Tennessee, Kentucky, Virginia, West Virginia, Ohio, and Pennsylvania, save for the Anthracite Region in the Ridge and Valley of northeastern Pennsylvania (Averitt 1975).

Although heavily logged during the nineteenth and early-to-mid-twentieth centuries following European settlement, second- and third-growth forests represent 65% of the current land base (Yarnell 1998); however, some counties in the region have landcover that exceed 90% forested (Boettner et al. 2014). Highly variable elevational and climatic conditions found in the region result in diverse growing conditions and habitats that support a wide range of forest types: montane red spruce (*Picea rubens*)-eastern hemlock (*Tsuga canadensis*) and northern hardwood beech (*Fagus grandifolia*)-birch (*Betula* spp.)-maple (*Acer* spp.)-cherry (*Prunus serotina*) types in the northern and higher elevation portions of the Plateau. Appalachian cove hardwoods and oak forests, dominated by yellow-poplar (*Liriodendron tulipifera*) and a variety of oaks (*Quercus* spp.), respectively, exist throughout the region. The botanically rich mixed-mesophytic hardwood forests occur in eastern Kentucky, southwestern Virginia, and southern West Virginia whereas oak-pine (*Pinus* spp.) types become common in Tennessee and Alabama (Braun 1950). The amount of early-successional lands present prior to European settlement is open for debate, particularly those shaped by Native American use of fire, but invariably these areas occurred throughout the landscape (Harper et al. 2016). For example, early accounts by explorers and surveyors in the mid to late eighteenth century described vast fields of native grasses and clovers (*Trifolium* spp.) in the region (Ford et al. 2003). Accordingly, the region supports tremendous habitat diversity that contributes to the high faunal richness found therein. Pickering et al. (2002) estimated that the Appalachian Region, including the Appalachian Plateau, contained 76 amphibian species, 58 reptile species, 255 bird species, and 78 mammals.

Wildlife communities and their habitats in the Appalachian Plateau have been modified and influenced by humans for thousands of years, beginning with late Pleistocene-early Holocene arrival of Native Americans through European settlement (Van Lear and Harlow 2002). Habitat change, especially the early conversion of river-bottom forests and switch cane (*Arundinaria gigantea*) to agriculture (level to moderately sloping tillable land is uncommon locally) by the early nineteenth century and then industrial logging and early coal mining before the twentieth century, combined with unregulated subsistence hunting profoundly affected wildlife abundance. Similarly, by 1940, the loss of the American chestnut (*Castanea dentata*) from the chestnut blight (*Cryphonectria parasitica*) removed a keystone mast source for regional wildlife (Hepting 1974). American bison (*Bison bison*) were extirpated by the 1820s and the eastern elk (*Cervus canadensis*) by the 1870s. The passenger pigeon (*Ectopistes migratorius*), whose sheer abundance constituted an important component of forest disturbance, was largely absent on the landscape shortly before the twentieth century (Ellsworth and McComb 2003). Predators such as the eastern mountain lion (*Puma concolor cougar*) and eastern wolf (*Canis lupus lycaon*) were gone by the 1890s, persisting longer than bison and elk due to the introduction of free-range livestock production. Species considered common today, following decades of closed hunting seasons, localized re-introduction, and regulated harvest, were typically present only in areas of low human population density. These species included: wild turkey (*Meleagris gallopavo*), beaver (*Castor canadensis canadensis*), white-tailed deer (*Odocoileus virginianus*), river otter (*Lutra canadensis*), and black bear (*Ursus americanus*) (Trefethn 1975). The remote, large post-mined landscape has facilitated the establishment of feral swine (*Sus scrofa*) populations in the region (Gipson et al. 1998). The broad-spectrum ecosystem-level impacts of swine, particularly to native wildlife, vegetation, and water quality, is a concern for managers in the Appalachian Plateau (Campbell and Long 2009).

Other than impacts from locally high human population densities and associated development from initial coal production that provided an impetus for subsistence hunting, wildfire, and logging, deep mining processes historically had few direct terrestrial wildlife impacts per se. However, throughout much of the region, surface mining became a prominent mining method following World War II with the development of larger, earth-moving equipment (Skousen and Zipper 2014) that have substantively altered the Appalachian landscape and impacted its wildlife (Wickham et al. 2013). Larger machinery and increased mechanization allowed large areas to be economically surface mined, with little to no legal requirements for post-mining reclamation (Skousen and Zipper 2014). These unreclaimed sites often contained exposed high walls and large expanses of overburden or spoil. Prior to 1977, if reclamation was performed, it generally included some back filling of excavated areas and the planting of trees, shrubs, or grasses on disturbed areas. These efforts were often only minimally successful, with even hardy volunteer species such as black locust (*Robinia pseudoacacia*) having difficulty establishing in mine soils. Nevertheless, depending on drainage and water quality, borrow pits at the base of highwalls often created permanent standing water in a region with little lentic conditions, creating habitat for Anurans, eastern newts (*Notophthalmus viridescens*), and occasionally shorebirds and waterfowl (Denmon 1998).

Passage of the Surface Mining Control and Reclamation Act (SMCRA) of 1977 established federal control over coal mining, reclamation, and environmental standards (Skousen and Zipper 2014). Among the requirements for a company to receive a mining permit are the following considerations relevant to post-mining wildlife use: (1) regrading (i.e., to approximate original contour) or, by approved variance, an alternate post-mining landscape; (2) establishment of a mine soil suitable for vegetation and selection of appropriate plant species for revegetation; and (3) development of the designated post-mining land use. Federal regulations for meeting these requirements are very explicit and Skousen and Zipper (2014) provide an excellent review of all permit requirements. The SMCRA allows for post-mining land that includes (1) prime farmland, (2) hay land and pasture, (3) biofuel crops, (4) forestry, (5) wildlife habitat, (6) building site development, (7) other suitable uses; and requires that mined lands be returned to a condition that will support their pre-mining land use or higher and better use. What constitutes "higher and better" use is subject to regulatory interpretation, but in economically-depressed Appalachia, economic development often is considered most desirable, followed by managed lands (e.g., farmland, hay, and pasture), and lastly forestry and wildlife habitat. Most mine sites in the Appalachian Region were forested prior to mining but were reclaimed as hay land or pasture (estimated at 80–90%) during the 1980s and 1990s (Skousen and Zipper 2014).

Part of the reason for reclaiming to pastoral use was the initial limited success in reforestation efforts. Implementation of SMCRA regulations inadvertently created physical and environmental conditions (e.g., soil compaction, dense ground cover) that led to poor tree growth on reclaimed mines (Simmons et al. 2008; Burger and Evans 2010). Following decades of reclamation research, the Forestry Reclamation Approach (FRA) was developed as a set of practices to enhance restoration of forest vegetation and ecosystem services (Burger et al. 2005; Zipper et al. 2011b), including: (1) create a suitable rooting medium, (2) loosely graded soil to reduce compaction, (3) use of less competitive ground covers, (4) planting both early successional tree species for wildlife and soil stability, and commercially valuable crop trees, and (5) use of proper tree planting techniques. The goal for adoption of FRA practices is to meet SMCRA guidelines whereby promoting a successional trajectory towards forest vegetation and a faster return of native flora and fauna and their associated ecosystem services (Zipper et al. 2011b).

Nonetheless, surface mining disturbance and reclamation, whether permitted before 1977 or following the SMCRA, has created somewhat non-analog early successional conditions in the Appalachian Plateau that strongly influence wildlife in the post-mined environment. Research on how reclaimed habitats are used by various taxa of wildlife shows varied levels of response: negative, neutral, and positive, depending on the species and landscape/temporal context (Buehler and Percy 2012). Although many of the negative surface mining impacts to forest-associated wildlife have been demonstrated (Wickham et al. 2013; Wood et al 2013), post-mining reclamation can benefit grassland- or early successional shrub-scrub wildlife species as well as generalist species that require a wide diversity of habitats across the large landscape (Zipper et al. 2011a). Of the numerous wildlife species found in the Appalachian Plateau, many rely to some extent on hard mast, particularly that produced by oaks. Within the region, the seasonal availability of oak mast strongly influences the annual survival and recruitment of species ranging from small mammals to white-tailed deer and black bear. Because of the importance of oak mast, the conversion of mature oak forest types to predominantly reclaimed grass-legume and shrub landcover will influence the diversity and abundance of wildlife species occupying the site for potentially decades. Species dependent on forest structure and hard mast, such as tree squirrels, will be absent from the reclaimed site until mature, mast-bearing forest returns. The habitat quality and carrying capacity for other less obligate species will depend on reclamation practices and their success in reaching successional milestones. Early efforts to re-establish oaks and other hard mast-producing species (e.g., black walnut *Juglans nigra*, hickory *Carya* spp.) on reclaimed mine sites were mostly unsuccessful. FRA practices and further research offer optimism that the reestablishment of native mast-bearing species is feasible as part of standard reclamation efforts (Davis et al. 2012). Information on occupancy and use of reclaimed mine sites among taxa varies because of regulatory requirements and recreational and economic development objectives. The amount of information available or lacking for a species group may reflect their regulatory status (Indiana bat *Myotis sodalis*), ecological role as an environmental indicator (amphibians) or as highly adaptable generalist (white-tailed deer). The following is not intended to be exhaustive and focuses mostly on post-SMCRA wildlife status.

2 Herpetofauna

Among all temperate ecosystems, the herpetofauna of the Appalachian region, including the Appalachian Plateau, is characterized by high salamander diversity, particularly the lungless salamanders in the family Plethodontidae (Petranka 1998). Conversely, diversity of other amphibian and reptile groups is low compared to other areas of North America. Anurans are primarily characterized by generalist species within the families Bufonidae, Hylidae, and Ranidae, although some members of these families are more specialized, such as wood frogs (*Lithobates sylvatica*) and mountain chorus frogs (*Pseudacris brachyphona*). Snake species in the region represent a broad array of ecological strategies, although most species are often more abundant in either open areas or forest patches containing suitable basking habitat (i.e., canopy gaps). These species span common generalists such as eastern garter snakes (*Thamnophis sirtalis*) and racers (*Coluber constrictor*), semi-aquatic species such as northern water snakes (*Nerodia sipedon*), and species with more restricted resource needs such as timber rattlesnakes (*Crotalus horridus*). There are also a number of small snakes typically found in forests and edges such as eastern worm snakes (*Carphophis amoenus*) and smooth earthsnakes (*Virginia valeriae*). The turtle species in the region are primarily semi-aquatic, with the exception of the eastern box turtle (*Carolina terrapene*). Finally, lizards are the least diverse of Appalachian taxa and, such as the eastern fence lizard (*Sceloporus undulatus*), generally are associated with edge habitat between forest and open areas. There are three major types of mining impacts that have been investigated with respect to amphibians and reptiles: loss of forest cover and the frequent conversion to grassland or shrubland; creation of ponds or wetlands post-mining; and stream occupancy particularly in valley fill areas following mountaintop removal mining.

2.1 Salamanders

The majority of Appalachian salamanders rely on forests for some aspects of their life history. As a result, minelands reclaimed as grasslands may not be suitable for salamander persistence and/or serve as barriers to a movement among forest patches (Fig. 1). For instance, spotted salamanders (*Ambystoma maculatum*), a widespread Appalachian species, will not move across grassland patches (Rittenhouse and Semlitsch 2006). In southwest Virginia, Carrozzino (2009) found only one salamander detection in reclaimed minelands as compared to 64 observations in either reference or pre-SMCRA mineland forest. Eastern newts successfully reproduced in wetlands at an abandoned contour mine in West Virginia, but there was no evidence of spotted salamander recruitment, which was attributed to reduced water quality (Loughman 2005). Terrestrial salamanders were primarily limited to intact forest or forest fragments around reclaimed mountaintop removal mines in West Virginia, and habitat type was the main factor affecting salamander occupancy (Wood and Williams 2013a, Williams et al. 2017). Similar patterns of low salamander richness and abundance on reclaimed mine sites also have been noted in the Illinois Basin (Lannoo et al. 2009; Terrell et al. 2014; Stiles et al. 2016). In West Virginia, wood frogs only were found in forested areas on a reclaimed mine (Williams et al. 2017) and were absent from an abandoned contour mine (Loughman 2005). However, three species of wetland breeding salamander (e.g., marbled salamander *Ambystoma opacum;* small-mouthed salamander *Ambystoma texanum;* and eastern newt) have successfully recruited in wetlands embedded in reclaimed mines adjacent to unmined forests.

Fig. 1 Slimy salamander (*Plethodon glutinosus*) found in a post-mined landscape. Woodland salamanders in the family Plethodontidae require forested, cool, and moist environments. Surface mining negatively impacts these species in the Appalachian Plateau. Given time, vegetation in-growth, and immigration from surrounding forest habitats, some recovery can occur. (Photo by S.F. Spears)

Unlike post-SMCRA sites, many pre-SMCRA minelands in the Appalachian Plateau reforested to some extent and likely provide a greater opportunity for salamander recolonization than more recently reclaimed areas. Brady (2016) compared pre-SMCRA sites to reference forests in eastern Ohio and found a higher species richness of salamanders in reference sites. However, this difference was due to the complete absence of stream breeding salamanders (e.g., northern dusky salamander *Desmognathus fuscus*), northern two-lined salamander *Eurycea bislineata,* long-tailed salamander *E. longicauda,* and red salamander *Pseudotriton ruber*) from pre-SMCRA sites. Other studies have also found stream-dependent salamanders responding negatively to mining (Merovich et al. 2021).

When only comparing strictly terrestrial salamanders, such as the northern ravine salamander (*Plethodon electromorphus*), eastern red-backed salamander (*Plethodon cinereus*), and northern slimy salamander (*Plethodon glutinosus*), there was no difference in abundance between pre-SMCRA sites and reference forests. Many pre-SMCRA surface mines are also characterized by rock highwalls that could potentially provide habitat for crevice dwelling species. These rock highwalls also typically have to borrow pits with water that likely provide habitat for several pond-breeding amphibian species. With respect to the potential for highwall rock crevices to provide habitat, Hinkle et al. (2018) examined the distribution of the outcrop associated green salamander (*Aneides aeneus*) across sites in three categories in Virginia: highwalls created by surface mining, remnant forested outcrops within a surface mine matrix, and unmined outcrop sites within a national forest. Although they did not find any salamanders on highwalls, they did find presence at 72% of remnant outcrops within mine sites. Furthermore, they found evidence of reproduction at both remnant and unmined outcrops. Highwalls were characterized by fewer crevices within an outcrop and lower forest cover but there were no differences between remnant and unmined outcrops in these categories. As a result, while mining activities likely exclude this species directly, green salamanders appear to persist in a mining landscape if remnant outcrops are preserved.

2.2 *Reptiles*

Terrestrial reptiles are often found on grasslands of reclaimed mines, but little data exist beyond observations. Snakes tend to occur primarily in the shrubland of post-SMCRA mines, although they do not respond strongly to habitat type (Williams et al. 2017). Species such as racers can be abundant in reclaimed minelands (Myers and Klimstra 1963; Williams et al. 2017), especially in rocky highwall areas (Loughman 2005). The federally threatened eastern massasauga (*Sistrurus catenatus*) has been observed on reclaimed minelands in Pennsylvania, though not believed to be common there (Brenner 2007). Results from research on box turtle use and abundance on reclaimed minelands versus surrounding forests were inconclusive in the Appalachian Plateau as well as the Illinois Basin (Lannoo et al. 2009; Stiles et al. 2016), whereas eastern fence lizards were common (Myers and Klimstra 1963).

2.3 Relationships to Aquatic Habitats

The abundance and heterogeneity of wetlands created from surface mining likely have benefited species such as frogs adapted to breed in small wetlands with a variety of hydroperiods (Stiles et al. 2016). In general, frogs common to the Appalachian Plateau tend to occur on reclaimed minelands when water is present (Lacki et al. 1992; Timm and Meretsky 2004; Turner and Fowler 1981; Loughman 2005; Carrozzino 2009; Lannoo et al. 2009; Terrell et al. 2014; Stiles et al. 2016; Williams et al. 2017). Reptile response is poorly studied in the Appalachian Plateau, but in the Illinois Basin, northern water snakes and painted turtles (*Chrysemys picta*) were common on pre-SMCRA mines (Myers and Klimstra 1963) and post-SMCRA sites (Lannoo et al. 2009; Stiles et al. 2016). Copperbelly water snakes (*Nerodia erythrogaster neglecta*) were observed over a wide array of post-SMCRA conditions with no apparent population impacts (Lacki et al. 2005).

Gore (1983) found the presence of northern dusky salamanders in 16 of 78 streams in surface-mined areas in eastern Kentucky and suggested that increased dissolved solids and lack of shading were reducing populations. Lower occupancy, abundance, and rates of colonization and persistence were observed for a community of stream amphibians (e.g., northern dusky salamander, seal salamander *Desmognathus monticola*, southern two-lined salamanders *Eurycea cirrigera*, spring salamanders *Gyrinophilus porphyriticus*, and red salamanders) in streams within or downstream of reclaimed mountain top removal mines versus streams in unmined second-growth forest reference sites in Kentucky (Muncy et al. 2014; Price et al. 2016; Price et al. 2018). The streams through the reclaimed mine sites were characterized by increased conductivity, fewer in-stream cover rocks, and less surrounding canopy cover, which was correlated with the reduced presence and abundance of salamander species. Bourne (2015) also detected increased selenium levels in salamander tail tips from valley-fill streams relative to those from reference streams. In eastern Tennessee, stream salamander numbers were negatively affected by low pH and high conductivity in streams draining minelands (Schorr et al. 2013). Comparing two reference streams with three valley-fill streams in West Virginia, Hamilton (2002) found reduced salamander abundance at two of the three valley-fill streams, although the oldest (18 years since reclamation) had similar abundances to reference sites, indicating the potential for recovery post-mining. Williams and Wood (2004) noted high rock density and abundant cover objects mitigated some of the impacts to stream salamanders in post-mined valley-fill streams. Further work by Wood and Williams (2013b) found similar salamander species richness between reference and valley-fill streams, but approximately double the number of salamander individuals was observed in reference streams relative to valley fill. Sweeten and Ford (2016)

Fig. 2 Post-mined landscape restored stream. Aquatic and semi-aquatic salamander species in the genus *Desmognanthus* may require > 20 years post-mining and restoration to approximate community composition and population numbers in pre-mined or reference streams in the Appalachian Plateau (Photo by S. Sweeten, Virginia Tech)

assessed occupancy and abundance of stream salamanders with stream habitat and landscape-level covariates in southwest Virginia and found that both occupancy and abundance of *Desmognathus* spp. was best explained by site variables such as stream canopy cover (Fig. 2). In contrast, occupancy of *Eurycea* spp. was related to overall percent mining and forest loss across the watershed but abundance was explained by stream sediments and embeddedness. Based on these results, they recommended that post-mining reforestation may especially benefit *Desmognathus* spp. due to their reduced dispersal compared to *Eurycea* spp. (Sweeten and Ford 2015). Accordingly, successful restoration of mined lands with streams using FRA practices would likely benefit the entire stream salamander community, as species tend to have similar ecological responses to mine disturbance (Price et al. 2018). Although a reduction in stream salamander abundance is seen in forests with timber harvest, the magnitude of this effect tends to be smaller compared to studies from minelands, and stand age is not always the strongest covariate influencing abundance (Moseley et al. 2008).

There are many gaps in knowledge regarding the influence of surface mining on herpetofauna species. Comparative studies examining the abundance and population trends on mined and unmined lands are generally needed for all reptile groups and pond-breeding amphibians in general. Terrestrial salamanders of the genus *Plethodon* can occur in reclaimed forests, but the full suite of factors that influence the ability of this group to recolonize and persist post-mining largely is unknown. Lastly, although stream salamander response is well documented, further work to better understand recovery times and optimal restoration practices to increase populations would be useful.

3 Avifauna

Researchers have long recognized that unprecedented landscape-level changes created by surface mines and resulting reclamation efforts affect avian communities (Riley 1952). Forest bird assemblages are initially altered by the complete clearing of forests for mining, but then effects can extend post-mining if reclamation does not involve reforestation. Clearing of forested areas changes the forest bird community into a grassland-early successional bird community. When large areas of forest are removed it also creates forest edges for the remaining forested areas which affect bird species reliant on core forest areas. However, reclaimed minelands, especially those with residual or created water features and wetlands can add to local avian diversity. For example, over 130 species of birds have been recorded using a large reclaimed mine site in southeastern Kentucky, including waterfowl and wading shorebirds such as the American avocet (*Recurvirosta americana*) that are otherwise rare to absent in the Appalachian Plateau (Cornell Lab of Ornithology 2020).

3.1 Forest Obligate Birds

Generally, forest songbirds are sensitive to changes resulting from surface mining and reclamation. Forest bird species occupancy and abundance decline following tree removal and mining, and core-forest bird species are affected disproportionately. For example, from 1992–2006 in 19 counties in the central portion of the Appalachian Plateau, ~92,500 ha of mature forest was changed to non-forest cover (McDermott et al. 2013). Becker et al. (2015) found that in Kentucky and West Virginia, a landcover transition away from forest after mining activities elicited more negative responses than positive ones for birds and that negative effects on avian species abundance occurred at thresholds lower than (or before) other species responded positively. Specifically, the forest interior guild was most sensitive to landscape changes wherein abundance negatively responded to even small (11%) landscape losses in forest cover (Becker et al. 2015).

Research that focuses on the effects of surface mine reclamation on forest songbirds is limited and usually species-specific (Becker et al. 2015). Cerulean warblers (*Setophaga cerulea*) avoid large-scale disturbances and edges such as large open fields and power lines (McDermott et al. 2013). In West Virginia, there were no clear differences in red-eyed vireo (*Vireo olivaceous*) nest success between forests on reclaimed minelands and non-mined areas (Mizel 2011). However, tree species composition contributed to guild differences, especially on heavily compacted mine benches where American redstart (*Setophaga ruticilla*), rose-breasted grosbeak (*Pheucticus ludovicianus*), and worm-eating warbler (*Helmitheros vermivorum*) were using subcanopy dominated by yellow-poplar for foraging (Mizel 2011). Also, black-and-white warbler (*Mniotilta varia*), ovenbird (*Seiurus aurocapilla*), and worm-eating warbler avoided maple stands because of reduced forest floor leaf litter as compared to oak-hickory (*Carya* sp.) dominant stands (Mizel 2011).

In the Appalachian Plateau, some advocate for minelands to be reclaimed as forests because the contiguous forest is critical for core forest songbirds (McDermott et al. 2013; Wood et al. 2013; Becker et al. 2015; Wood and Ammer 2015). The FRA advocates for science guided reforestation of minelands that outlines an explicit approach dependent on region-specific target species and goals (Wood et al. 2013). Highlighted is the potential for reforested areas to initially provide young-forest habitat for ruffed grouse (*Bonasa umbellus*), golden-winged warbler (*Vermivora chrysoptera*), and American woodcock (*Scolopax minor*; Wood et al. 2013). In West Virginia, American woodcock use reclaimed minelands, which may be the best available habitat in some landscapes, though soil conditions and earthworm biomass is best on reclaimed mines > 5 yr old (Gregg et al. 2001). Also in West Virginia, ruffed grouse utilized reclaimed surface mines, but non-mined areas had greater plant-feeding rates and provided better food resources (Kimmel and Samuel 1984). Overall, there remains a significant lack of information regarding forest reclamation effects on forest birds, in part because it requires decades, at minimum, for forests on reclaimed mines to adequately regenerate.

3.2 Grassland and Early-Successional Birds

In the Appalachian Plateau, reclaimed minelands are dominated by non-native invasive grass and herbaceous vegetation and offer a unique heavily impacted cover type that would otherwise not occur (Stauffer et al. 2011; Sena et al. 2021). Despite the clear disparity in plant species composition and habitat structure between reclaimed mines and surrounding Appalachian Plateau forests, reclaimed minelands provide unique areas for an assemblage of otherwise absent grassland and shrubland bird species (Graves et al. 2010; Ingold et al. 2010; Fig. 3). Reclaimed minelands generally provide adequate habitat for grassland-dependent birds (DeVault et al. 2002), and in limited species-specific examples, do not act as population sinks. Though largely from work in the Illinois Basin, red-winged blackbird (*Agelaius phoeniceus*), grasshopper sparrow (*Ammodramus savannarum*), and eastern meadowlark (*Sturnella magna*) occurred on >95% of reclaimed mineland survey points. Dickcissel (*Spiza americana*; 67%), common yellowthroat (*Geothlypis trichas*; 65%) occurred less frequently, and Henslow's sparrow (*Centronyx henslowii;* 59%) were patchily distributed (Scott et al. 2002). Landscape factors generally did not affect Henslow's sparrow abundance on reclaimed minelands, and Henslow's sparrows were virtually absent from pasture and hayfields in areas around reclaimed mines (Bajema and Lima 2001). Patch-level relationships showed red-winged blackbird abundance was negatively correlated with litter and canopy cover, whereas eastern meadowlark and grasshopper sparrow abundance was negatively correlated with visual obstruction preferring less dense vegetation (Scott et al. 2002). Henslow's sparrows preferred areas dominated by grasses, specifically smooth brome (*Bromus inermis*) and broom

Fig. 3 Common yellowthroat (*Geothlypis trichas*) is an example of an early-successional shrub-scrub obligate bird species that can respond favorably to the post-mined landscape in the Appalachian Plateau (Photo by J. Cox, University of Kentucky)

sedge bluestem (*Andropogon virginicus*). In Tennessee, golden-winged warbler, a critically declining shrubland bird, nest on reclaimed minelands (Bulluck and Buehler 2008). In southeastern Ohio, populations of grassland birds (grasshopper sparrows and Henslow's sparrows) are increasing, whereas in other parts of the state they are decreasing (Ingold 2002). In the Appalachian Plateau, population estimates confirm that reclaimed minelands provide critical habitat in maintaining population densities for grassland birds, specifically Henslow's sparrow (Mattice et al. 2005). In Pennsylvania, the distribution of Henslow's sparrows closely matches the distribution of reclaimed surface mines (Hill 2012). Moreover, there was no support that landscape characteristics affected Henslow's sparrow occupancy; rather local vegetation structure was the most influential factor (Hill and Diefenbach 2014). In fact, typically there are species-specific responses to vegetation structure that are comparable to non-reclaimed mine sites.

Species abundance and nest success are influenced by vegetation structure and disturbances on reclaimed mines, analogous to patterns seen on non-reclaimed mine habitats. Grasshopper sparrow, Henslow's sparrow, savannah sparrow (*Passerculus sandwichensis*), and bobolink (*Dolichonyx oryzivorus*) abundance was negatively correlated with increasing shrub cover on reclaimed mines in southeastern Ohio. Alternatively, eastern meadowlark and dickcissel abundances were not correlated to shrub cover (Graves et al. 2010). In West Virginia, grasshopper sparrows on reclaimed minelands require grass structure similar to that found in native prairie, with ~25% bare ground in their territories (Whitmore 1979; Wood and Ammer 2015). In Pennsylvania, Henslow's sparrow abundance on reclaimed minelands was positively related to increasing grass cover, but the inverse was true for herbaceous cover (Hill and Diefenbach 2014).

A fundamental question for reclaimed minelands is if they act as ecological traps by attracting nesting birds that ultimately are not reproductively successful (Wray et al. 1982). For golden-winged warbler on reclaimed mines, nest sites tended to have more grass and forb cover and less woody cover than random sites, and no micro-habitat variables investigated were significantly correlated with daily nest survival (Bulluck and Buehler 2008). Reclaimed minelands in Pennsylvania provide brood habitat and food resources for imprinted wild turkey poults when compared to unmined areas (Anderson 1980). In southeastern Ohio, random locations had ~2.5 times the amount of woody vegetation, but nest placement was not associated with the number of woody patches or distance to an edge (Graves et al. 2010). Grasshopper sparrow and Henslow's sparrow daily nest survival was negatively correlated with the amount of woody vegetation surrounding nests (Graves et al. 2010). Henslow's sparrow nest survival on reclaimed minelands that are small in size (<100 ha) was comparable to other values reported in the literature; and in Pennsylvania, Henslow's sparrow nest almost always occurred on reclaimed minelands (Stauffer et al. 2011). However, at these sites, Henslow's sparrow produced fewer young, as increasing fledgling production was correlated with decreasing woody vegetation, decreasing plot-level woody vegetation, and decreasing bare ground around the nest (Hill 2012). Another consistent finding is that nesting songbirds on reclaimed minelands have low brood parasitism rates by brown-headed cowbirds (*Moluthrus ater*; Wray et al. 1982; DeVault et al. 2002; Scott and Lima 2004; Hill 2012; Wood and Ammer 2015).

Woody vegetation removal, mowing, prescribed burning (Fig. 4), and grazing are some disturbance tools recommended to improve grassland and shrubland bird habitat on reclaimed sites (Whitmore 1981; Ingold et al. 2009; Hill and Diefenbach 2014; Brooke et al. 2015). Woody vegetation removal on reclaimed minelands correlated with a three-fold decline in Henslow's sparrow survival, but grasshopper sparrow population increased by 15% on woody removal treatments (Hill and Diefenbach 2014). On reclaimed sites in Ohio, Henslow's sparrows and bobolink avoided recently mowed areas for nesting, but savannah sparrow and grasshopper sparrow nested equally on mowed and unmowed areas (Ingold 2002). Site fidelity of Henslow's sparrows was low (~13%) and individuals that returned were only individuals that were banded in unmowed areas (Ingold et al. 2009); no individuals banded

Fig. 4 Prescribed fire being used to maintain and improve grassland and shrub habitat conditions for northern bobwhite (*Colinus virginianus*), white-tailed deer (*Odocoileus virginianus*), and elk (*Cervus canadensis*) in the post-mined landscape in the Appalachian Plateau (Photo by D. Ledford, Appalachian Wildlife Foundation)

in mowed areas returned. Average survival probabilities of returning Henslow's sparrows were greater on mowed than unmowed areas (Ingold et al. 2010) and mowing did not affect grasshopper sparrow or savannah sparrow returns. Prescribed burning reduces litter and encourages pioneer plants, positively affecting northern bobwhite (*Colinus virginianus*) brood habitat (Brown 1981), though prescribed burning is believed to have negatively affected winter survival by reducing the quality of shrub cover in some mined sites (Peters et al. 2015). Northern bobwhite on reclaimed minelands is most limited by the wide interspersion of shrub cover (Brooke et al. 2015).

In the absence of reclamation that focuses on reforestation, grassland and shrubland cover is created and can provide habitat for a community of grassland and shrubland birds that continues to experience dramatic population declines in the eastern United States. Many grassland and shrub bird species not only occupy reclaimed minelands but occur in abundances comparable to un-mined areas. Additionally, these species can nest and reproduce at rates comparable to unmined areas, suggesting that reclaimed minelands are usually not acting as ecological traps. Specifically, for northern bobwhite, reclaimed mines managed or otherwise offer novel opportunities for management directed at maintaining populations that are otherwise lacking in most of the Appalachian Plateau (Brown 1981; Wood et al. 2013; Brooke et al. 2015; Peters et al. 2015).

3.3 Raptors

Post-SMCRA mines can also provide habitat for a variety of raptor species. In northwestern Pennsylvania, reclaimed mines had greater counts of spring raptors than surrounding agricultural areas (Yahner and Rohrbaugh 1998), and the most frequently counted species were red-tailed hawk (*Buteo jamaicensis*; 58%), American kestrel (*Falco sparverius*; 36%), and northern harrier (*Circus cyaneus*; 6%). Red-shouldered hawks (*Buteo lineatus*) tolerated fragmentation created by reclaimed minelands in West Virginia, but were still more likely to occur close to wetland areas (Balcerzak and Wood 2003). On reclaimed mines, these areas were often represented by fill ponds created to control mine erosion. Short-eared owls (*Asio flammeus*) and northern harriers (*Circus hudsonius*) have been documented to nest on reclaimed sites in the Appalachian Plateau, and northern harriers tended to nest in drier areas with denser vegetation with a greater proportion of subadult females comprising the nesting population than expected (Vukovich and Ritchison 2006, 2008).

4 Mammals

4.1 Bats

Bats comprise an ecologically important faunal component with at least 16 species potentially occurring in all or portions of the Appalachian Plateau Coalfields. During spring through fall, outside of hibernation and migratory seasons, all regionally extant species forage in both forested and open upland areas and along all orders of riparian corridors. Most species in the region day-roost in foliage, exfoliating bark, or cavities of live trees or snags including Rafinesque's big-eared bat (*Corynorhinus rafinesquii*), big brown bat (*Eptesicus fuscus*), eastern red bat (*Lasiurus borealis*), hoary bat (*Lasiurus cinereus*), Seminole bat (*Lasiurus seminolus*), silver-haired bat (*Lasionycteris noctivagans*), little brown bat (*Myotis lucifugus*), the threatened northern long-eared bat (*Myotis septentrionalis*), the endangered Indiana bat, evening bat (*Nycticeius humeralis*), and tri-colored bat (*Perimyotis subflavus*). Exceptions to the use of trees as day-roosts are the cave-obligate and endangered Virginia big-eared bat (*Corynorhinus townsendii virginianus*) and endangered gray bat (*Myotis grisescens*). Rafinesque's big-eared bat, big brown bat, little brown bat, and Brazilian free-tailed bat (*Tadarida brasiliensis*) will commonly use anthropogenic structures that mimic hollow trees or snags as day-roosts (Ellison et al. 2007; Fagan et al. 2018). Additionally, Indiana bats and northern long-eared bats will readily use artificial roost-structures designed specifically for bats (Adams et al. 2015; De La Cruz et al. 2018; Hoeh et al. 2018). The eastern small-footed bat (*Myotis leibii*) is the exception in that it day-roosts primarily in emergent rock (i.e., cliff-faces) and talus slopes (Moosman et al. 2015). Many of the cave- or mine-hibernating species (i.e., eastern small-footed bat, little brown bat, northern long-eared bat, tri-colored bat, and Indiana bat) have suffered precipitous declines approaching or exceeding 90% due to White Nose Syndrome (WNS), caused by the novel fungal pathogen *Pseudogymnoascus destructans* (Francl et al. 2012; Powers et al. 2015). Another additive stressor to non-hibernating bats, such as the eastern red bat and hoary bat, in the region has come from additive mortality associated with wind-energy development in Pennsylvania and West Virginia, some occurring on reclaimed surface mines (Arnett et al. 2008).

4.2 Bat Day-Roosting

Surface-mining deforestation removes both day-roosting and foraging bat habitat. Although diurnal roosts are likely not a limiting factor in this largely forested region, the size (i.e., 25–2000 ha) of most surface mines in the Appalachian Plateau increases the likelihood that roosting networks used by active maternity colonies of the social northern long-eared and/or the Indiana bat are altered or destroyed (Menzel et al. 2001; Silvis et al. 2015, 2016). Throughout the 1970s to present, prior to mining, operators are required to survey for Indiana bats and locate day-roosts to determine minimization and mitigation measures (e.g., seasonal tree-clearing) for continued species conservation (US FWS 2020). Migratory and foliage-roosting bats such as the eastern red bat, hoary bat, and Seminole bat do not form large colonies and presumably are not overly impacted by pre-mining tree removal (Menzel et al. 2000). Despite the occurrence of northern long-eared bats and Indiana bats in the Appalachian Plateau, particularly prior to WNS, little or no research has examined post-mining day-roost use in forests surrounding mined lands or implications of colony formation within mined areas themselves. Presumably from research conducted in managed and unmanaged forest landscapes, some pre-SMCRA woody regrowth, particularly cavity bearing black locust trees or snags, may provide quality northern long-eared day-roost habitat (Johnson et al. 2009; Silvis et al. 2016).

Application of the FRA holds promise for improved day-roost habitat for bats as trees mature, particularly species that exhibit exfoliating bark or are prone to cavity formation. Wildlife-friendly tree species provided in FRA guidelines (Wood et al. 2013) such as sugar maple (*Acer saccharum*), shagbark hickory (*Carya ovata*), and American elm (*Ulmus americana*) are preferred by Indiana bats (Menzel et al. 2001; Johnson et al. 2010; Jachowski et al. 2016), whereas black locust and sassafras (*Sassafras albidum*) is preferred by northern long-eared bats in the region (Silvis et al. 2016). Presumably, the lower tree densities and higher snag creation rates than surrounding forests and increased solar radiation will improve future day-roosting conditions (Johnson et al. 2009; Ford et al. 2016). In the interim, young forests on reclamation sites provide bat above-canopy foraging substrate and a source of arthropod prey (Sheets et al. 2013). Similarly, if strategically placed during planting (Wood et al. 2013), these young forest patches create connective corridors between the unmined forest surrounding the mine site as Indiana bats and other forest-adapted foraging species are less likely to traverse open landscapes (Menzel et al. 2005). Additions of artificial day-roost structures which have been used in the region by northern long-eared bats where natural day-roost loss has occurred due to deforestation (De La Cruz et al. 2018) could be used before FRA-planted stands achieve sufficient structure to serve as day-roost habitat. Emergent rock features destroyed by surface mining have impacted eastern small-footed bats, however, this species has been found to readily use crevices in residual highwalls and waste rock, and also will utilized "engineered" rock structures whether designed expressly for the species (Tompkins 2014) or associated with other activities such as rock placement for stream restoration or boulder-lined dam surfaces (Moosman et al. 2013).

4.3 Bat Foraging

Post-reclamation foraging activity by bats in the region also is poorly studied. Research in the Appalachian region generally shows neutral to positive responses from bats following forest harvesting or prescribed burning that reduces forest clutter, particularly among larger-bodied species with lower echolocation call characteristics adapted to forage in more open conditions (Ford et al. 2005; Austin et al. 2018).

Whether these findings are directly transferable to reclaimed minelands is unknown considering the differences in disturbance size between forest management activities (i.e., < 30 ha) versus surface mining, or between natural forest regeneration versus either grass/forb reclamation or FRA-planting practices. Whether in a closed forest or open conditions, overall bat foraging activity is related to proximity to water in the Appalachian Plateau (Ford et al. 2005). Assuming good water quality, creation and maintenance of wildlife-friendly water retention areas (Wood et al. 2013) will benefit bats by providing a source for drinking water and potential foraging areas that mitigate for loss of streams and stream function post-mining. Waterhole creation on xeric ridgetops distant from lentic or lotic habitats to benefit bats has been demonstrated in the Appalachian Plateau (Huie 2002; Maslonek 2010; Johnson et al. 2010; 2013). Research regarding bat response to stream reclamation and restoration is limited (Ciechanowski et al. 2011); however, analogous to the FRA process, as vegetation matures and stream processes stabilize, bats will likely utilize these redeveloping foraging and commuting habitats.

4.4 Bat Hibernacula

As a somewhat positive legacy of coal mining in the Appalachian Plateau, in areas that lack limestone solution caves in karst geology, abandoned underground mines and associated structures such as vent shafts often serve as maternity and bachelor roosts, migratory stopover sites, and hibernacula for bats where prior to mining few had existed before (Krusac and Mighton 2002 Johnson et al. 2005; 2006; Buehler and Percy 2012; Furey and Racey 2015; Fig. 5). In the Appalachians generally, the functional extirpation of bat species such as the northern long-eared bat occurred rapidly with the fungal infection of caves, particularly in settings where long hibernation periods occurred (Johnson et al. 2013; Ford et al. 2016; Austin et al. 2018).

Fig. 5 Legacy deep mines have provided day-roosting and hibernation habitat for bats such as these endangered Indiana bats (*Myotis sodalis*) in the Appalachian Plateau (Photo by Tomas Nocera, Virginia Tech)

However, in the Appalachian Plateau of West Virginia and eastern Ohio, anecdotal evidence suggests some residual populations of bats normally affected by WNS persist perhaps due to lower exposure to WNS in pre-law mines than would occur in karst formation caves. Analogous to persistence being reported on the Coastal Plain to the east where bats overwinter in forests or aberrant hibernacula (Grider et al. 2016; Dowling and O'Dell 2018), the same is observed for northern long-eared and Indiana bats overwintering in small legacy coal adits or emergent rock features on pre-law mines in the Appalachian Plateau (De La Cruz and Schroder 2015). These legacy mines may minimize other bat species' exposure to WNS-vectoring little brown bats, may be utilized only by local bats, and/or provide unsuitable substrates (i.e., substrate pH < 5) for *Pseudogymnoascus destructans,* persistence and growth (Wilder et al. 2011; Raudabaugh and Miller 2013). Underground mines often have microclimates suitable for bat hibernacula or if not, maybe modified in a number of ways to increase habitat suitability including venting, entrance and interior stabilization and modification, gating, and installation of monitoring devices (Carter et al. 2010). Exclusion structures such as gates may help to stabilize the entrance of an underground mine but more importantly will protect remaining bats, that may be somewhat WNS resistant, from unnecessary and costly human-induced arousal (Kunz et al. 2010). Impacts to non-bat cave-obligate biota, including native fungi, have limited the use of antifungal treatments of karst caves, but because mines often lack such species the potential treatment of these areas may enhance use by WNS-susceptible bats and subsequently their use of post-mined landscapes (Carter et al. 2010; Sewall et al. 2016).

4.5 Small and Meso-Mammals

Pre- and post-SMCRA mine restoration that included replanting with a grass-legume mixture created sites that lacked vegetational and structural diversity (McGowan and Bookhout 1986), often dominated by grasses or combination of exotic species (e.g., fescue; sericea lespedeza, *Lespedeza cuneata*; Zipper et al. 2011a, Sena et al. 2021). The persistence of this vegetation can have a decadal impact on successional progression (Holl 2002; Skousen et al. 2006). Small mammal diversity on such sites is low (Lacki et al. 1982; McGowan and Bookhout 1986; Larkin et al. 2008) and the lack of woody cover is a limiting factor in the presence of woodland mice (*Peromyscus* spp.; Sly 1976). Whereas results from these early studies provide information on species richness and relative abundance, their objective was not to compare among reclamation practices or between reclaimed habitat and pre-mining conditions. Chamblin (2002) compared mammal species richness among four treatments (intact forest- pre-mining; grassland—7–21 years post reclamation; shrub/pole—18–28 years; fragmented forest– streamside buffer surrounded by reclaimed lands) and reported no differences in species richness among treatments and no correlation between time since reclamation and species richness on mountaintop-removal mines in southern West Virginia. Use of reclaimed mine sites by fossorial (ground-dwelling) and semi-fossorial mammals depends on depth of soil and degree of substrate compaction

(Lawer et al. 2019). Species requiring exposed rock outcrops such as the Allegheny woodrat (*Neotoma magister*) may decline or be absent from post-SMCRA minelands unless an artificial structure can serve as an alternative to native habitat (Chamblin et al. 2004). Meso-mammals either captured or observed on reclaimed sites include Virginia opossum (*Didelphis virginiana*), eastern cottontail rabbit (*Sylvilagus floridanus*), woodchuck (*Marmota monax*), raccoon (*Procyon lotor*), striped skunk (*Mephitis mephitis*), gray fox (*Urocyon cinereoargenteus*), and red fox (*Vulpes vulpes*) (Yearsley and Samuel 1980; Brenner et al. 1982; Lacki et al. 1982). Species with large home ranges with wide habitat tolerances, such as bobcat (*Lynx rufus*) and coyote (*Canis latrans*) would be expected to opportunistically use portions of reclaimed minelands to meet their habitat requirements, if available.

4.6 White-Tailed Deer and Black Bear

White-tailed deer are found throughout coal-bearing areas of the Appalachian Plateau, from high-elevation forests to bottomland agriculture/urbanized areas, though often at lower densities than elsewhere in the Appalachian region (Kniowski and Ford 2017; but see Campbell et al. 2005). As highly adaptable herbivores, they occupy all suitable habitats including those created during mining and reclamation. However, limited research has examined white-tailed deer use of mined lands (Brenner et al. 1975; Brenner et al. 1977; Knotts and Samuel 1982). White-tailed deer herbivory impacts on tree restoration have been noted (Jacobs et al. 2004; Burney and Jacobs 2018). It should be assumed that white-tailed deer occupied forested lands cleared during mining activities and that they returned to these sites post-mining, although research is needed on changes to carrying capacity, whether positive or negative, resulting from habitat alterations.

Reclaimed surface mines, particularly mountain-top removal sites that encompass thousands of hectares, can provide habitat for black bears within Appalachia, particularly along ecotones where soft mast-producing plants (e.g., *Rubus* spp.) thrive. Oak mast is a primary food source for black bears in the Appalachians, and consequently its availability influences spatial and temporal habitat use by bears (Vaughan 2002), including the relative value of interspersed minelands within an otherwise forested landscape. Because of their large scale and conversion of mature hardwood forest to grasslands, however, reclaimed minelands often contain lower habitat quality for black bears because of reduced mast production (Ryan 2009), thereby representing a net loss in overall habitat. Mineland areas with dense cover by the soft-mast-producing exotic autumn olive (*Elaeagnus umbellata*; as per Oliphant et al. 2017) are readily used by black bears (Allan and Steiner 1972). Regionally, black bears have been frequently observed to den in brushpiles, clearcuts, and even under discarded or idle machinery on or near minelands (J. Cox, University of Kentucky, pers. observation). Bears may also have access to anthropogenic food (e.g., large dumpsters) on operational mines, and limited public access to mines can create refugia from harvest because of reduced hunter access (Ryan 2009); however, these same attractants can condition bears to human foods and lead to poaching or increased risk-taking behavior (e.g. frequently crossing roads) whereas other bears seem to totally avoid mines where disturbance is high.

4.7 Elk

Although early accounts from European settlers indicate the presence of elk throughout the Appalachian Plateau, elk were not found at high densities in more forested areas. Elk most often occurred in more open environments in mixed-species aggregations that included bison and white-tailed deer (Bryant and Maser 1982). Early European settlers, and Native Americans before them, did not primarily rely on elk for food and hides, possibly indicating their relatively low abundance compared to other big game species in the region (McCabe 1982). Nonetheless, in the early twentieth century, a few eastern states, including some in the Appalachian Plateau such as Pennsylvania and Virginia, attempted to reestablish elk, though usually with limited success. Regionally, human development and agriculture had displaced formerly suitable elk habitat on floodplains. Accordingly, in the 1990s, managers turned to reclaimed surface coal mines with mosaics of herbaceous/grass and shrub-scrub habitat on mine benches and surrounding forests as potential analogs to occupied elk habitat in the West. Elk are mixed feeders, spending most of their time in open areas grazing grasses, forbs, and other low-growing herbaceous plants; however, they also spend considerable time within closed-canopy forests and shrub-scrub ecotonal communities that provide browse foods, as well as thermal and security cover. On the Appalachian Plateau, surface mines are often located in relatively remote areas away from human development, and therefore, it was suggested that elk placed in these areas would minimize human-elk conflict. Additionally, it was believed that the vast, grassy "minescapes" would enhance release site fidelity and large herd cohesion, thereby decreasing the possibility of multiple founder effects resulting from scattered, small, low-density populations that overtime be non-viable (Maehr et al. 1999).

Over the past 25 years, elk have been reintroduced to portions of Kentucky, Tennessee, Virginia, and West Virginia within the Appalachian Plateau, with most release sites occurring at post-SMCRA reclaimed surface mines (Fig. 6). The reintroduction of elk into Kentucky provides perhaps the most well-studied example of the relationship between elk and this novel landscape. From 1997 to 2002, ~1550 elk were translocated from western states to a 14-county area "elk zone" in southeastern Kentucky. Elk was released at eight different sites, seven of which were on coal surface mines (some active) with varying levels of ongoing human activity (i.e., off-road vehicle use, blasting, and coal transport). This area was the only region of Kentucky that could support large elk herds while minimizing the potential for human-elk interactions. Human activity, along with poaching, meningeal worm infection (*Parelaphostrongylus tenuis*) that causes neurological lesions in elk, and

Fig. 6 Elk (*Cervus canadensis*) originally from western population sources have thrived on surface mines reintroduction sites in the Appalachian Plateau. Note radio-telemetry collars on left- and right-most cows that allows researchers and managers the ability to track movement patterns and habitat use (Photo by D. Ledford, Appalachian Wildlife Foundation)

poor forage conditions, had the potential to reduce survival, decrease site fidelity and cause human-elk conflict in the surrounding landscape (Lankester 2001; Larkin et al. 2001). Several western states provided what was considered the Rocky Mountain elk subspecies (*Cervus canadensis nelsoni*; Larkin et al. 2001) to Kentucky; however, whether western elk would be a suitable surrogate or not for the extinct eastern elk subspecies was a concern (Polziehn et al. 2000).

In eastern Kentucky, the first few hundred elk released were outfitted with VHF radio-collars, allowing initial assessments of elk habitat use and movement. Reintroduced elk exhibited high rates of annual survival (≥90%; Larkin et al. 2003) and release site fidelity (≥53% within 10 km after one year; Larkin et al. 2004). Release site fidelity was higher where public access was lower and edge habitat was higher (Larkin et al. 2004). Indeed, both survival and natality were very high during the first few years post-translocation and may have reflected a brief period of irruptive growth in a region that, unlike western elk range, lacked harsh winters and large predators such as wolves and mountain lions (Larkin et al. 2003). Despite this early success, research indicated elk could be locally impacted by human-caused disturbances leading to changes in temporal activity and movement patterns, including disruption of herd units (Wichrowski et al. 2005; Olsson et al. 2007). Currently, regulated hunting, where allowed, is the primary mortality agent in adult elk. Meningeal worm infections have occurred, however, parasitic and other stressors have minimally impacted survival, reproduction, and recruitment (Kentucky Department of Fish and Wildlife Resources 2015; Slabach et al. 2018).

With an estimated population of ~ 3,500 (G. Jenkins, Kentucky Department of Fish and Wildlife Resources, pers. communication), elk in Kentucky are now well established, with high densities still occurring at or near some of the original surface-mine release sites. Despite high site fidelity of the Kentucky herd, some natural immigration from Kentucky into Tennessee, Virginia, and West Virginia has occurred over the past 20 years. These reintroductions occurred in Tennessee beginning in 2000 (Kindall et al. 2011), Virginia in 2012 (Virginia Department of Game and Inland

Fisheries 2020), and lastly in West Virginia in 2018 (West Virginia Division of Natural Resources 2020). Combined with reintroduction efforts in the Blue Ridge portion of the Appalachians in the Great Smoky Mountains National Park (Murrow et al. 2009), the larger region has the potential to harbor a much greater interstate population, albeit probably always centered in the Appalachian Plateau mined landscape.

Although the reintroduction of elk in Kentucky and surrounding states has so far been a program success, there may be long-term ecological and economic consequences of having high densities of a large, herd-forming herbivore in the post-mined landscape. These consequences could lead to severe ecological degradation in terms of the composition, structure, and successional trajectories of native plant communities (Bradshaw and Waller 2016), which in turn, affect other wildlife species; however, this early in the reintroduction saga, research on the potential ecological impact of reintroduced elk in the Appalachian Plateau is lacking. Shortly after their reintroduction into Kentucky on post-SMCRA sites, elk established wide (~0.3–0.6 m) trails with visible soil erosion leading from bedding grounds within the forest to reclaimed minelands where they forage during crepuscular and nocturnal periods. Heavily used bedding areas have sparse or no leaf litter, trampled/browsed vegetation (Fig. 7), and large deposits of elk feces and urine-saturated soils (TerBeest 2005). Consequently, these elk-disturbed ridgetops and ravine bottoms had higher soil ammonium, lower soil carbon, and lower soil moisture than non-elk reference forest sites. Although not quantified, Maigret et al. (2019) suggested that elk impacts to unmined forest "islands" could reduce diversity and ecological integrity of native flora and fauna and/or reduced opportunities for natural colonization of reclaimed mine sites by these taxa.

Fig. 7 Pine (*Pinus spp.*) damaged by elk (*Cervus canadensis*) antler rubbing on a reclaimed surface mine in the Appalachian Plateau (Photo by J. Cox, University of Kentucky)

Within the Appalachian Plateau, elk inhabit a very different forest ecosystem than that which existed even a century ago (Yarnell 1998). In southwest Virginia, elk monitored with GPS-collars from 2014 to 2019 were found to use all forest types equitably with their proportion in the landscape; however, their use of reclaimed mine sites was greater than expected based on availability (Quinlan et al. 2020), a finding similar to elk in Kentucky (Cox 2003). The fragmentation of forests by surface mining and the extensive use of invasive exotic plant species to reclaim these areas has created an entirely new, non-analog plant-herbivore regional dynamic. Elk rapidly takes advantage of early plant germination following hydroseeding, frequently consuming the new-growth of both forbs and grasses preventing successful establishment or arrested growth. In a large mountaintop removal mine matrix landscape, Schneider et al. (2006) found that 51% of elk diet in Kentucky consisted of native and exotic grasses and forbs commonly planted on surface mines including Brome (*Bromus* spp.), Kentucky 31 fescue (*Lolium arundinaceum*), broom sedge bluestem, orchard grass (*Dactylis glomerata*), Chinese silver grass (*Miscanthus sinensis*), crown vetch (*Coronilla varia*), Chinese-bush clover (*Lespedeza cuneata*), and red clover (*Trifolium pretense*). Many of these grasses and forbs were used by elk in Tennessee (Lupardus et al. 2011) and Pennsylvania (Heffernan 2009) as well. In Tennessee, the summer diet was dominated by forbs with jewelweed (*Impatiens* spp.) being the most selected. Schneider et al. (2006) also observed that woody browse in elk diet was dominated by black locust and autumn olive (*Elaeagnus umbellate*), two species commonly planted or invasive on surface mines (Oliphant et al 2017). In Pennsylvania, Heffernan (2009) found elk heavily used autumn olive because of the high nutritional value of the fruit in the fall. Yearling females used autumn olive and other browse species more during summer compared to males or other age groups (Heffernan 2009). In Tennessee, within an area that only contained about 12% open habitat from pasture and legacy bench and strip mines, Lupardus et al. (2011) found a far greater percentage of native plant species in elk diets; however, non-native autumn olive, tall fescue (*Festuca arundinacea*), and exotic legumes still were important diet items.

Regionally, elk appear to be seed vectors of invasive species such as multi-flora rose (*Rosa multiflora*) that can become prolific in both mined and non-mined forest sties (Schneider et al. 2006; Lupardus et al. 2011). Hackworth et al. (2018) found that elk were responsible for much of the damage to planted and volunteer tree seedlings in an active area of mine reclamation. Black locust seems particularly vulnerable to elk browsing and girdling from elk antler rubbing. Accordingly, high elk densities may complicate FRA efforts through intense herbivory, trampling, and antler rubbing whereby succession patterns along forest-mine ecotones are arrested or altered. Throughout the Appalachian Plateau, mining and post-mining action create residual overburden and structures such as sediment ponds that contain high concentrations of sodium, sulfur, and in some cases heavy metals. These sites are attractive to elk and white-tailed deer as mineral licks encouraging geophagy and creating other localized areas of high-elk impact to vegetation and soils (Campbell et al. 2004). Whether these legacy environmental contamination effects are detrimental or beneficial to elk and white-tailed deer in the region is unknown.

Nonetheless, the return of elk to the Appalachian Plateau appears to be a restoration success story. Elk were and are once again an important component of the mixed-mesophytic ecosystem of central Appalachia. As large, gregarious herbivores, they perform vital ecological roles as a prey species (neonates) for black bear and coyote, food for scavengers such as ravens (*Corax corax*), plant seed vectors, and physical modifiers of soils and plant communities. In the absence of wolves, mountain lions, or harsh winters, however, elk have demonstrated a remarkable ability to irruptively grow from founding populations and expand their range in the Appalachian Plateau, in large part to the habitat provided by reclaimed mines. Managers are just now beginning to understand the ecological implications of returning elk to a region increasingly fragmented and denatured by surface mining for coal, mineral extraction, logging, and the post-extraction land practices that ensue. Hunting will likely remain the primary mechanism of elk population control in the immediate future, and we suggest long-term studies are needed to better characterize the role of elk in these novel landscapes.

5 Conclusion

Coal mining, particularly surface mining, has left an enduring legacy in Appalachia's landscapes; its impacts on wildlife communities are complex. Overall biodiversity has declined and some species have been negatively impacted, while other species have responded positively to the novel habitats on post-mining areas. How to manage these lands going forward remains a question not fully answered. Reaching consensus between reforestation to restore lost forest habitat or maintaining large, high-quality early-successional grasslands and shrub-scrub with both native and exotic plants will require an assessment of not only wildlife needs at both the local and landscape scales but also the logistical and financial constraints for managers and stake-holders. Nonetheless, opportunities for using mined lands of the Appalachian Plateau both as a demonstration for large-scale ecological restoration and to explore approaches for managing non-analog environments with novel plant communities that have conservation value for adaptive wildlife will abound in the coming decades.

References

Adams J, Roby P, Sewell P, Schwierjohann J, Gumbert M, Brandenburg M (2015) Success of Brandenbark, an artificial roost structure designed for use by Indiana bats (*Myotis sodalis*). J Amer Soc Mining Reclam 4:1–15

Allan PF, Steiner WW (1972) Autumn olive for wildlife and other conservation used. U.S, Soil Conservation Service Resource Note, p 458

Anderson RJ (1980) Evaluation of Reclaimed Surface Mines as Wild Turkey Brood Range. M.S. Thesis, West Virginia University, Morgantown

Arnett EB, Brown WK, Erickson WP, Fielder JK, Hamilton BL et al (2008) Patterns of bat fatalities at wind energy facilities in North America. J Wild Manage 72:61–78
Austin LV, Silvis A, Ford WM, Muthersbaugh M, Powers KE (2018) Bat activity following restoration prescribed burning in the central Appalachian upland and riparian habitats. Nat Areas J 38:183–195
Averitt P (1975) Coal resources of the United States. U.S. Geol Surv Bull 1412, 131 p
Bajema RA, Lima SL (2001) Landscape-level analyses of Henslow's sparrow (*Ammodramus henslowii*) abundance in reclaimed coal mine grasslands. Amer Midl Nat 145:288–298
Balcerzak MJ, Wood PB (2003) Red-shouldered hawk (*Buteo lineatus*) abundance and habitat in a reclaimed mine landscape. J Raptor Res 37:188–197
Becker DA, Wood PB, Strager MP, Mazzarella C (2015) Impacts of mountaintop mining on terrestrial ecosystem integrity: identifying landscape thresholds for avian species in the central Appalachians, United States. Lands Ecol 30:339–356
Boettner F, Clingerman J, McIlmoil R, Hansen E, Hartz L, Hereford A, Vanderberg M, Arano K, Deng J, Strager J, Strager M, Donahue C (2014) An assessment of natural assets in the Appalachian Region: forest resources. Appalachian Regional Commission Report, Washington DC, p 97
Bourne JC (2015) Examining the Impacts of Valley Fills In Stream Ecosystems on Amphibian and Aquatic Insect Communities in Southeastern Kentucky. M.S. Thesis, Eastern Kentucky University, Richmond
Bradshaw L, Waller DM (2016) Impacts of white-tailed deer on regional patterns of forest tree recruitment. For Ecol Manage 375:1–11
Brady JK (2016) Salamander diversity of reforested abandoned surface coal mines in the Appalachian Region, USA. Restor Ecol 24:398–405
Braun EL (1950) Deciduous forests of eastern North America. Blakiston Company, Philadelphia
Brenner FJ (2007) Impact of the surface mining control and reclamation act on species of special concern in Pennsylvania. In: Proc., Am Soc Min Reclam, p.:79–86
Brenner FJ, Crowley RH, Musaus MJ, Goth JH (1975) Evaluation and recommendations of strip mine reclamation procedures for maximum sediment erosion control and wildlife potential. Pp 2–23 in: Third Symposium on Surface Mining and Reclamation Volume II, NCA/BCR Coal Conf, Louisville KY
Brenner FJ, Kelly RB, Kelly j (1982) Mammalian community characteristics on surface mine lands in Pennsylvania. Environ Manage 6:241–249
Brenner FJ, Musaus MJ, Granger W (1977) Selectivity of browse species by white-tailed deer on strip-mined lands in Mercer County, Pennsylvania. Proc Penn Acad Sci 51:105–108
Brooke JM, Peters DC, Unger AM, Tanner EP, Harper CA et al (2015) Habitat manipulation influences northern bobwhite resource selection on a reclaimed surface mine. J Wild Manage 79:1264–1276
Brown SL (1981) The Effects of Controlled Burning on Potential Bob White Quail Brood Habitat on Surface Mines In West Virginia. M.S. Thesis, West Virginia University, Morgantown
Bryant LD, Maser C (1982) Classification and distribution. Pp 1–59 in: J.W. Thomas JW, Toweill DE (Eds.), Elk of North America: Ecology and Management. Wildlife Management Institute. Washington DC
Buehler DA, Percy K (2012) Coal mining and wildlife in the eastern United States: A literature review. Final Report to the Appalachian Wildlife Habitat Foundation, University of Tennessee, Knoxville TN
Bulluck LP, Buehler DA (2008) Factors influencing golden-winged warbler (*Vermivora chrysoptera*) nest-site selection and nest survival in the Cumberland Mountains of Tennessee. Auk 125:551–559
Burger JA, Evans DM (2010) Ripping compacted mine soils improved tree growth 18 years after planting, In: Barnhisel RI (ed) 27th Annual National Conference of the American Society of Mining and Reclamation, Lexington KY, pp 55–69
Burger JA, Graves D, Angel P, Davis V, Zipper C (2005) The forestry reclamation approach. For Reclam Advis 2:1–4

Burney OT, Jacobs DF (2018) Species selection– a fundamental silvicultural tool to promote forest regeneration under high animal browsing pressure. For Ecol Manage 408:67–74

Campbell TA, Laseter BR, Ford WM, Miller KV (2004) Unusual white-tailed deer movements to a gas well in the central Appalachians. Wild Soc Bull 32:983–986

Campbell TA, Laseter BR, Ford WM, Miller KV (2005) Population characteristics of a central Appalachian white-tailed deer herd. Wild Soc Bull 33:212–221

Campbell TA, Long DB (2009) Feral swine damage and damage management in forested ecosystems. For Ecol Manage 257:2319–2326

Carrozzino AL (2009) Evaluating Wildlife Response to Vegetation Restoration on Reclaimed Mine Lands in Southwestern Virginia. Ph.D. Dissertation, Virginia Tech, Blacksburg

Carter TC, Steffen BJ, Feldhamer GA (2010) Converting abandoned mines to suitable hibernacula for endangered Indiana bats (*Myotis sodalis*). In: Vories KC, Caswell AK, Price TM (eds) Proc, Protecting Threatened Bats at Coal Mines: A Technical Interactive Forum. US Office of Surface Mining, Alton IL, pp 205–213

Chamblin HD (2002) Small Mammal Communities on a Reclaimed Mountaintop Mine/Valley Fill Landscape in Southern West Virginia. MS Thesis, West Virginia University, Morgantown

Chamblin HD, Wood PB, Edwards JW (2004) Allegheny woodrat (*Neotoma magister*) use of rock drainage channels on reclaimed mines in southern West Virginia. Amer Mid Nat 151:346–354

Ciechanowsk M, Kubic W, Rynkiewicz A, Zwolicki A (2011) Reintroduction of beavers *Castor fiber* may improve habitat quality for vespertilionid bats foraging in small river valleys. Euro J Wild Res 57:737–747

Cornell Lab of Ornithology (2020) eBird. https://ebird.org/hotspot/L2887476

Cox JJ (2003) Community Dynamics among Reintroduced Elk, White-Tailed Deer, and Coyotes in Southeastern Kentucky. Ph.D. Dissertation. University of Kentucky, Lexington

Davis V, Burger JA, Rathfon R, Zipper CE, Miller CR (2012) Selecting tree species for reforestation of Appalachian mined land. US Office of Surface Mining, Appalachian Regional Reforestation Initiative, p 9

De La Cruz JL, Schroder ES (2015) Kanawha State Forest hibernaculums suitability determination and spring emergence survey report, Kanawha County. West Virginia - Final Report to the Mary Ingles Trail Blazers, Charleston, WV, p 52

De La Cruz J, Ward RL, Schroder ES (2018) Landscape characteristics related to use of artificial roosts by northern long-eared bats in north-central West Virginia. Northeast Nat 25:487–501

Denmon P (1998) Early Successional Habitat Use by Nongame Wildlife Species in American Woodcock Breeding Habitat in West Virginia. M.S. Thesis, West Virginia University, Morgantown

DeVault TL, Scott PE, Bajema RA, Lima SL (2002) Breeding bird communities of reclaimed coal-mine grasslands in the American Midwest. J Field Ornith 73:268–275

Dowling ZR, O'Dell D (2018) Bat use of an island off the coast of Massachusetts. Northeast Nat 25:362–382

Ellison LE, O'Shea TJ, Neubaum DJ, Bowen RA (2007) Factors influencing movement probabilities of big brown bats (*Eptesicus fuscus*) in buildings. Ecol Appl 17:620–627

Ellsworth JW, McComb BC (2003) Potential effects of passenger pigeon flocks on the structure and composition of presettlement forests of eastern North America. Cons Biol 17:1548–1558

Fagan KE, Willcox EV, Tran LT, Bernard RF, Stiver WH (2018) Roost selection by bats in buildings, Great Smoky Mountains National Park. J Wild Manage 82:424–434

Ford WM, Mardarish D, Schuler TM, Castleberry SB (2003) Influence of white-tailed deer digestion on running buffalo clover (*Trifolium stonoiferum*: Fabaceae Muhl. ex. A Eaton) germination. Amer Midl Nat 149:425–428

Ford WM, Menzel MA, Rodrigue JL Menzel JM, Johnson JB (2005) Relating bat species presence to simple habitat measures in a central Appalachian forest. Biol Cons 126:528–539

Ford WM, Silvis A, Johnson JB, Edwards JW, Karp M (2016) Northern long-eared bat day-roosting and prescribed fire in the central Appalachians, USA. Fire Ecol 12:13–27

Francl KE, Ford WM, Parks DW, Brack V (2012) Capture and reproductive trends in summer bat communities in West Virginia: Assessing the impact of white-nose syndrome. J Fish Wild Manage 3:33–42

Furey NM, Racey PA (2015) Conservation ecology of cave bats. In: Voigt CC, Kingston T (eds) Bats in the Anthropocene: Conservation of Bats in a Changing World. Springer International Publishing, Cham, Switzerland, pp 463–500

Gipson PS, Hlavachik B, Bergerl T (1998) Range expansion by wild hogs across the central United States. Wild Soc Bull 26:279–286

Gore JA (1983) The distribution of desmognathine larvae (Amphibia: Plethodontidae) in coal surface mine impacted streams of the Cumberland Plateau, USA. J Freshw Ecol 2:13–23

Graves BM, Rodewald AD, Hull SD (2010) Influence of woody vegetation on grassland birds within reclaimed surface mines. Wilson J Ornith 122:646–654

Gregg ID, Wood PB, Steketee AK (2001) Comparison of reclaimed and unmined Woodcock summer diurnal habitat in West Virginia. Proc, Ann Conf Southeast Assoc Fish Wild Agencies 55:488–497

Grider JF, Larsen AL, Homyack JA, Kalcounis-Rueppell MC (2016) Winter activity of coastal plain populations of bat species affected by white-nose syndrome and wind energy facilities. PLoS ONE 11:1–14

Hackworth Z, Lhotka J, Cox JJ, Barton CD, Springer MT (2018) First-year vitality of reforestation plantings in response to herbivore exclusion on reclaimed Appalachian surface-mined land. Forests 9(4). https://doi.org/doi.org/10.3390/f9040222

Hamilton MS (2002) Effects of Developmental Activities on Streamside Salamander Communities in Boone County, West Virginia. M.S. Thesis, Marshall University, Huntingdon

Harper CA, Ford WM, Lashley MA, Moorman CE, Stanbaugh M (2016) Fire effects on wildlife in Central Hardwoods and Appalachian regions. Fire Ecol 12:127–159

Hepting GH (1974) Death of the American chestnut. J For Hist 18:60–67

Hill JM (2012) Population Ecology of Grassland Sparrows on Reclaimed Surface Mine Grasslands in Pennsylvania. Ph.D. Dissertation, The Pennsylvania State University, State College

Hill JM, Diefenbach DR (2014) Occupancy patterns of regionally declining grassland sparrow populations in a forested Pennsylvania landscape. Cons Biol 28:735–744

Hinkle MP, Gardner LA, White E, Smith WH, VanGundy RD (2018) Remnant habitat patches support Green salamanders (*Aneides aeneus*) on active and former Appalachian surface mines. Herpetol Conserv Biol 13:634–641

Hoeh JPS, Bakken GS, Mitchell WA, O'Keefe JM (2018) In artificial roost comparison, bats show preference for rocket box style. PLoS ONE 13:1–16

Holl KD (2002) The effect of coal surface mine revegetation practices on long-term vegetation recovery. J Appl Ecol 39:960–970

Huie K (2002) Use of Constructed Woodland Ponds by Bats in the Daniel Boone National Forest. MS Thesis, Eastern Kentucky University, Richmond

Ingold DJ (2002) Use of a reclaimed stripmine by grassland nesting birds in east-central Ohio. Ohio J Sci 102:56–62

Ingold DJ, Dooley JL, Cavender N (2009) Return rates of breeding Henslow's sparrows on mowed versus unmowed areas on a reclaimed surface mine. Wilson J Ornith 121:194–197

Ingold DJ, Dooley JL, Cavender N (2010) Nest-site fidelity in grassland birds on mowed versus unmowed areas on a reclaimed surface mine. Northeast Nat 17:125–134

Jachowski DS, Rota CT, Dobony CA, Ford WM, Edwards JW (2016) Seeing the forest through the trees: Considering roost-site selection at multiple spatial scales. PLoS ONE 11:1–19

Jacobs DF, Ross-Davis AL, Davis AS (2004) Establishment success of conservation tree plantations in relation to silvicultural practices in Indiana, USA. New Forest 28:23–36

Johnson JB, Ford WM, Rodrigue JL, Edwards JW, Johnson C (2010) Roost selection by male Indiana Myotis following forest fires in central Appalachian hardwoods forests. J Fish Wild Manage 1:111–121

Johnson JB, Edwards JW, Ford WM, Gates JE (2009) Roost tree selection by northern myotis (*Myotis septentrionalis*) maternity colonies following prescribed fire in a central Appalachian Mountains hardwood forest. Forest Ecol Manage 258:233–242

Johnson JB, Edwards JW, Wood PB (2005) Virginia big-Eared bats (*Corynorhinus townsendii virginianus*) roosting in abandoned coal. Northeast Nat 12:233–240

Johnson JB, Rodrigue JL, Ford WM (2013) Nightly and yearly bat activity before and after white-nose syndrome on the Fernow Experimental Forest in West Virginia. U.S. Forest Service, Res Pap NRS-24

Johnson J, Wood PB, Edwards JW (2006) Are external mine entrance characteristics related to bat use? Wild Soc Bull 34:1368–1375

Kentucky Department of Fish and Wildlife Resources (2015) 2015–2030 Kentucky elk management plan. Kentucky Dept of Fish and Wildlife Resources Report, Frankfort, KY, p 12

Kimmel RO, Samuel DE (1984) Ruffed grouse brood habitat on reclaimed surface mines in West Virginia. Minerals Environ 6:150–153

Kindall JL, Muller LI, Clark JD, Lupardus JL, Murrow JL (2011) Population viability analysis to identify management priorities for reintroduced elk in the Cumberland Mountains, Tennessee. J Wild Manage 75:1745–1752

Kniowski AB, Ford WM (2017) Predicting intensity of white-tailed deer herbivory in the Central Appalachian Mountains. J For Res. https://doi.org/10.1007/s11676-017-0476-6

Knotts R, Samuel DE (1982) White-tailed deer movement and distribution about surface coal mines in West Virginia. Minerals Environ 4:17–25

Krusac DL, Mighton SR (2002) Conservation of the Indiana bat in national forests: where we have been and where we should be going. In: Kurta A, Kennedy J (eds) The Indiana Bat: Biology and Management of an Endangered Species. Bat Conservation International. Austin, TX, pp 55–65

Kunz TH, Foster JT, Frick WF, Kilpatrick AM, McCracken GF, et al. (2010) White-Nose Syndrome: An overview of ongoing and future research news. Pages189–203 in: Vories KC, Caswell AK, Price TM (Eds.). Proceedings of Protecting Threatened Bats at Coal Mines: A Technical Interactive Forum. Office of Surface Mining, Alton IL

Lacki MJ, Hummer JW, Fitzgerald JL (2005) Population patterns of copperbelly water snakes (*Nerodia erythrogaster neglecta*) in a riparian corridor impacted by mining and reclamation. Amer Midl Nat 153:357–369

Lacki MJ, Hummer JW, Webster HJ (1982) Effect of reclamation technique on mammal communities inhabiting wetlands on mined lands in east-central Ohio. Ohio J Sci 91:154–158

Lacki MJ, Hummer JW, Webster HJ (1992) Mine-drainage treatment wetland as habitat for herptofaunal wildlife. Environl Manage 16:513. https://doi.org/10.1007/BF02394127

Lankester MW (2001) Extrapulmonary lungworms of cervids. In: Samuel WM, Pybus MJ, Kocan AA (eds) Parasitic Disease of Wild Mammals, 2nd edn. Iowa State University Press, Ames, IA

Lannoo M, Kinney V, Heemeyer J, Engbrecht NJ, Gallant AL, Klaver RWl (2009) Mine spoil prairies expand critical habitat for endangered and threatened amphibian and reptile species. Diversity 1:118–132

Larkin JL, Cox JJ, Wichrowski MW, Dzialak MR, Maehr DS (2004) Influences of release site fidelity of translocated elk. Restor Ecol 12:97–106

Larkin JL, Grimes RA, Cornicelli Cox JJ, Maehr DS (2001) Returning elk to Appalachia: foiling Murphy's Law. In: Maehr DS, Noss RF, Larkin JL (eds) Large Mammal Restoration. Island Press, Washington, D.C, pp 101–117

Larkin JL, Maehr DS, Cox JJ, Bolin DC, Wichrowski MW (2003) Demographic characteristics of a reintroduced elk population in Kentucky. J Wild Manage 67:467–476

Larkin JL, Maehr DS, Krupa JJ, Cox JJ, Alexy K et al (2008) Small mammal response to vegetation and spoil conditions on a reclaimed surface mine in eastern Kentucky. Southeast Nat 7:401–412

Lawer EA, Mupepele AC, Klein AM (2019) Responses of small mammals to land restoration after mining. Land Ecol 34:473–485

Loughman ZJ (2005) Natural History and Conservation Biology of a Southern West Virginia Contour Surface Mine Reptile and Amphibian Community. M.S. Thesis, Marshall University, Huntington

Lupardus JL, Muller LI, Kindall JL (2011) Seasonal forage availability and diet for reintroduced elk in the Cumberland Mountains, Tennessee. Southeast Nat 10:53–74

Maehr DS, Grimes R, Larkin JL (1999) Initiating elk restoration: the Kentucky case study. Proc Annu Conf Southeast Fish Wild Agenc 53:350–363

Maigret TJ, Cox JJ, Yang J (2019) Persistent geophysical effects of mining threaten ridgetop biota of Appalachian forests. Front Ecol Environ. https://doi.org/10.1002/fee.1992

Maslonek ML (2010) Bat Use of Created and Natural Wetlands. M.S. Thesis, West Virginia University, Morgantown

Mattice J A, Brauning DW, Diefenbach DR (2005) Abundance of grassland sparrows on reclaimed surface mines in western Pennsylvania. Pages 504–510 in: Ralph JC, John RC, Rich TD (Eds) Bird Conservation Implementation and Integration in the Americas: Proceedings of the Third International Partners in Flight Conference. U.S. Forest Service, Gen Tech Rep PSW-191

McCabe RE (1982) Elk and Indians: historical values and perspectives. In: Thomas JW, Toweill DE (eds) Elk of North America: Ecology and Management. Wildlife Management Institute. Washington, D.C, pp 61–123

McDermot ME, Shumar MM, Wood PB (2013) Case study: prioritization strategies for reforestation of minelands to benefit Cerulean warblers. J Am Soc Min Reclam 2:80–98

McGowan KJ, Bookhout TA (1986) Small mammal populations on Ohio strip-mined lands reclaimed with herbaceous vegetation under old and new reclamation laws. Ohio J Sci 86:29–32

Menzel MA, Carter TC, Ford WM, Chapman BR, Ozier J (2000) Summer roost tree selection by eastern red (Lasiurus borealis), Seminole (*L. seminolus*), and evening (*Nycticeius humeralis*) bats in the Upper Coastal Plain of South Carolina. Proc Annul Conf Southeast Assoc Fish Wild Agenc 54:304–313

Menzel JM, Ford WM, Menzel MA, Carter TC, Gardner JE et al (2005) Summer habitat use and home-range analysis of the endangered Indiana bat. J Wild Manage 69:430–436

Menzel MA, Menzel JM Carter TC, Ford WM, Edwards JW (2001). Review of the forest habitat relationships of the Indiana bat (*Myotis sodalis*). U.S. Forest Service Gen Tech Rep NE-284

Merovich GT, Hitt NP, Merriam ER, Jones JW (2021) Response of aquatic life to coal mining in Appalachia. Chapter 10 in: Zipper CE, Skousen JG (eds) Appalachia's Coal-Mined Landscapes. Springer

Mizel J (2011) Avian Assemblages and Red-Eyed Vireo Nest Survival within Mineland Forest. M.S. Thesis, West Virginia University, Morgantown

Moosman PR, Warner DP, Hendren RH, Hosle MJ (2015) Potential for monitoring eastern small-footed bats on talus slopes. Northeast Nat 22:N1–N13

Moosman PR, Veilleux JP, Pelton GW, Thomas HH (2013) Changes in capture rates in a community of bats in New Hampshire during the progression of white-nose syndrome. Northeast Nat 20:552–558

Moseley KR, Ford WM, Edwards JW, Schuler TM (2008) Long-term partial cutting impacts on *Desmognathus* salamander abundance in West Virginia headwater streams. Forest Ecol Manage 254:300–307

Muncy B, Price SJ, Bonner SJ, Barton CD (2014) Mountaintop removal mining reduces stream salamander occupancy and richness in southeastern Kentucky (USA). Bioll Cons 180:115–121

Murrow JL, Clark JD, Delozier EK (2009) Demographics of an experimentally released population of elk in Great Smoky Mountains National Park. J Wild Manage 73:1261–1268

Myers CW, Klimstra WD (1963) Amphibians and reptiles of an ecologically disturbed (strip-mined) area in southern Illinois. Amer Midl Nat 70:126–132

Oliphant AJ, Wynne RH, Zipper CE, Ford WM, Donovan PF, Li J (2017) Autumn olive (*Elaeagnus umbellate*) presence and proliferation on former surface coal mines in Eastern USA. Biol Invas 19:179–195

Olsson PMO, Cox JJ, Larkin JL, Maehr DS, Widen P, Wichrowski MW (2007) Movement and activity patterns of translocated elk (*Cervus elaphus*) on an active coal mine in Kentucky. Wild Biol Pract 1:1–8

Peters DC, Brooke JM, Tanner EP, Unger AM, Keyser PD et al (2015) Impact of experimental habitat manipulation on Northern bobwhite surivial. J Wild Manage 79:605–617

Petranka JW (1998) Salamanders of the United States and Canada. Smithsonian Press, Washington, D.C.

Pickering JR, Kays R, Meier A, Andrew S, Yatskievych K (2002) The Appalachians. In: Gil PR, Mittermeier RA, Mittermeier CG et al (eds) Wilderness–Earth's Last Wild Places. Conservation International. Devon, United Kingdom, pp 458–467

Polziehn RO, Hamr J, Mallry FF, Strobeck C (2000) Microsatellite analysis of North American wapiti (*Cervus elaphus*) populations. Molecular Ecol 9:1561–1576

Powers KE, Reynolds RJ, Orndorff W, Ford WM, Hobson CS (2015) Post-White-nose syndrome trends in Virginias cave bats, 2008-2013. J Ecol Nat Environ 7:113–123

Price SJ, Freytag FB, Bonner SJ, Muncy B, Drayer AN et al (2018) Mountaintop removal mining alters stream salamander population dynamics. Divers Distrib 24:1242–1251

Price SJ, Muncy B, Bonner SJ, Drayer A, Barton CD (2016) Effects of mountaintop removal mining and valley filling on the occupancy and abundance of stream salamanders. J Appl Ecol 53:459–468

Quinlan B, Abernathy H, Kalb D, Thorne ED, Ford WM, Cherry, MJ (2020) Elk in the Old Dominion: preliminary assessment of habitat selection. Abstracts of the 27th Annual Meeting of the Wildlife Society. Louisville, KY

Raudabaugh DB, Miller AN (2013) Nutritional capability of and substrate suitability for *Pseudogymnoascus destructans*, the causal agent of bat white-nose syndrome. PLoS ONE 8:e78300

Riley CV (1952) An Evaluation of Reclaimed Coal Strip Mined Lands as Wildlife Habitat. Ph.D. Dissertation, The Ohio State University, Columbus

Rittenhouse TAG, Semlitsch RD (2006) Grasslands as movement barriers for a forest-associated salamander: Migration behavior of adult and juvenile salamanders at a distinct habitat edge. Biol Cons 131:14–22

Ryan CW (2009) Population Ecology, Residents' Attitudes, Hunter Success, Economic Impact, Modeling Management Options and Retention Time of Telazol of West Virginia Black Bears. Ph.D. Dissertation, West Virginia University, Morgantown

Schorr MS, Dyson MC, Nelson CH, Van Horn GS, Collins DE, Richards SM (2013) Effects of stream acidification on lotic salamander assemblages in a coal-mined watershed in the Cumberland Plateau. J Freshwater Ecol 28:339–353

Scott PE, DeVault TL, Bajema RA, Lima SL (2002) Grassland vegetation and bird abundances on reclaimed midwestern coal mines. Wild Soc Bull 30:1006–1014

Scott PE, Lima SL (2004) Exotic grasslands on reclaimed midwestern coal mines: an ornithological perspective. Weed Tech 18:1518–1521

Sena K, Franklin JA, Swab RM, Hall SL (2021) Plant communities on Appalachian mined lands. Chapter 5 in: Zipper CE, Skousen JG (eds) Appalachia's Coal-Mined Landscapes. Springer

Sheets JJ, Whitaker JO, Brack V, Sparks DW (2013) Bats of the hardwood ecosystem experiment before timber harvest. Pages 191–202 in: Swihart RK, Saunders MR, Kalb RA, Haulton GS, Michler CH (Eds). The Hardwood Ecosystem Experiment: a Framework for Studying Responses to Forest Management. U.S. Forest Service, Gen Tech Rep NRS-P-108

Silvis A, Kniowski AB, Gehrt SD, Ford WM (2015) Roosting and foraging social structure of the endangered Indiana bat (*Myotis sodalis*). PLoS ONE 9:e96937

Silvis A, Perry RW, Ford WM (2016) Relationships of three species of bats impacted by white-nose syndrome to forest condition and management. U.S. Forest Service Gen Tech Rep SRS–214

Simmons J, Currie W, Eshleman KN, Kuers K, Monteleone S et al (2008) Forest to reclaimed land use change leads to altered ecosystem structure and function. Ecol Appl 18:104–18

Skousen J, Ziemkiewicz P, Venable C (2006) Tree recruitment and growth on 20-year-old, unreclaimed surface mined lands in West Virginia. Inter J Min Reclam Environ 20:142–154

Skousen J, Zipper CE (2014) Post-mining policies and practices in the Eastern USA coal region. Inter J Coal Sci Tech 1:135–151

Slabach BL, Hast, JT, Murphy SM, Bowling WE, Crank, RD, Jenkins G, Johannsen KL, Cox JJ (2018) Survival and cause-specific mortality of elk Cervus canadensis in Kentucky, USA. Wildl Biol 2018(1). https://doi.org/10.2981/wlb.00459

Sly GR (1976) Small mammal succession on strip-mined land in Vigo County, Indiana. The Amer Midl Nat 95:257–267

Stauffer GE, Diefenbach DR, Marshall MR, Brauning DW (2011) Nest success of grassland sparrows on reclaimed surface mines. J Wild Manage 75:548–557

Stiles RM, Swan JW, Klemish JL, Lannoo MJ (2016) Amphibian habitat creation on postindustrial landscapes: a case study in a reclaimed coal strip-mine area. Can J Zool 95:67–73

Sweeten SE, Ford WM (2015) Effects of microhabitat and land use on stream salamander abundance in the southwest Virginia coalfields. Pages 122–140 in: Craynon JR (Ed) 2nd Proceedings of Environmental Considerations in Energy Production. Soc Min Metallur Explor, Englewood CO

Sweeten SE, Ford WM (2016) Effects of microhabitat and large-scale land use on stream salamander occupancy in the coalfields of Central Appalachia. J Ecol Nat Environ 8:129–141

TerBeest J (2005) Effects of a Restored Elk Population on Soils, Vegetation, and Water Quality in Eastern Kentucky. M.S. Thesis. University of Kentucky, Lexington

Terrell VCK, Klemish JL, Engrecht NJ, May JA, Lannoo PJ et al (2014) Amphibian and reptile colonization of reclaimed coal spoil grasslands. J N Am Herpetology 2014:59–68

Timm A, Meretsky V (2004) Anuran habitat use on abandoned and reclaimed mining areas of southwestern Indiana. Proc Indiana Acad Sci 113:140–146

Tompkins D (2014) Utilizing geophysics and biology to identify bat habitat on a proposed right of way. Abstracts of the (2014) Northeastern Bat Working Group. Clinton, NJ

Trefethen JB (1975) An American crusade for wildlife. Winchester Press, Winchester, Ontario

Turner LJ, Fowler DK (1981) Utilization of surface mine ponds in East Tennessee by breeding amphibians. Tennessee Valley Authority, Division of Land and Forest Resources, Norris, TN

United States Fish and Wildlife Service (USFWS) (2020) Range-wide Indiana bat summer survey guidance. https://www.fws.gov/midwest/endangered/mammals/inba/surveys/pdf/2019_Rangewide_IBat_Survey_Guidelines.pdf

Van Lear DH, Harlow RE (2002) Fire in the eastern United States: Influence on wildlife habitat. Pages 2–10 in: Ford WM, Russell KE, Moorman CE (Eds). The Role of Fire in Nongame Wildlife Management and Community Restoration: Traditional Uses and New Direction. U.S. Forest Service, Gen Tech Rep NE-288

Vaughan R (2002) Oaks trees, acorns and bears. In: McShea WJ, Healy WH (eds) Oak Forest Ecosystems: Ecology and Management for Wildlife. John Hopkins University Press, Baltimore MD, pp 224–240

Virginia Department of Game and Inland Fisheries. (2020) Virginia elk reintroduction. https://www.dgif.virginia.gov/wildlife/elk/restoration/

Vukovich M, Ritchison G (2006) Nesting success and behavior of Northern harriers on a reclaimed surface mine grassland in Kentucky. J Raptor Res 40:210–216

Vukovich M, Ritchison G (2008) Foraging behavior of Short-eared owls and Northern harriers on a reclaimed surface mine in Kentucky. Southeast Nat 7:1–10

West Virginia Division of Natural Resources (2020) Elk restoration program. www.wvdnr.gov/Wildlife/ElkRestorationProgram.shtm

Whitmore RC (1979) Short-term changes in vegetation structure and its effect on Grasshopper sparrows in West Virginia. Auk 96:621–625

Whitmore RC (1981) Structural characteristics of Grasshopper sparrow habitat. J Wild Manage 45:811–814

Wichrowski MW, Maehr DS, Larkin JL, Cox JJ, Olsson MPO (2005) Activity and movements of reintroduced elk in southeastern Kentucky. Southeast Nat 4:365–374

Wickham J, Wood PB, Nicholson MC, Jenkins W, Druckenbrod D et al (2013) The overlooked terrestrial impacts of mountaintop mining. BioSci 63:335–348

Wilder AP, Frick WF, Langwig KE, Kunz TH (2011) Risk factors associated with mortality from whitenose syndrome among hibernating bat colonies. Biol Letters 7:950–953
Williams JM, Brown DJ, Wood PB (2017) Responses of terrestrial herpetofauna to persistent, novel ecosystems resulting from mountaintop removal mining. J Fish Wild Manage 8:387–400
Williams JM, Wood PB (2004) Streamside salamanders in valley fill and reference streams in southern West Virginia. Proc, Am Soc Min Reclam, pp 2027–2041
Wood PB, Ammer FK (2015) Grasshopper sparrow reproductive success and habitat use on reclaimed surface mines varies by age of reclamation. Wild Soc Bull 39:512–520
Wood PB, Larkin JL, Mizel J, Zipper C, Angel P (2013) Reforestation to enhance Appalachian mined lands as habitat for terrestrial wildlife. U.S, Office of Surface Mining, Appalachian Regional Reforestation Initiative, p 10
Wood PB, Williams JM (2013a) Impact of valley fills on streamside salamanders in southern West Virginia. J Herp 47:119–125
Wood PB, Williams JM (2013b) Terrestrial salamander abundance on reclaimed mountaintop removal mines. Wild Soc Bull 37:815–823
Wray T, Strait KA, Whitmore RC (1982) Reproductive success of grassland sparrows on a reclaimed surface mine in West Virginia. Auk 99:157–164
Yahner RH, Rohrbaugh RW (1998) A comparison of raptor use of reclaimed surface mines and agricultural habitats in Pennsylvania. J Raptor Res 32:178–180
Yarnell SL (1998) The Southern Appalachians: a history of the landscape. U.S. Forest Service, Southern Research Station, General Technical Report SRS-18. Asheville NC. p 52
Yearsley EF, Samuel DE (1980) Use of reclaimed surface mines by foxes in West Virginia. J Wild Manage 44:729–734
Zipper CE, Burger JA, McGrath JM, Rodrigue JA, Holtzman GI (2011a) Forest restoration potentials of coal-mined lands in the eastern United States. J Environ Qual 40:1567–1577
Zipper CE, Burger JA, Skousen JG, Angel PN, Barton CD et al (2011b) Restoring forests and associated ecosystem services on Appalachian coal surface mines. Environ Manage 47:751–765

Conversion Options for Mining-Affected Lands and Waters in Appalachia

Carl E. Zipper, Carmen T. Agouridis, Christopher D. Barton, and Jeff Skousen

Abstract More than 10,000 km^2 of Appalachia have been surface mined for coal. Surface-mined lands represent an underutilized resource and can be converted to more productive uses that support ecological and human needs. Seeding, fertilization, and vegetation management can convert degraded mine sites to livestock pasture. Mined lands occupied by invasive plants can be made more similar to native ecosystems by controlling non-native species, mitigating soil limitations, and planting native trees. Degraded mine-site streams can be rebuilt to restore hydrologic function and aquatic habitat. If mine soils have suitable properties, appropriate cultural techniques can be applied to produce biomass crops. If geotechnically stable, mined lands can support housing projects, industrial sites, and other large-scale building structures. Due to their wide availability and lack of competing uses, mined lands may prove suitable as locations for renewable energy projects. Mined-land conversions require technical expertise and often require significant expense; when such resources are available, degraded mined lands of the types that are common throughout Appalachia's coalfield can be converted to beneficial uses.

Keywords Biomass crops · Land development · Photovoltaic · Reforestation · Stream restoration

1 Introduction

Extensive areas of former Appalachian coal surface mines are in an unmanaged condition. Although generally vegetated and geotechnically stable, such lands are typically occupied by unmanaged mixtures of non-native and native plants (Sena

C. E. Zipper (✉)
Virginia Tech, Blacksburg, Virginia, USA
e-mail: czip@vt.edu

C. T. Agouridis · C. D. Barton
University of Kentucky, Lexington, Kentucky, USA

J. Skousen
West Virginia University, Morgantown, West Virginia, USA

C. E. Zipper and J. Skousen (eds.), *Appalachia's Coal-Mined Landscapes*,
https://doi.org/10.1007/978-3-030-57780-3_7

et al. 2021). Hence, their conditions are far from optimal from a human-needs standpoint. Such lands and their waters have potential to better serve the human condition by providing environmental services and, in some cases, by supporting economic activity.

No scientific data are available to describe Appalachia's mining-affected lands and waters in aggregate but several studies provide insight. It is clear that such lands and waters are extensive. Since the late 1970s, approximately 10,000 km^2 in Appalachia have been disturbed by coal surface mining (data from federal and state mining agencies; Zipper 2020), while additional acreages were disturbed prior to that time (Johnson and Paone 1981). For central Appalachia, research suggests that the majority of coal-mined lands are occupied by non-forest and unmanaged vegetation, much of which is exotic and invasive (Zipper et al. 2011a). Corresponding data are not available for northern Appalachia but similar non-managed land conditions with exotic vegetation occur (Cavender et al. 2014). Although some miners have been employing reclamation methods intended to restore native forest trees on Appalachian mines in recent years (Burger et al. 2005), no regional assessments of these methods' success have been conducted. Central Appalachia also lost 2% or more of its streams due to valley-fill construction and other forms of mining disturbance (USEPA 2011); and reconstructed streams are altered from a natural condition, especially on older mine sites (Fritz et al. 2010).

Mined areas in the Appalachian coalfield present significant opportunities for conversion for improved environmental quality and to better serve the human condition. This chapter reviews methods for converting Appalachian mined lands and affected waters to alternative conditions and uses.

2 Hayland and Pasture

Mined-land haylands and pastures are common throughout Appalachia. Active mine permits commonly define hayland and/or pasture as post-mining land uses for reclamation purposes (GAO 2009). However, lands so reclaimed are not always further developed and managed for agriculture following mining (Angel et al. 2005).

When reclaimed and revegetated appropriately, mined lands can be converted for use as haylands and pastures (Fig. 1). Essential characteristics are land slope, soil properties, and, for pasture uses, access to suitable water. Pastures can be established on relatively steep slopes while hay production requires slopes suitable for agricultural equipment, typically less than 15%. Moderately acidic or circumneutral soil pHs are suitable for livestock pasture and hay crops while soils lacking excessive stoniness will generally be more productive than those with high rock contents. Livestock require adequate quality and quantities of water. Cattle, for example, require 50 gallons of water per day per head. Most livestock prefer waters with pH in the range of 6–8, with dissolved solids <1000 mg/L, and lacking toxic concentrations of trace elements (Whittier et al. 2010). While some mined lands discharge waters of

Fig. 1 Cattle grazing on a reclaimed-mine pasture in Wise County, Virginia. Note the undesirable non-native shrubs (autumn olive) in the background. In years following the photo, these non-native shrubs invaded the pasture area further, leaving less area producing forage for grazing. (Photo by C. E. Zipper)

that quality, others do not (Clark et al. 2021). Pasture areas will be used most effectively by livestock if they are within a reasonable distance of water supply; cattle, for example, utilize lands most effectively within 250 m of suitable water (Ditsch et al. 2006). The terrain will also influence the cost and feasibility of site fencing, thus limiting conversion opportunities on mountainous lands. Fencing can direct animals to more effectively utilize the forage base and can influence pasture composition. If terrain is too steep for agricultural equipment, management practices such as liming, fertilizing, overseeding, and invasive species control will be limited; hence, grazing operations on such lands require larger areas per animal unit than where management can be applied more effectively.

The mined-land conversion process generally begins by clearing woody vegetation. Even when revegetated solely with grasses and legumes, mined lands are vulnerable to invasions by non-native species with limited or no utility as livestock feed, such as autumn olive (*Elaeagnus umbellata*) and multiflora rose (*Rosa multiforum*). Such species tend to expand over time, thus further limiting forage availability unless controlled. Hayland-pasture conversion will be aided by seeding with suitable grasses and legumes (Ditsch and Collins 2000). Mined-land sites often require sustained fertilization to maintain plant productivity, especially on soils created from rock spoils that lack the organic-matter and nutrient-holding capacities of natural soils (Ditsch et al. 2006). Because of altered mineral profiles in vegetation, animals grazing animals on mined lands may require mineral supplements to a greater extent than on natural soils (Ditsch et al. 2006). Because of productivity limitations, grazing systems established on mine soils created from rock spoils sustain lower animal densities than would be typical on natural soils (Teutsch et al. 2008).

Mined-lands typically offer lower rental or ownership costs than natural-soil sites in agricultural areas, thus providing some level of potential economic advantage. But those advantages may be offset by higher operating costs required by lower animal densities, repeated fertilizations, costs of dealing with invasions by unwanted non-native plants, and management and fencing difficulties imposed by terrain. These characteristics vary widely among sites. Many small-scale livestock producers operate successfully on mined lands in the Appalachian coalfield.

3 Native Forest

Appalachian forests are some of the most biodiverse terrestrial ecosystems of the non-tropical world but they are also among the most threatened. Appalachian forests face challenges from invasive pests and pathogens, urbanization, and climate-change stresses as well as surface mining (Aukema et al. 2010; Drummond and Loveland 2010; Butler et al. 2015). Forest loss due to surface mining has been extensive in Appalachia (see above).

Reforestation of Appalachian mined lands can aid in reversing those losses. Mined-land reforestation techniques, although costly, have been developed through research and are widely applied. Through their application, the non-productive and invasive vegetation that is widespread on mined lands can be converted to plant communities with closer resemblance to Appalachian forests.

3.1 Reforestation Methods

The non-forest vegetation occupying mined lands is typically persistent and well-adapted to the mine sites' altered soil conditions. Hence, successful conversion to native forest requires skillful application of cultural techniques that are based on soil and vegetation science and silvicultural management principles (Burger et al. 2013).

A usual first step for any mine-site reforestation project is site assessment and planning. Following that, fieldwork typically begins with removal of unwanted vegetation. This can be achieved manually in some cases but often requires more forceful intervention such as careful application of an herbicide or, in extreme cases where invasive plants are dominant, use of a dozer to scrape off the upper soil and vegetation layer (Adams et al. 2019). Removal of above-ground plant material will be only a first step, however, given that many plants will leave seeds in the soil and some of the most problematic species can resprout from living roots if not killed using an herbicide.

Since most Appalachian mine soils are compacted, soil loosening will improve survival and growth of planted trees and is often necessary for successful reforestation. Soil loosening is typically achieved by dragging a deep-tillage device such as a ½-to-1 m long steel ripping tooth through the soil over the mine site using heavy equipment such as a dozer. Soil amendments, fertilization, and in some cases liming for pH-remediation are sometimes applied, but such treatments are best conducted strategically so as to avoid stimulating undesirable vegetation that will compete with the planted trees (Burger et al. 2013). Forest trees are planted as bare-root seedlings, typically during the winter or early spring season following site preparation and in the loosened channels produced by deep tillage. Most mine reforestation projects select species that are prominent in local forests and suited to the site conditions (Davis et al. 2012). Multiple tree species are usually intermixed and planted to recreate diverse forest-tree communities. Planting mixes often include tree species that produce mast for birds and other wildlife, anticipating that attracting animals for feeding will enable site occupation by additional native species deposited as live seed. Assurance of successful planted-tree establishment generally requires control of competing vegetation for several years following, which can be accomplished by manual spot-spraying with herbicides (Burger et al. 2013).

The above method has been executed successfully throughout the Appalachian coalfield. For example, the green jobs and green economy program called Green Forests Work (GFW) has treated nearly 2000 hectares of previously reclaimed mines using these techniques accompanied by the planting of more than 3 million trees (GFW 2019). However, reforestation success requires careful execution for each step of the process; in the absence of such, prior vegetation can regenerate from residual seeds and roots to proliferate (Evans et al. 2013).

Research has demonstrated successful reestablishment of forest vegetation in mine sites, and that such vegetation can be both productive and diverse (Fig. 2). When properly prepared, mine soils can provide a deep rooting medium rich in geologically

Fig. 2 A forest re-established on a former coal surface mine in Perry County, Kentucky. (Photo by C. D. Barton)

derived nutrients such as Ca, Mg, and K and produce forests with similar growth characteristics as those on unmined areas. Rodrigue et al. (2002) reported forest growth on 12 of 14 selected older coal mine sites in the eastern and midwestern US that was similar to local unmined forests. Casselman et al. (2007) measured a 50-year site index for eastern white pines (*Pinus strobus*) stand growing on an uncompacted mine site in Virginia that was considerably greater than the average site index for the southern Appalachians. Similarly, site index for yellow poplar (*Liriodendron tulipfera*) growing on 40-to-50-year-old Tennessee mine sites with uncompacted soils averaged 32.3 m, which was greater than the 26.5-m regional average (Franklin and Frouz 2007). Cotton et al. (2012) showed that 10-year-old yellow-poplar and white oak growing on loose-dumped and partially weathered Kentucky spoils exhibited similar stem diameters to those of regenerating non-mined stands of the same age. These are best-case scenarios, however; as other examples that were less successful have been noted (Evans et al. 2013). Skillful intervention is required in order for forest re-establishment on mined areas reclaimed and revegetated previously to be successful.

Also, few comparisons of the full plant communities on reforested mine sites to those of native forests have been conducted. Studies of that type have found that although some native-forest plant species do, in fact, colonize and establish on the mine sites via natural processes, others do not or are much slower to do so (Holl 2002). Hence, the extent to which mine reforestation practices restore the full complement of forest plants remains unclear. Nonetheless, skillful interventions can improve the potential for degraded mined-land plant communities to support human needs.

3.2 Societal Benefits

Although challenging and costly, successful conversion of non-forested mined lands to native-forest-like ecosystems can be extremely beneficial. Appalachian forests aid in maintenance of ecosystem services and values that are commonly lost in intensively mined areas, including high-quality water resources (Neary et al. 2009), flood attenuation (Negley and Eshleman 2006), and carbon storage (Littlefield et al. 2013). Therefore, re-establishing forests on degraded mined landscapes can help to restore services and values that are typically degraded by mining (Simmons et al. 2008; Zipper et al. 2011b). Targeted reforestation can reduce forest fragmentation from mining to restore habitat for wildlife species that depend on large expanses of unbroken forest (Wickham et al. 2013). Reforestation with native species can also improve landscape aesthetics, thus enhancing the capacity of mining-area communities to serve as tourist destinations and to support tourism-related businesses and jobs. Forests also clean air and sequester atmospheric carbon; thus, re-establishing forests on mined landscapes can offset in small measure some of the climate-warming effects of the coal that was mined to create them (Amichev et al. 2008). Mined-land reforestation can also restore plant communities that produced valued products, including non-timber products such as forest herbs and high-quality Appalachian timber that is in demand by global markets.

4 Headwater Stream Restoration

Headwater streams are commonly identified as first-, second-, and third-order reaches. Ephemeral, intermittent, and perennial headwater streams account for nearly 80% of the total stream length in central Appalachia (Shreve 1969) (Fig. 3). In non-mined landscapes, these streams provide many ecosystem services related to water, sediment, and organic matter transport; nutrient cycling; support of aquatic and terrestrial wildlife; peak flow attenuation; and water storage (Gomi et al. 2002).

Surface coal mining impacts headwater streams by causing physical, chemical, and biological degradation. Physical degradation occurs when parts of the landscape surface that include stream channels are removed to extract coal, and when excess spoil is placed in valleys resulting in stream burial and loss (USEPA 2011). Chemical impacts occur when dissolved elements, both major ions and trace elements, are released from fractured mine spoils and transported into stream waters (Clark et al. 2021). Physical and chemical modifications to streams degrade biological conditions resulting in altered aquatic communities (USEPA 2011; Merovich et al. 2021).

Unavoidable impacts to headwater streams from coal mining activities require compensatory mitigation via Section 404 of the Clean Water Act. Loss of stream form and function (i.e., physical, chemical, and biological degradation) are compensated by restoring, reconstructing, enhancing, or preserving streams (USCFR 2020). Coal mine operators can offset stream losses through permittee-responsible mitigation, in-lieu fee programs, or mitigation banks. Compensatory mitigation can occur on-site at the reclaimed mine or off-site in a nearby watershed. Restoration of lost form and function for streams located on mined lands is challenging due to factors such

Fig. 3 Central Appalachia has a dense drainage network consisting of small headwater streams such as this one at Robinson Forest in eastern Kentucky. (Photo by M. Barton, University of Kentucky, Agricultural Communications Service)

as variations in soil quality, loss of streambed geologic integrity and groundwater connections, and presence of invasive plants and large ungulates (Agouridis et al. 2018).

4.1 Restoring Stream Structure and Function

Stream restoration involves re-establishing the structure and function of stream ecosystems, thereby attempting to create stream systems in human-altered landscapes that resemble pre-disturbance or reference conditions. Structure refers to a stream's dimension, pattern, and profile while stream functions are processes such as water and sediment transport, carbon and nutrient cycling, energy production, habitat provision, and other processes (Cummins 1974). Harman et al. (2012) grouped stream functions are into five categories represented as a pyramid with attainment of upper levels dependent on the successful attainment of lower levels. From bottom to top, these functional categories are hydrology, hydraulic, geomorphology, physiochemical, and biology.

While early stream restoration designs largely focused on restoring structure, recent efforts have shifted towards restoring stream function (Hill et al. 2013; Rubin et al. 2017). Restoration of stream function requires a watershed-based approach; practitioners seek understanding of how ecosystem processes are influenced by upland and riparian conditions in reference systems and utilize that knowledge in stream restoration design (Bohn and Kershner 2002).

4.2 Stream Restoration Methodologies

Each stream restoration project has unique characteristics though most designs share main components: reconnection of stream to its floodplain, reconstruction of stream morphology, streambank stabilization, incorporation of instream structures, establishment of a riparian vegetation, and addition of instream and riparian habitat enhancement features (e.g., snags and vernal pools) (Doll et al. 2003). Stream restoration design methodologies are broadly grouped according to their applicability to alluvial or threshold channels (NRCS 2007). Alluvial channels are those whose bed and banks are formed by water-transported sediments; hence, geomorphic stability or dynamic equilibrium is the main design goal for these streams. Threshold channels have bed and banks comprised of relatively immobile material meaning their boundaries experience little to no adjustment; for these channels, ensuring sufficient capacity and maintaining the shear stress appropriate for bank materials are primary design goals. These methods, discussed briefly below, are detailed in Copeland et al. (2001) and NRCS (2007).

4.2.1 Alluvial Channel Design

Alluvial channel design approaches include using an analog or reference reach, hydraulic geometry relationships, and natural channel design (NCD) techniques.

Analog Approach. With this approach, morphology (e.g., dimension, pattern, and profile) for the design stream is determined using one or more reference reaches. A reference reach is a stable section of stream located in a similar physiographic and climatic region as the impacted stream (Rosgen 1998; Hey 2006). The difficulty in identifying suitable reference reaches for streams on mined lands is a major challenge for this approach.

Hydraulic Geometry Relationships. Streams tend to develop in predictable patterns that create an equilibrium between discharge and sediment inputs (Leopold and Maddock 1953). Termed hydraulic geometry curves, these relationships predict channel characteristics from a representative condition such as bankful discharge, the maximum discharge a channel can carry before spilling out onto the floodplain (Leopold et al. 1964; Agouridis et al. 2011). Bankfull discharge data, however, are seldom available; hence, designers often use bankful regional curves, formulas which relate channel dimensions (e.g., discharge, cross-sectional area, width, and depth) to drainage area, a more readily obtained parameter (Agouridis 2014). As with the analog approach, hydraulic geometry relationships are developed for a given physiographic region (Blackburn-Lynch et al. 2017). Challenges with this approach include identification of enough study sites to allow for statistical validity given the stream-systems' inherent variability, and accounting for physiographic changes resulting from mining and reclamation within stream watersheds (NRCS 2007).

Natural Channel Design. Natural channel design (NCD) methodology incorporates aspects of the analog and hydraulic geometry approaches, such as reference reaches and regional relationships, but also includes watershed and geomorphic assessments, hydraulic and sediment transport modeling, design of instream structures, and development of a streamside vegetation plan (Doll et al. 2003; Hey 2006; NRCS 2007). Design validation is performed using hydraulic modeling software. The NCD approach faces challenges similar to those of other methods, including reliance on the correct identification of bankful elevation and the hydrologic variability of mined landscapes.

4.2.2 Threshold Channel Design

Threshold channel design approaches include allowable velocity and allowable shear stress, discussed below, and tractive force. These methods seek to compute streamflow characteristics such as flow velocity and/or the shear stresses that will be exerted by flowing waters for the purpose of ensuring that materials used to construct the stream bank and stream bottom (boundary materials) will remain stable.

Allowable Velocity. The premise of the allowable or permissible velocity method is that computed average channel velocities for design flows are compared to permissible velocities for boundary materials (NRCS 2007; Baird et al. 2015). This technique is applicable to meandering channels and those with depths of 1 m (3 ft) or less. Consideration is also given to the effects of suspended sediment in the water column (Haan et al. 1994).

Allowable Shear Stress. This approach is largely used on gravel, cobble, or rock-lined channels. As with the allowable velocity approach, the applied shear stress on the channel boundary is computed and compared to the allowable shear stress for that boundary material (NRCS 2007; Baird et al. 2015). This technique is applicable to meandering streams but is not recommended for systems that experience high bedloads.

4.3 Special Considerations for Mined Lands

The altered nature of mined landscapes, relative to unmined conditions with intact bedrock and native soils, creates challenges for all stream restoration approaches.

Hydrologic Alteration: Restoring streams on mined lands requires a focus on restoring hydrologic functions at the watershed scale (Fig. 4). Mining has major effects on watershed hydrology due to fracturing of bedrock and replacing of natural soils with mine soils. Although landscape hydrologic responses to mining vary widely due to variations of landscape reconstruction and reclamation methods, common responses include lower infiltration rates, increased peak flows, and increased runoff volumes (Evans et al. 2015).

Reforestation: Appalachian forests influence watershed hydrology (e.g., storm volume and peak reductions, baseflow provision), water quality (by enabling rainwater infiltration, reducing sediment transport), and aquatic habitat (e.g., organic matter inputs, modifying water temperature via shading) of headwater streams in unmined areas. Mining typically replaces forest with non-forest vegetation. The Forestry Reclamation Approach (FRA), a method for re-establishing forest trees on mine sites (Burger et al. 2005), has proven effective at reducing mined-land peak flows and runoff volumes (Taylor et al. 2009; Sena et al. 2014). Reforestation of mined areas also has the potential to improve water quality depending on the strata used for soil construction (Sena et al. 2014). Zipper et al. (2018) provide guidance on using FRA to establish woody vegetation on streams in mined lands, but restoration of mature woody canopy requires decades.

Water Quality: Restoration of stream function is dependent on good quality water. Poor water quality can limit the presence of benthic macroinvertebrates and other biota in mine-impacted waters (USEPA 2011; Merovich et al. 2021). When restoring streams on mined land, identifying high-conductivity spoils within the disturbance area and avoiding their use in and around constructed streams may improve water quality and will aid riparian reforestation (Zipper et al. 2018). However, the water quality of restored streams in mined watersheds will also be influenced by watershed

Fig. 4 The Guy Cove stream restoration project, located in eastern Kentucky, considered the entire watershed when developing a stream restoration plan. The mined-area project site prior to restoration in 2005 (lower left), immediately after construction in 2009 (lower right), and in 2017, about 8 years after construction and tree planting (upper). The stream is stable, tree growth and survival are good, and water quality and habitat metrics show improvements (Agouridis et al. 2018). Photos by C. T. Agouridis (lower left, lower right), and C. D. Barton (upper)

spoil materials; hence, water-quality parameters such as major-ion concentrations and conductivity may remain elevated for decades following post-mining landscape reconstruction (Evans et al. 2014).

Groundwater Connection: Surface coal mining alters ephemeral and intermittent stream watersheds by removing bedrock, soils, and forests, thus drastically modifying surface and groundwater connections (Evans et al. 2015). Intermittent and perennial headwater streams experience lateral and vertical exchanges between surface and ground waters in the hyporheic zone (Cardenas 2009). Recreating hyporheic exchange during stream reconstruction is challenging in mined landscapes due in part to highly altered subsurface geologic and soil conditions and groundwater regimes. The unconsolidated and porous nature of overburden also presents challenges for sustaining baseflow in reconstructed streams. Practices such as compacting or rock-lining the streambed can aid in maintaining water flows but often do not re-establish hyporheic exchange.

Organic Matter: In unmined watersheds, allochthonous organic matter such as leaves and branches is introduced to headwater streams from the surrounding riparian forest. Re-establishing a forested canopy on constructed streams can improve habitat quality and functional processes by providing a source of organic matter, moderating sunlight in a manner that restores more-reference-like primary production processes, and moderating stream temperatures (Krenz et al. 2016, 2018). Stream restoration practitioners can add organic matter to stream beds (e.g., woody debris in riffles, snags in pools, leaf packs along banks) and can ensure more reference-like stream conditions over longer terms by reforesting riparian areas as well as watersheds using the FRA.

Niche Habitats: The rich diversity of species in Appalachia provides opportunities to create niche habitats in stream restoration projects. Such habitats may include vernal pools, which can be constructed as depressional areas within the stream's floodplain but outside of the stream channel. Vernal pools serve as critical breeding and nursery habitat for salamanders and frogs and a water source and feeding station for bats, snakes, and waterfowl (Biebighauser 2003). Practitioners may construct stream channels to include niche habitats similar to those occurring in natural streams, such as periodic pools where loose sediments, rocks, and cobbles can accumulate. In the absence of such features, the compacted and rock-lined surfaces required to maintain water flows within the otherwise porous mine-spoil materials may inhibit occupation by aquatic organisms that require a more varied and looser substrate.

4.4 *Stream Restoration Effectiveness on Appalachian Mined Lands*

Stream restoration and creation efforts on Appalachian mined lands have yielded mixed results. Because monitoring data are limited, assessment is difficult (Palmer and Hondula 2014; Agouridis et al. 2018). For instance, compensatory mitigation projects largely focus on assessing stream geomorphic stability with limited assessment emphasis on the evaluation of hydrology, water quality, and aquatic ecosystem health. Such data are needed to understand what techniques work well, which require modification or elimination, and where innovation is needed (Palmer et al. 2007). The few studies of function conducted in mined-land stream-restoration projects to date indicate that some level of ecosystem function may be restored, but that those levels vary widely among projects and from functional measures in unmined reference streams (Krenz et al. 2016, 2018).

A major challenge in executing mined-land stream restoration is the shift in knowledge and skill required, as these efforts differ substantially from the design and construction of features such as rock-lined channels that are common on Appalachian mines. Restoration projects conducted with the intent of restoring stream structure and function require knowledge of hydrology, geomorphology, biology, ecology, and forestry in addition to engineering. Also, the extent to which ecosystem processes have been restored within a stream's watershed will influence functional restoration in a reconstructed stream (Wohl et al. 2015). When achieved, successful restoration of headwater stream structure and function on Appalachian mines can be expected to yield benefits, including improved ecosystem health both within the mined landscape and in larger river systems located downstream.

5 Biomass Crops for Bioenergy

Bioenergy systems utilize biomass from herbaceous and woody crops as substitutes for fossil-fuel energy sources. Research demonstrates that many mined lands have potential to produce biomass crops for bioenergy. Surface mined landscapes with low to moderate slopes and with suitable soil properties can be planted with herbaceous biomass crops that require agricultural management, while steeper lands have potential to produce woody biomass. Producing biomass on mined lands will displace existing vegetation, but this may represent significant ecosystem improvement where mined lands are degraded with invasive species. Use of mined lands for bioenergy cropping may also reduce the potential competition of bioenergy crops for agricultural land that could otherwise be used for food production. Developing biomass for bioenergy production on Appalachian mined lands also would have potential to provide jobs in areas where mining jobs have been lost.

Uses of plant products as energy sources range from household heating with wood to the fueling of electric power plants, to the conversion of plant products to liquid or gaseous transportation fuels. The majority of current liquid biofuel production in the USA is based on processing corn (*Zea mays*)—typically grown on high-quality farmland – to produce ethanol. But concerns associated with use of quality farm land to produce fuel has led to research on mined-land production options.

5.1 Herbaceous Crops

Many perennial herbaceous plants can produce sufficient quantity and quality of cellulosic feedstocks for conversion to either solid or liquid biofuel (Ballesteros and Manzaneres 2019). Leading feedstock candidates include switchgrass (*Panicum virgatum*) and *Miscanthus* × *giganteus* (miscanthus) (Fig. 5). Switchgrass, a warm

Fig. 5 Switchgrass (left) and miscanthus (right) are two species of grass which are being tested for biomass production on reclaimed mine lands in West Virginia. (Photos by J. G. Skousen)

season perennial grass native to North America, has commonly been used for conservation plantings (e.g., for wildlife habitat or stream buffers). It is the most extensively-studied species for use as a bioenergy feedstock in North America because of its high biomass production potential, adaptability and tolerance to adverse growing conditions, and low input requirements (Parrish and Fike 2005). Miscanthus, a sterile hybrid from Asia, may have even greater production potential in the humid east, although it must be established vegetatively because it does not produce viable seed. (*Miscanthus* × *giganteus* should not be confused with other fertile *Miscanthus* species which have proven invasive on mined lands.)

To be suitable for herbaceous biofuel crops mined lands should be relatively level, with slopes that do not hinder tillage and harvesting machinery. Soils should not be excessively stony nor compacted but with suitable nutrient- and water-holding capacity. Soil chemical properties should be assessed via soil testing and compared to what is required for productive growth by the crop in question. Switchgrass, for example, can grow across a range of soil pH conditions, from moderately acidic to slightly alkaline. Phosphorus and potassium in the moderate range of most soil tests will be adequate. One of the advantages of growing switchgrass and miscanthus on mined lands is their ability to thrive in less fertile and marginal soils, such as are found on mined areas. Once established, they can persist for 10–15 years with annual cutting and few inputs.

Where site conditions are suitable, a first step in preparing lands for biomass cropping will be to remove existing vegetation. This operation may require physical removal if woody trees or shrubs are present, although herbicide application may be adequate for initial control. If the dead vegetation is quite thick, it will need to be removed either by burning or by incorporating vigorous disking or other tillage. If tillage is utilized, lime and fertilizer can be applied in accord with soil test recommendations prior to tilling. Nitrogen applications should be limited and used with caution as the nitrogen can stimulate vigorous growth of annual grasses which in turn out-compete the emerging switchgrass.

Switchgrass seeding rates generally are in the 7–11 kg/ha (6–10 lb/acre) range, although the higher end may be appropriate for challenging planting environments. Switchgrass seed is small and establishment success is greatest when the soil is even and firm, which prevents the seed getting placed too deep. Once the seedbed is prepared, switchgrass can be planted with standard or no-till drills or by broadcasting the seed, followed by an operation intended to improve seed-soil contact such as over-raking or chain-dragging.

Miscanthus is planted from rhizomes or vegetative plugs. Although establishment may benefit from tillage, miscanthus can be established by directly planting the propagules with the assistance of a commercial tree-planting crew. Planting with human labor may limit scale, however, as miscanthus typically is planted at 12,000 plants/ha (4800 plants/acre); large-scale mechanized systems for miscanthus establishment have not been tried on mined lands.

Once established, switchgrass and miscanthus have the ability to persist as long-lived perennials. Two or three years are required for these grasses to be well established with stands close to their productive potential. Once established, both grasses stabilize soil and reduce potential erosion losses. Neither species requires high levels of nutrients. In agricultural settings, productivity can be supported by low levels (45–55 kg/ha, 50–60 lb/acre) of nitrogen fertilizer application. Response to nitrogen may be greater on mine soils, given their lower organic matter (and thus less nitrogen available in the organic pool). Regardless, the lower nutrient requirements of these species relative to traditional row crops represent a particularly important advantage for production on Appalachian mine soils which generally have higher stone and gravel contents and lower fertility than native agricultural soils. Harvesting can be done with conventional haying equipment but generally will require keeping the cutting equipment slightly higher from the ground surface to reduce risk of hitting surface-protruding rocks.

Research demonstrates that some mined lands are capable of producing herbaceous biofuels in quantities adequate for commercial activity. Switchgrass yield was more than 5 Mg/ha (2.2 tons/ac) after the third year on a fertile mine soil in West Virginia (Brown et al. 2015). Miscanthus achieved 20 Mg/ha (9 tons/ac) on a reclaimed mine site in West Virginia after four years of growth (Scagline-Mellor et al. 2018). These results demonstrate the potential opportunity of high yielding biomass crops on well-managed mined lands.

5.2 *Woody Crops*

Mined lands can be prepared for woody biofuel crops using procedures described for forest establishment, as stated above. These procedures include removing hindering and competing woody vegetation with equipment, manually or with herbicides. Ripping can also be helpful to disturb the soil; woody stems, rhizomes, or other propagation materials can then be planted in the rips. However, the land surface must be suitable for powerful wood-harvesting equipment and in some cases ripping the soil may make the surface too rough for this equipment. Liming and fertilization may be necessary. Hybrid poplars, for example, prefer soil pH in the range of 6.0–7.0, which is higher than the optimal range for native hardwoods.

Productivity of woody biofuel crops under commercial-scale management has not been well studied. Species that have performed well on Appalachian mined areas include willow, hybrid poplar, American sycamore, and black locust (Brinks et al. 2011; Caterino et al. 2020; Zipper et al. 2011c). Woody biofuel crops may prove advantageous, relative to herbaceous crops, on mined land sites with slopes slightly greater than those suitable for agricultural biofuel crop harvesting. However, little research has attempted to develop the cultural, management, and harvesting systems that would enable cost-effective woody biofuel production.

5.3 Markets

Current biofuel markets are restricted to direct combustion uses, such as firing of heating systems or electrical generation, either as a sole fuel or via co-firing with coal. These markets, however, are limited geographically, typically demand feedstocks of low cost, and do not support biomass transport over extended distances.

Advanced technologies such as production of liquid biofuels for transportation uses from cellulosic materials have potential to provide access to more extended and high-paying markets. Such markets are also very limited as of this writing (Lynd 2017), and do not occur in the coalfield region. The major constraint in the current economy is market competition by lower-cost fossil fuels such as petroleum-derived gasoline.

If climate-change mitigation policies were to be enacted, results would likely include expanded markets for commercial-scale biofuel products generally and perhaps in the coalfields. Increased costs and/or usage restrictions for petroleum-based liquids would likely increase demands for liquid biofuels. Climate-change mitigation pathways commonly project increased usage of biofuels for multiple purposes including production of liquid fuels and direct combustion in association with carbon-capture and sequestration technology as a means of removing carbon dioxide from the atmosphere (IPCC 2018).

6 Renewable Energy Facilities

A global energy transition is affecting Appalachia's coalfield (Zipper et al. 2021). As coal mining declines in many parts of the world, renewable energy sources including solar photovoltaics and wind are becoming cost-competitive with fossil fuels for electrical generation (NREL 2019). These changing economics are causing landscape transformations outside of Appalachia's coalfield as solar photovoltaic and wind-energy "farms" are proposed in response to changing energy demands. Such facilities have been proposed to displace agricultural businesses by occupying farmland, occur adjacent to residential areas, require clearing of environmentally-valued forests, or create visual contrasts to surrounding landscapes and therefore are not always welcomed by residents or communities.

As this transition unfolds, Appalachia's coalfield finds itself with thousands of hectares of unused mined lands, many with characteristics conducive to solar energy development. Unobscured exposure to the sun's path across the sky is a critical site feature. The rocky soils, settlement-prone subsurfaces, and limited water access that hinder agriculture and some other forms of development on many Appalachia mines would not hinder solar energy development where other essential characteristics were present.

Mine site competitive advantages for solar generation facilities can include large contiguous acreages without competing uses and often under single ownership;

low land ownership costs; and the absence of sun-obscuring buildings and vegetation. Also, infrastructure necessary for solar development, including roadways and electric-power transmission connections, are sometimes in place because of mining. Proximity and access to transmission with adequate capacity to power markets, however, can be a constraint, especially for larger projects. Two recent studies have assessed commercial-scale solar conversion opportunities for mined lands in West Virginia (Campoli et al. 2019; James and Hansen 2017). Site selection criteria included adequate size (>16 ha), favorable slopes (<8% as ideal, <15% as adequate) (Campoli et al. 2019), proximity to the electrical grid, and lack of competing managed uses such as buildings or agriculture (both studies). Both studies found multiple sites appearing as suitable for solar development. Development of former coal mines in Appalachia and elsewhere for photovoltaic generation is already underway (Burger 2019).

Wind-development opportunities are also present but more constrained. Critical site conditions include wind exposure and electrical transmission access to a market. Multiple ridgetops with wind exposure are present in the coalfield, but suitable wind-energy locations appear as less extensive than potential solar development sites. Mountaintop removal mining has lowered the heights of many former highland areas, and thus their wind exposure. For mountaintop mines whose contours have been reconstructed, the replacement of intact geologic structures with loose mine spoils can create geotechnical stability concerns for the large wind-generating units that are most economical in today's markets. While wind turbines have been developed on high ridges of the Appalachian mountains, most such development to date has occurred outside of the coalfield.

Even under the best circumstances, solar- and wind-energy development is unlikely to fully replace coal mining as a source of well-paying jobs. However, such development can provide some economic input while producing renewable carbon-emission-free energy for American homes and businesses. Solar and wind development can have additional economic benefits by drawing businesses that desire renewable-energy generated electric power, such as data centers, to the coalfield (Dodson 2019). Parties within Appalachia are not alone in recognizing the potential for mine sites to support renewable energy generation, as mined lands in other parts of the world are also being converted renewable energy facilities (EC 2018; Bodis et al. 2019; Fig. 6).

7 Developed Land Uses

Reclaimed mines are often discussed as potential sites for economic development, commercial and industrial businesses that are much desired in the coalfield. Modern reclamation, however, rarely prepares mined areas for building support. With added expense, many former mine sites can be made suitable for residences and for development that will support businesses and jobs (Fig. 7).

Fig. 6 Photovoltaic energy facilities, "solar farms", constructed on mined land in Europe (Left) An 8.4 MW solar array constructed on approximately 20 ha of coal-mine spoils near Goettelborn in western Germany; (right) One segment of a 168 MW photovoltaic facility constructed on lignite mine spoils near Senftenberg in eastern Germany. (Photos from Google Earth)

7.1 Factors Influencing Mined Lands' Development Potentials

Several factors are critical to successful industrial and commercial development of mined areas. Location and access to infrastructures such as roads and utilities must be considered when essential to the development's success (Zipper and Yates 2018). It is also important to assess influence by mining-related hazards. Geotechnical stability of sites located on or immediately above slopes, proximity to potential hazards such as dangerous highwalls, the presence of underlying mine voids subject to subsidence or with potential to produce toxic gases, for example, can be considered when evaluating mined lands for development projects.

One factor unique to surface-mine land development is the underlying mine spoil and its potential for physical consolidation. As filled lands, virtually all reclaimed mine surfaces will be subject to some settlement over time. Rates of settlement will be dependent upon certain factors:

- Depth of underlying spoil: The greater the depth or thickness of the fill layer, the greater the potential for downward movement of the land surface.
- Mine site age: Mined land spoil typically settles most rapidly in years immediately following placement. As sites age, rates of settlement typically decline. As a rule of thumb: Most mines sites experience the majority of potential settlement within the first decade following spoil placement (Zipper and Winter 2018). Depending on the nature of underlying material, however, measurable settlement may continue for decades.
- The nature of the underlying mine spoil: Sandstones spoils will typically experience less overall settlement than more finely-textured rocks such as siltstones and

Fig. 7 (Above) Residential housing under development in 2015 on an area that was mined during the early 2000s in Buchanan County, Virginia. (Below) A segment of Lonesome Pine Business Development Park on an area that was mined prior to 1984 in Wise County, Virginia. (Photos from Google Earth)

shales; while durable-rock sandstone with quartzitic cement will experience less settlement than non-durable rock materials that disintegrate with time. If spoils have been compacted during placement, settlement will be less than if spoils have been loose-dumped; such compaction, however, rarely occurs.

- The presence of groundwater: Saturation of previously unsaturated mine spoils can stimulate additional settlement, even if the unsaturated mine spoils have been in place for years. Such water can accelerate geochemical reactions of mine-spoil minerals, causing them to lose physical rigidity and enabling additional settlement.

Although it can damage utility connections, a uniform, even ground settlement of 5 cm or less will generally cause little or no damage to most building types (Zipper and Winter 2018). "Differential settlement"—where the depth of settlement varies spatially under different parts of a structure—can damage buildings by causing structural distortion. Differential settlement occurs when significant variations in depth, composition, or compaction are present in spoils underlying a building site.

The best way to minimize settlement is to construct a mine-spoil fill area as a building-support pad during mining. Such construction will utilize relatively uniform materials and will employ thin lifts and spoil compaction (Zipper and Winter 2018). On existing mined lands with loose-dumped spoil, however, such construction procedures are not an option.

7.2 *Development of Settlement-Prone Mine Sites*

When developing land areas to support buildings and structures on a mine site, a geotechnical professional should be engaged to make recommendations for reducing site-development risks. If investigations determine that a site is subject to problematic settlement, several options are available.

One option is to employ piers extending through the mine spoil to solid ground for building support. This option will be feasible economically only on sites capped with relatively thin coverings of mine spoil.

Another option is to adopt building and site-development procedures that can tolerate the amount of settlement that is expected. Such techniques could include:

- Employ gravel or other flexible surfaces for roadways, parking lots, and sidewalks.
- Use flexible piping and/or piping connectors for underground utilities
- Use building techniques with sufficient tolerance to potential settlement. Factory-built, modular structures that can be trucked to the site, for example, typically have sufficient structural flexibility to accommodate some level of differential settlement or foundation shifting. The degree of structural distortion that small buildings of this type will experience can be reduced by employing adjustable foundation supports (Krebs and Zipper 2018). Post-and-beam structures are able to tolerate some level of structural distortion; although internal wall coverings and external cladding must also have some distortion tolerance if such methods are to be successful. Unless constructed with modular and pre-stressed components, masonry buildings are intolerant of even minimal differential settlement.
- Direct water to flow off-site and away from buildings to avoid saturation of the building-support pad, as such saturation may induce additional settlement.

Finally, it is also possible to employ methods intended to accelerate mine-spoil settlement so as to create a relatively stable surface for use in construction (Zipper and Winter 2018). At least two methods are available:

- Pre-Loading: This method consists of placing additional earth material on the building-site surface, and leaving it in place for sufficient time to enable compression of the underlying material. The amount of time needed will likely extend over multiple months, and in some cases for years. Greater thicknesses of pre-loading material relative to underlying spoil depth will enable more rapid settlement. The pre-loading material would then be removed prior to building construction.

- Dynamic compaction: This method employs a large crane which lifts and drops a weight onto the land surface. The lift-and-drop procedure is conducted repeatedly at individual locations, and systematically such that the entire building-support surface is compressed by the dropped weight.

These two procedures may be employed separately or in combination. For example, the Red Onion Prison was developed on a mine site in Wise County, Virginia, using these techniques (Zipper and Winter 2018). It was essential to minimize settlement of underlying fill materials to enable the prison's predominantly masonry construction.

Using these techniques, it is possible to develop mine sites for improved uses. Multiple sites have been developed on mined areas throughout the Appalachian coalfields. These include Lonesome Pine Business Development Park in Wise County, Virginia. Most mined lands throughout Appalachia, however, remain in an unmanaged and undeveloped state (Zipper et al. 2011a).

8 Mined-Land Conversion Progress

Despite reclamation requirements, many former Appalachian mines are unmanaged in their present form. They commonly are undeveloped, have unmanaged vegetation with exotic invasive components (Sena et al. 2021) that contrasts visually with surrounding land, release water pollutants (Clark et al. 2021), and provide minimal contributions to societal welfare. Even with available procedures, conversions of degraded mined lands to more beneficial conditions are infrequent because of the many constraints discussed herein. Site locations and cost barriers further restrict conversion.

Essential to any conversion are the people with interest and resources necessary to make it happen. Conversion of mined lands to hayland or pasture use, for example, is often easily accomplished and for relatively low cost; but makes sense only when an interested party has the know-how and willingness to operate an agricultural business. Similarly, land development for commercial purposes generally requires entrepreneurial initiative as well as capital investment.

The potential for profitability is another limitation. Conversions to biomass production, for example, can often be achieved for relatively low cost but markets for biomass products are limited. While local timber markets are available, reforestation costs typically exceed what can be justified based on projected economic returns due to eventual timber sales alone. Development for industrial uses can be hindered if a site with similar suitability can be developed on non-mined land for less cost. Another factor limiting conversion opportunities is ownership, as many former Appalachian mines are under ownership by corporate interests with limited local presence.

While society at large can benefit from restoration of ecosystem services and productivity to mined lands through reforestation, stream restoration, and wildlife habitat enhancement (Lituma et al. 2021), the investment capital necessary to support such non-revenue-generating conversions can be difficult to obtain. The solar- and wind-energy businesses are being transformed by improving technologies, but how the resulting prospects will play out for the coalfield is yet to be seen.

Degraded mined lands represent both an economic burden and opportunity. Investments to support businesses and jobs on unused mine sites can support economic diversification in a region where such is badly needed, especially during a time of declining coal-related activity. Such investments can produce additional societal benefits when displacing invasive-plant-dominated ecosystems. Investments in ecological restoration can support jobs in restoration work and monitoring (Holl and Howarth 2000) while improving long-term ecosystem productivity, values, and services (Tallis et al. 2008). Investment in ecological restoration of degraded lands may be considered profitable if non-market and indirect benefits are tallied against costs (De Groot et al. 2013). Recent studies indicate that stream restoration contributes over a \$1 billion annually to the US economy (Bernhardt et al. 2005), while reforestation of mined lands generates associated economic impacts and jobs to the nursery, tree planting, and geotechnical industries.

Despite limitations, some Appalachian mined lands are being converted to alternative conditions and uses. No data are available to quantify those conversions. Our experience indicates, however, that the majority of Appalachian mined lands and waters remain in the condition created by mine reclamation and often in a neglected state; and that numerous opportunities for conversion to more socially beneficial conditions remain available.

References

Adams MB, Sanderson T, Sena K, Barton C, Agouridis C, Angel P, Zipper CE (2019) Managing invasive exotic plant species on legacy mine lands. US Office of Surface Mining, Appalachian Regional Reforestation Initiative Advisory 16

Agouridis CT (2014) Bankfull frequency in rivers. In: Eslamian S (ed) Handbook of engineering hydrology, Modeling climate changes and variability, vol 2. CRC Press, Boca Raton, FL

Agouridis CT, Barton CD, Warner RC (2018) Recreating a headwater stream system on a valley fill in the Appalachian coal field, USA. In: Bolan NS, Kirkham MB, Ok YS (eds) Spoil to soil: mine site rehabilitation and revegetation. CRC Press, Boca Raton, FL, pp 147–174

Agouridis CT; Brockman RA, Workman SR, Ormsbee L, Fogle AW (2011) Bankfull hydraulic geometry relationships for the Bluegrass Region of Kentucky. Water 3:923–948

Amichev BY, Burger JA, Rodrigue JA (2008) Carbon sequestration by forests and soils on mined land in the Midwestern and Appalachian coalfields of the U.S. Forest Ecol Manage 256:1949–1959

Angel P, Davis V, Burger J, Graves D, Zipper C (2005) The Appalachian regional reforestation initiative. US Office of Surface Mining, Appalachian Regional Reforestation Initiative Advisory 1

Aukema JE, McCullough DG, Von Holle B, Liebhold AM, Britton K, Frankel SJ (2010) Historical accumulation of nonindigenous forest pests in the continental United States. Bioscience 60:886–897

Baird DC, Fotherby L, CC Klumpp, SM Sculock (2015) Bank stabilization guidelines. Report No. SRH-2015–25. U.S. Bureau of Reclamation, Technical Service Center, Sedimentation and Hydraulics Group

Ballesteros M, Manzaneres P (2019) Liquid biofuels. Chapter 3 in: Lago C, Caldes N, Lechon Y (eds) The role of bioenergy in the emerging bioeconomy. Academic Press

Bernhardt ES, Palmer M, Allan J et al (2005) Synthesizing U S. river restoration efforts. Science 308:636–637

Biebighauser TR (2003) A guide to creating vernal ponds. USDA Forest Service

Blackburn-Lynch W, Agouridis C, Barton C (2017) Development of regional curves for hydrologic landscape regions (HLR) in the contiguous United States. J Am Water Resourc Assoc 53:903–928

Bódis K, Kougias I, Taylor N, Jäger-Waldau A (2019) Solar photovoltaic electricity generation: a lifeline for the European coal regions in transition. Sustainability 11:3703

Bohn BA, Kershner JL (2002) Establishing aquatic restoration priorities using a watershed approach. J Enviro Manage 64:355–363

Brinks JS, Lhotka JM, Barton CD, Warner RC, Agouridis CT (2011) Effects of fertilization and irrigation on American sycamore and black locust planted on a reclaimed surface mine in Appalachia. Forest Ecol Manage 261:640–648

Brown C, Griggs T, Keene T, Marra M, Skousen J (2015) Switchgrass biofuel production on reclaimed surface mines: I. Soil quality and dry matter yield. Bioenergy Res 9:31–39

Burger A (2019) The switch to solar at coal power plants and mines is on. Solar Mag, 25 April https://solarmagazine.com/switch-to-solar-at-coal-power-plants-and-mines-is-on/

Burger JA, Graves D, Angel P, Davis V, Zipper C (2005) The Forestry reclamation approach. US Office of Surface Mining, Appalachian Regional Reforestation Initiative Advisory 2

Burger JA, Zipper CE, Angel P, Hall N, Skousen J, Barton C, Eggerud S (2013) Establishing native trees on legacy surface mines. US Office of Surface Mining, Appalachian Regional Reforestation Initiative Advisory 11

Butler PR, Iverson L, Thompson FR et al (2015) Central Appalachians forest ecosystem vulnerability assessment and synthesis: a report from the Central Appalachians climate change response framework project. Gen. Tech. Rep. NRS-146. U.S. Department of Agriculture, Forest Service, Northern Research Station

Campoli AA, Gardner JS, Mynear DK, Willis DA, Mittle S (2019) Adaptive reuse of mined lands for solar energy. Min Eng, 18–31

Cardenas MB (2009) Stream-aquifer interactions and hyporheic exchange in gaining and losing sinuous streams. Water Resourc Res 45:W06429

Casselman CN, Fox TR, Burger JA (2007) Thinning response of a white pine stand on a reclaimed surface mine in southwest Virginia. North J Appl Forest 24:9–13

Caterino B, Schuler J, Grushecky S, Skousen J (2020) Early growth and survival of shrub willow on newly reclaimed mine soil. New For. https://doi.org/10.1007/s11056-020-09776-4

Cavender N, Byrd S, Bechtoldt CL, Bauman JM (2014) Vegetation communities of a coal reclamation site in southeastern Ohio. Northeast Nat 21:31–47

Clark EV, Zipper CE, Soucek DJ, Daniels WL (2021) Contaminants in Appalachian water resources generated by non-acid-forming coal-mining materials. Chapter 9 in: Zipper CE, Skousen JG (eds) Appalachia's coal-mined landscapes. Springer

Copeland RR, McComas DN, Thorne CR, Soar PJ, Jonas MM, Fripp JB (2001) Hydraulic design of stream restoration projects. ERDC/CHL TR-01–28, U.S. Army Corps of Engineers, Coastal and Hydraulics Laboratory

Cotton C, Barton C, Lhotka J, Angel P, Graves D (2012) Evaluating reforestation success on a surface mine in Eastern Kentucky In: Proceedings of forest and conservation nursery associations RMRS-P-65. USDA Forest Service, Rocky Mountain Research Station, pp 16–23

Cummins KW (1974) Structure and function of stream ecosystems. Bioscience 24:631–641

Davis V, Burger JA, Rathfon R, Zipper CE, Miller CR (2012) Selecting tree species for reforestation of Appalachian mined lands. US Office of Surface Mining, Appalachian Regional Reforestation Initiative Advisory 9

De Groot RS, Blignaut J, Ploeg S, Aronson J, Elmqvist T, Farley J (2013) Benefits of investing in ecosystem restoration. Conserv Biol 27:1286–1293

Ditsch DC, Collins M (2000) Reclamation considerations for pasture and hay lands receiving sixty-six or more centimeters of precipitation annually. In Barnhisel RI, Darmody RG, Daniels WL (eds.) Reclamation of drastically disturbed lands. ASA, CSSA, and SSSA, Madison, WI. pp. 241–272

Ditsch DC, Teutsch C, Collins M, Whittier W, Rockett J, Zipper CE, Johns J (2006) Managing livestock forage for beef cattle production on reclaimed surface-mined land. University of Kentucky Cooperative Extension Publication ID-157

Dodson T (2019) $500K grant will open solar project in Wise County. Bristol Herald Courier

Doll BA, Grabow GL, Hall KR, Halley J, Harman WA, Jennings GD, Wise DE (2003) Stream restoration: a natural channel design handbook. Stream Restoration Institute, North Carolina State University, Raleigh, NC

Drummond MA, Loveland TR (2010) Land-use pressure and a transition to forest-cover loss in the eastern United States. Bioscience 60:286–298

European Commission (EC). (2018) EU coal regions: opportunities and challenges ahead. https://ec.europa.eu/jrc/en/publication/eur-scientific-and-technical-research-reports/eu-coal-regions-opportunities-and-challenges-ahead

Evans DM, Zipper CE, Burger JA, Strahm B, Villamagna A (2013) Reforestation practice for enhancement of ecosystem services on a compacted surface mine: Path toward ecosystem recovery. Ecol Eng 51:16–23

Evans DM, Zipper CE, Donovan PF, Daniels WL (2014) Long-term trends of specific conductance in waters discharged by coal-mine valley fills in central Appalachia, USA. J Am Water Resourc Assoc 50:1449–1460

Evans DM, Zipper CE, Hester ET, Schoenholtz SH (2015) Hydrologic effects of surface coal mining in Appalachia (U.S.). J Am Water Resourc Assoc 51:1436–1452

Franklin JA, Frouz J (2007) Restoration of soil function on coal mine sites in eastern Tennessee 50 years after mining. In: Proceedings of Ecological Restoration in a Changing World. Ecological Society of America and Society for Ecological Restoration joint conference, San Jose, CA

Fritz KM, Fulton S, Johnson BR, Barton CD, Jack JD, Word DA, Burke RA (2010) Structural and functional characteristics of natural and constructed channels draining a reclaimed mountaintop removal and valley fill coal mine. J N Am Benthol Soc 29:673–689

Gomi T, Sidle RC, Richardson JS (2002) Understanding processes and downstream linkages of headwater systems. Bioscience 52:906–916

Green Forests Work (GFW) (2019) https://www.greenforestswork.org/

Haan CT, Barfield BJ, Hayes JC (1994) Design hydrology and sedimentology for small catchments. Academic Press, New York, NY

Harman W, Starr R, Carter M, Tweedy K, Clemmons M et al (2012) A functional-based framework for stream assessment and restoration projects. U.S. Environmental Protection Agency, Washington, D.C. EPA 843-K-12-006

Hey R (2006) Fluvial geomorphological methodology for natural stable channel design. J Am Water Resourc Assoc 42:357–374

Hill T, Kulz E, Munoz B, Dorney JR (2013) Compensatory stream and wetland mitigation in North Carolina: an evaluation of regulatory success. Environ Manage 51:1077–1091

Holl KD (2002) The effect of coal surface mine revegetation practices on long-term vegetation recovery. J Appl Ecol 39:960–970

Holl KD, Howarth RB (2000) Paying for restoration. Restor Ecol 8:260–267

Intergovernmental Panel on Climate Change (IPCC) (2018) Global warming of 1.5 °C. https://www.ipcc.ch/sr15/

James J, Hansen E (2017) Prospects for large-scale solar on degraded land in West Virginia. Downstream Strategies, Morgantown WV

Johnson W, Paone J (1981) Land utilization and reclamation in the mining industry, 1930–80. US Bureau of Mines Information Circular 8862, Pittsburgh, PA

Krebs RD, Zipper CE. (2018) Foundations for housing on reclaimed mined lands. Virginia Coop Extension Pub 460–115

Krenz RJ, Schoenholtz SH, Zipper CE (2016) Riparian subsidies and hierarchical effects of ecosystem structure on leaf breakdown in Appalachian coalfield constructed streams. Ecol Eng 97:389–399

Krenz RJ, Zipper CE, Schoenholtz SH (2018) Periphyton structure and function in constructed headwater streams of the Appalachian coalfield. Freshw Sci 37(4):780–791

Leopold LB, Maddock T (1953) The hydraulic geometry of stream channels and some physiographic implications. U.S. Geological Survey Prof Paper 252

Leopold LB, Wolman MG, Miller JP (1964) Fluvial processes in geomorphology. Dover Publications, New York

Littlefield T, Barton C, Arthur M, Coyne M (2013) Factors controlling carbon distribution on reforested minelands and regenerating clearcuts in Appalachia, USA. Sci Total Environ 465:240–247

Lituma CM, Cox JJ, Spear S, Edwards JW, De La Cruz JL, Muller LI, Ford WM (2021) Terrestrial wildlife in the Appalachian post-mined landscape: status and opportunities. Chapter 6 in: Zipper CE, Skousen JG (eds) Appalachia's coal-mined landscapes. Springer

Lynd LR (2017) The grand challenge of cellulosic biofuels. Nat Biotechnol 35:912–915

Merovich GT, Hitt NP, Merriam ER, Jones JW (2021) Response of aquatic life to coal mining in Appalachia. Chapter 10 in: Zipper CE, Skousen JG (eds) Appalachia's coal-mined landscapes. Springer

National Renewable Energy Laboratory (NREL) (2019) Annual technology baseline: electricity 2019 ATB cost and performance summary. https://atb.nrel.gov/electricity/2019/summary.html

Neary DG, Ice GG, Jackson CR (2009) Linkages between forest soils and water quality and quantity. Forest Ecol Manage 258:2269–2281

Negley TL, Eshleman KN (2006) Comparison of stormflow responses of surface-mined and forested watersheds in the Appalachian mountains, USA. Hydrol Proc 20:3467–3483

Palmer MA, Allan JD, Meyer J, Barnhardt ES (2007) River restoration in the twenty-first century: data and experimental knowledge to inform future efforts. Restor Ecol 15:472–481

Palmer MA, Hondula KL (2014) Restoration as mitigation: analysis of stream mitigation for coal mining impacts in southern Appalachia. Environ Sci Technol 48:10552–10560

Parrish DJ, Fike J (2005) The biology and agronomy of switchgrass for biofuels. Crit Rev Plant Sci 24:423–459

Rodrigue JA, Burger JA, Oderwald RG (2002) Forest productivity and commercial value of pre-law reclaimed mined land in the eastern United States. North J Appl Forest 19:106–114

Rosgen D (1998) The reference reach—a blueprint for natural channel design. In: Proceedings, Wetlands and Restoration conference, Denver, Co.

Rubin Z, Kondolf CM, Rios-Touma B (2017) Evaluating stream restoration projects: what do we learn from monitoring? Water 9:174

Scagline-Mellor S, Griggs T, Skousen J, Wolfrum E, Holaskova I (2018) Switchgrass and giant miscanthus biomass and theoretical ethanol production from reclaimed mine lands. BioEnergy Res 11:562–573

Sena K, Barton C, Angel P, Agouridis C, Warner R (2014) Influence of spoil type on chemistry and hydrology of interflow on a surface coal mine in the eastern US Coalfield. Water Air Soil Poll 225:2171

Sena K, Franklin JA, Swab RM, Hall SL (2021) Plant communities on Appalachian mined lands. Chapter 5 in: Zipper CE, Skousen JG (eds) Appalachia's coal-mined landscapes. Springer

Shreve R (1969) Stream lengths and basin areas in topologically random channel networks. J Geol 77:397–414

Simmons J, Currie W, Eshleman KN, Kuers K, Monteleone S, Negley TL, Pohlad B, Thomas C (2008) Forest to reclaimed land use change leads to altered ecosystem structure and function. Ecol Appl 18:104–118

Tallis H, Kareiva P, Marvier M, Chang A (2008) An ecosystem services framework to support both practical conservation and economic development. Proc Natl Acad Sci USA 105:9457–9464

Taylor TJ, Agouridis CT, Warner RC, Barton CD, Angel PN (2009) Hydrologic characteristics of Appalachian loose-dumped spoil in the Cumberland Plateau of eastern Kentucky. Hydrol Proc 23:3372–3381

Teutsch CD, Collins M, Ditsch DC (2008) Cow-calf production on reclaimed surface mined pastures in Appalachia. In: Barnhisel RD (ed) Proceedings, American Society of Mining and Reclamation, pp 1252–1266

US Code of Federal Regulations (USCFR) (2020) 33 CFR Part 332—compensatory mitigation for losses of aquatic resources

US Environmental Protection Agency (USEPA) (2011) The effects of mountaintop mines and valley fills on aquatic ecosystems of the Central Appalachian Coalfields. EPA/600/R-09/138F

US Government Accounting Office (GAO) (2009) Surface coal mining characteristics of mining in mountainous areas of Kentucky and West Virginia. GAO-10–21

US Natural Resources Conservation Service (NRCS) (2007) Stream restoration design (National engineering handbook 654). Available at https://www.nrcs.usda.gov/wps/portal/nrcs/detail/national/water/manage/restoration/?cid=stelprdb1044707

Whittier WD, Gerken HJ, Zipper CE (2010) Management of cow-calf production on reclaimed surface-mined land. Virginia Cooperative Extension Publication 460–128 (Revised)

Wickham J, Wood PB, Nicholson MC et al (2013) The overlooked terrestrial impacts of mountaintop mining. Bioscience 63:335–348

Wohl E, Lane SN, Wilcox AC (2015) The science and practice of river restoration. Water Resourc Res 51:5974–5997

Zipper CE (2020) Appalachian coal data. University Libraries, Virginia Tech. https://doi.org/10.7294/BDA6-7R15

Zipper CE, Burger JA, McGrath J, Rodrigue JA, Holtzman GI (2011a) Forest restoration potentials of coal mined lands in the eastern United States. J Environ Qual 40:1567–1577

Zipper CE, Burger JA, Skousen JG, Angel PN, Barton CD, Davis V, Franklin JA (2011b) Restoring forests and associated ecosystem services on Appalachian coal surface mines. Environ Manage 47:751–765

Zipper CE, Evans DM, Burger JA, Fields-Johnson CW, Brunner A, Stanton B (2011c) Woody biomass production on post-SMCRA mined lands over three years and comparisons with other studies. In: Proceedings, American Society of Mining and Reclamation, pp 768–786

Zipper CE, RJ Krenz, SE Sweeten, CT Agouridis, Barton CD, PN Angel (2018) Establishing riparian woody vegetation for constructed streams on mined lands using the Forestry Reclamation Approach. US Office of Surface Mining, Appalachian Regional Reforestation Initiative Advisory 15

Zipper CE, Skousen JG, Barton CD (2021) The Appalachian coalfield's energy transition and prospects. Chapter 13 in: Zipper CE, Skousen JG (eds) Appalachia's coal-mined landscapes. Springer

Zipper CE, Winter S (2018) Stabilizing reclaimed mines to support buildings and development. Virginia Cooperative Extension Publication 460–130

Zipper CE, Yates C (2018) Reclaiming mined lands as industrial sites. Virginia Cooperative Extension Publication 460–132

Acid Mine Drainage in Appalachia: Sources, Legacy, and Treatment

Natalie Kruse Daniels, Julie A. LaBar, and Louis M. McDonald

Abstract Metal sulfides in coal and associated rocks of the Appalachian coalfield weather in the presence of air and water to produce metalliferous, acidic mine drainage (AMD). When watersheds lack sufficient geologic buffering, AMD from surface and underground mines can impair surface waters. Appalachian coalfield AMD is similar to acidic mine waters found worldwide and typically contains elevated acidity; low or no alkalinity; elevated iron, aluminum, manganese, sulfate, and conductivity; and localized elevated trace metals and metalloids including copper, zinc, selenium, and arsenic. AMD has impaired thousands of kilometers of streams in Appalachia with metal-rich, acidic waters leading to metal-rich sediments, impaired aquatic habitat, and nutrient limitations. Treatment has often been undertaken with the goal of eliminating these impacts. Treatment strategies include active treatment, where alkaline chemicals are added to AMD or impaired streams, and passive treatment, where alkaline addition and metal removal are accomplished in pond or wetland-based systems through combinations of alkaline material, organic substrate, and hydraulic retention. Treatment of AMD in Appalachia has improved water quality and biological communities at thousands of sites. Waters treated for AMD typically remain high in conductivity and with biological communities less complex than those at unimpacted sites.

Keywords Alkalinity · Iron · Manganese · Remediation · Water treatment

N. Kruse Daniels (✉)
Ohio University, Athens, OH, USA
e-mail: krusen@ohio.edu

J. A. LaBar
Centenary University, Hackettstown, NJ, USA

L. M. McDonald
West Virginia University, Morgantown, WV, USA

C. E. Zipper and J. Skousen (eds.), *Appalachia's Coal-Mined Landscapes*,
https://doi.org/10.1007/978-3-030-57780-3_8

1 Introduction

Coals and associated rocks are often laden with pyrite and other metal sulfides, which upon exposure to air and water during mining or other land disturbance react to form protons (H^+), sulfate (SO_4^{2-}), dissolved iron (Fe), and other products. These low-pH waters can dissolve or otherwise react with other rocks and minerals to increase dissolved concentrations of other metals, especially aluminum (Al) and sometimes manganese (Mn). These acids and metals are released into mine drainage waters to contaminate local streams and waterways. Such conditions are especially prevalent in the northern part of the Appalachian coalfield but can also occur further south when sulfide-bearing strata are disturbed by mining (Eriksson and Daniels 2021). Acid mine drainage (AMD) is the term for drainage that is low in pH (<6.0) and contains excessive concentrations of metals and sulfate. Estimates of stream miles impacted by AMD in Appalachia range from 17,000 km (10,500 miles) (ARC 1969; USEPA 1995) to 62,700 km (39,000 miles) (Hansen et al. 2010). Not all drainages from coal mines in Appalachia are acidic as some mines produce non-acidic waters with elevated levels of other mining-origin constituents (Clark et al. 2021). The focus of this chapter is on the production, extent, impact, and treatment of AMD.

2 Acid Mine Drainage Formation and Properties

AMD is formed due to oxidative weathering of pyrite and other metal sulfides in the absence of buffering by water constituents released by the surrounding geology. The chemical reactions that describe pyrite oxidation have been understood for some time (Singer and Stumm 1970; Evangelou 1995; Nordstrom 2001, and references therein). As a solid, pyrite can occur with crystalline or framboidal structures; the oxidation rate is faster for the framboidal forms because of the larger surface areas. The first step in the process is the oxidation of reduced sulfur (S^{2-} or S_2^{2-}), as in Eq. 1 to ferrous iron, sulfate and protons.

$$FeS_{2(s)} + 3.5O_{2(g)} + H_2O \rightarrow Fe^{2+}_{(aq)} + 2SO_4{}^{2+}_{(aq)} + 2H^+_{(aq)} \quad (1)$$

The ferrous iron (Fe^{2+}) then reacts with oxygen and hydrogen to produce ferric iron (Fe^{3+}).

$$Fe^{2+}_{(aq)} + 0.25O_{2(g)} + H^+_{(aq)} \rightarrow Fe^{3+}_{(aq)} + 0.5H_2O \quad (2)$$

Equation 2 is pH-dependent, proceeding more rapidly at neutral pH, but is often quite slow except in the presence of iron-oxidizing archaea and bacteria. Ferric iron is unstable except at very low pH (<≈3) and often precipitates as a simple iron hydroxide, generating additional acidity.

$$Fe^{3+}_{(aq)} + 3H_2O \rightarrow Fe(OH)_{3(s)} + 3H^+_{(aq)} \quad (3)$$

Depending on solution characteristics such as pH and other dissolved-element concentrations, Fe^{3+} may also form other solid-phase minerals such as jarosite, lepidocrocite, goethite, or schwertmannite. The Fe^{3+} of Eq. 2 is often reduced at the pyrite surface to produce even more proton acidity,

$$FeS_{2(s)} + 14Fe^{3+}{}_{(aq)} + 8H_2O \rightarrow 15Fe^{2+}{}_{(aq)} + 2SO_4{}^{2+}{}_{(aq)} + 16H^+{}_{(aq)} \quad (4)$$

When reactive sulfides are present in disturbed mine rocks, the chemical, physical and biological properties of mine water are directly affected by sulfide oxidation and associated chemical reactions (Eqs. 1–4). The low pH and high dissolved metals and sulfates are a direct consequence of Eq. 4. The high concentrations of dissolved Fe^{2+} and Fe^{3+} are a result of Eqs. 1, 2, and 4. The high electrical conductivity is a result of Eq. 1, and the high turbidity/suspended solids concentration is a result of metal precipitates (Eq. 3). The solids produced in Eq. 3 are yellowish-orange in color and coat the bottom of streams and reduce populations of aquatic organisms. When AMD is produced by mineral forms that do not contain iron, the AMD waters lack that characteristic orange color and may be clear. The chemistry of Eqs. 1–4 is a simplified representation of what occurs in nature, since sulfidic minerals often include minor components of other elements including metals (e.g., Cu, Ni, Pb, Zn) that substitute for Fe, and elements such as As and Se that substitute for S. Sulfide oxidation often stimulates other geochemical reactions that release water contaminants; these vary based on the nature of sulfide-associated minerals. When affected by severe AMD, waters are not usable by plants, animals, or humans.

Throughout this chapter, we use the term acidity to describe a water's capacity to neutralize alkaline additions. A water's acidity is a product of both its pH and its content of metallic and other elements that are able to buffer alkaline additions (Kirby and Cravotta 2005). For example, a water's dissolved Fe^{3+} content is a source of acidity due to its production of protons (H^+) when neutralized and hydrolyzed (Eq. 3). Numerous other acid-soluble metallic elements also serve as sources of acidity when occurring as dissolved forms in AMD.

3 Impacts

Release of AMD to the environment can have a variety of impacts, including destruction of instream habitat due to water-quality impairment and clogging of sediments with metal precipitates such as iron oxyhydroxides (Eq. 3). Such effects can extend to disturbance of aquatic food chains, both locally and downstream of AMD inputs.

AMD in the Appalachian coalfield (Fig. 1) occurs in areas with high-sulfur geology without sufficient natural buffering capacity to neutralize acidic discharges. Such conditions occur primarily in central Pennsylvania, northern West Virginia, southeastern Ohio, and central Tennessee.

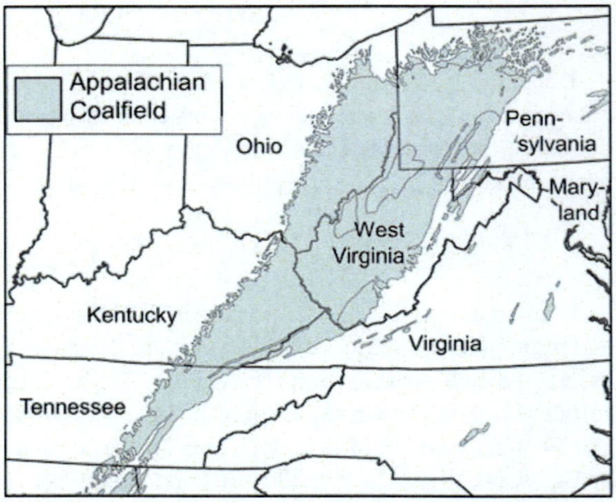

Fig. 1 The Appalachian coalfield extending from Tennessee northward

3.1 Impacts to Water and Biota

Water quality impacts in streams receiving AMD can have ecological consequences extending far downstream. Density and diversity of macroinvertebrates typically decline in response to AMD influences (Simmons et al. 2005; Freund and Petty 2007). Decomposition of organic material by macroinvertebrate shredders, filterers, and scrapers as well as microbes can slow significantly in AMD-impacted streams, slowing or even preventing the complete cycling of carbon and nutrients (Simmons et al. 2005; Bott et al. 2012). When comparing a stream receiving AMD, albeit with a near-neutral in-stream pH, and a slightly acidic reference stream, Bott et al. (2012) found total macroinvertebrate density to be 76% lower in the AMD-impacted stream. When looking at pollution-sensitive taxa (Ephemeroptera, Trichoptera, and Plecoptera; EPT), density was 96% lower. The same pattern was seen when looking at species richness, with total richness 58% lower and EPT richness 67% lower in the AMD-impacted stream than in the reference stream. Other studies have found similar results, with a large study in West Virginia demonstrating the significant impacts that AMD has on major and minor water quality constituents as well as biological structure (Table 1).

In extreme cases, AMD impacts to water resources can be far more severe. Untreated AMD in Pennsylvania, for example, has been found with pH values as low as 2.8, sulfate concentrations >10,000 mg/L, specific conductance values as high as 13,000 $\mu S\ cm^{-1}$, and multiple trace metals including Al, Cd, Cu, Ni, and Zn occurring at concentrations that would exceed in-stream water quality and/or drinking water standards if released without treatment (Cravotta and Brady 2015).

Clogging of stream sediments due to precipitation of metal hydroxides has physical, chemical, and biological repercussions. Settling of metal precipitates creates a blanket of solids on the stream bed that suffocates macroinvertebrates and destroys

Table 1 Comparison of physicochemical parameters and biological community metrics (mean ± standard deviation) in fourteen receiving moderate AMD impacts and four reference streams in West Virginia. Data from Watson et al. (2017)

Parameter/metric	AMD-impacted (n = 14)	References (n = 4)
pH	6.25 ± 0.74	7.29 ± 0.42
Specific conductance (μS/cm)	313.93 ± 69.59	118.8 ± 68.69
Alkalinity (mg/L)	4.03 ± 3.39	24.57 ± 13.75
Fe (mg/L)	0.51 ± 0.81	0.03 ± 0.00
Al (mg/L)	0.47 ± 0.82	0.01 ± 0.01
Mn (mg/L)	1.82 ± 1.15	0.01 ± 0.00
Ca (mg/L)	26.35 ± 6.21	11.55 ± 4.80
Mg (mg/L)	9.23 ± 1.94	2.35 ± 0.57
Co (mg/L)	0.06 ± 0.04	0.01 ± 0.00
Cr (mg/L)	0.02 ± 0.01	0.01 ± 0.01
Ni (mg/L)	0.05 ± 0.03	0.01 ± 0.00
Zn (mg/L)	0.10 ± 0.07	0.01 ± 0.00
Sulfate (mg/L)	113.11 ± 30.6	11.48 ± 1.50
Benthic macroinvertebrate richness	13.00 ± 7.68	27.40 ± 1.52
EPT richness	5.36 ± 4.75	16.80 ± 3.96
% EPT families	28.87 ± 25.28	61.75 ± 8.94
Fish species richness	0.73 ± 1.75	3.75 ± 2.43

habitats, creating embedded substrates. Stream sediments may serve as temporary sinks for metals associated with AMD, with metals being released slowly back into the water column as water quality conditions change over time or sediments move downstream (Carroll et al. 2003). Presence of metal hydroxyoxides in sediments has also been demonstrated to affect nutrient cycling in streams. Fe- and Al-hydroxyoxides have high capacities for adsorption of phosphorus, and their presence results in decreased potential for eutrophication in streams with high nutrient inputs. They may also cause streams to become phosphorus-limited by reducing biofilm primary production and impairing biological function (Simmons 2010; Drerup and Vis 2017).

3.2 *Historic Impacts*

Estimates of current AMD impact are not available, but surveys conducted in past years document extensive occurrences. By the late 1960s, an estimated 16,920 km

(10,575 stream-miles) in Appalachia were impacted by AMD and non-acidic mine drainages (Warner 1970). Streams impacted by AMD exhibit low pH, and contain elevated acidity, metals, and sulfate concentrations, while streams impacted by circum-neutral and alkaline mine drainage exhibit circumneutral or alkaline pH, may contain elevated trace elements and sulfate, and often contain elevated calcium, magnesium, and sodium, and conductivity (Clark et al. 2021). Past studies have indicated that water impacts by mining in northern Appalachia (northern West Virginia, and areas further north) were predominantly AMD; while AMD impacts were also present but not as prevalent in Appalachian coalfield areas further south (ARC 1969; Herlihy et al. 1990).

Some Appalachian states have more extensive inventories of AMD-impacted streams than others, making it difficult to estimate the total impacted stream lengths within Appalachia impaired by AMD. The Abandoned Mine Land Inventory System (AMLIS) (OSMRE 2019) inventories coal mines that were closed prior to passage of the Surface Mining Control and Reclamation Act in 1977, which are known as abandoned mined lands (AML). The AMLIS estimates over 7600 units in Appalachia are still considered to have water that does not meet standards for either human consumption or agricultural/industrial uses due to AML impacts (Table 2). These estimates include abandoned coal mine sites that existed prior to passage of the Surface Mining Control and Reclamation Act (SMCRA) and pose a risk to human health, safety, or welfare (Priority 1 or 2). These estimates do not include units that only pose a risk to aquatic life and ecological health unless directly associated with Priority 1 or 2 sites. Addition of these parameters can significantly raise the estimation of impacts. Streams throughout Appalachia's mining areas are also impacted by pollutants such as conductivity and TDS not directly related to AMD (Clark et al. 2021).

No comprehensive database of AMD-impacted streams in the Appalachian coalfield exists. The AMLIS data provided above are intended to give the reader an idea of the extent of the problem by using reclamation project numbers as a surrogate. However, given the constraints on these data, these values likely vastly underestimate

Table 2 Inventoried units impacted by coal mine drainage from AML in Appalachian states and their current reclamation status. Units are stream miles. Data from OSMRE (2019)

State	Completed	Funded	Unfunded	Total
Kentucky	19,146	978	990	21,114
Maryland	268	1.0	2067	2336
Ohio	396.1	1.0	22	419.1
Pennsylvania	3589.7	276.9	150	4016.6
Tennessee	217	0.0	95	312
Virginia	7554	74	536	8164
West Virginia	26,372.3	1175	1320	28,867.3
Total	57,543.1	2,505.9	5,180.0	65,229.0

the actual extent of AMD impacts in the region. These underestimates are likely to persist until a national inventory of AMD-impacted streams is developed.

3.3 Current Impacts

Although data on current impacts by AMD are not available in a form similar to past studies (ARC 1969; Warner 1970; Herlihy et al. 1990), several factors indicate such impacts today have been reduced although they are still quite evident in some areas.

First, new mining is not creating widespread AMD sources. The SMCRA was established in 1977 as the USA's federal law regulating coal mining and reclamation (Skousen and Zipper 2021) and has been a major factor in reducing AMD effects. The extent of AMD problems in Appalachia was a major factor that led to the law's passage. Hence, SMCRA and its regulations established controls for environmental effects of active mining including AMD. Miners are now required to test overburden prior to mining disturbance using Acid–Base Accounting or equivalent techniques to evaluate the AMD-production and neutralization potentials of each geologic stratum intended for disturbance (Skousen et al. 2002; Skousen 2017). When applying for a mining permit, miners must prepare an overburden handling plan as an effort to avoid production of acidic and/or metals-laden drainage. Methods for limiting AMD production can include blending potential acid-producing strata with other strata containing minerals that can neutralize acidity, such as carbonates, in the mine backfill; or constructing backfills using techniques intended to isolate highly acidic strata from atmospheric exposure and hydrologic flows (Skousen et al. 1999, 2019).

The SMCRA also requires miners to route all water runoff through sedimentation ponds, and mandates that water discharges from permitted areas be subjected to federal water-quality standards under the Clean Water Act. Federal regulations under the Clean Water Act require effluent waters to satisfy federal standards for pH, Fe and Mn (Skousen and Zipper 2021). Mine operators may treat AMD discharge waters as a means of meeting these water quality standards while the permit is active but are unable to achieve permit and performance-based bond release unless discharge waters meet federal standards without water treatment. If the mine permit application fails to demonstrate a likelihood that water quality standards will eventually be met without treatment, federal policy under SMCRA prevents issuance of the mine permit. If the mine permit is issued but persistent acid drainage is produced despite the permit plans, miners are required to continue treating that drainage until water-quality standards are met without treatment. Such continued treatment beyond the coal-production period can become costly, hence this requirement is a disincentive for mining in highly acidic strata. The SMCRA also enables citizens to petition the federal agency for designation of areas as "unsuitable for mining" and several large areas have been so designated because of extensive acidic overburden. Because of these legal and regulatory measures under SMCRA (Zipper 2000), the incidence of AMD from active and recently reclaimed coal surface mines in Appalachia has been reduced dramatically relative to pre-SMCRA mining.

Second, progressive depletion of pyritic minerals exposed by past mining is reducing AMD impacts from some past mined areas. Studies have shown that quality of waters discharging from mines changes over time. In surface coal mines, AMD has been shown to decrease over windows as short as 10–20 years as acid-generating minerals weather rapidly and become depleted (Meek 1996). Times of dissipation can vary, however, based on the extent and forms of sulfidic minerals present, particle size of the material, and leaching characteristics. In underground mines that are below drainage with limited oxygen exposure and limited through-flow, water quality has been shown to improve over 30–40 years, although the seminal studies were conducted in the United Kingdom, not Appalachia (Wood et al. 1999; Younger et al. 2002 and references therein). In above-drainage mines where the mine workings can drain freely to the surface, the pattern of water-quality change is site specific. In some above-drainage mines, AMD waters released exhibit clear improvement over time (Demchak et al. 2004; Burrows et al. 2015). For other Appalachian mines, little change has occurred over multiple decades of monitoring (e.g. Bowman et al. 2017), although there may be change from pre-monitoring conditions for which no data exist. Throughout northern Appalachia, many streams remain acidified due to AMD from mining conducted prior to federal regulation of mining which was established more than four decades ago (Simmons et al. 2005; Drerup and Vis 2017).

The site-specific nature of temporal change in AMD chemistry is shown in Fig. 2. Acidity concentrations from four underground mine discharges in southern Ohio that are partially above and partially below drainage are plotted for two decades. Two sites show apparent declining acidity over time while two show no trend. The two demonstrating no change over time drain different parts of a large (~4 km^2) mine pool. Truetown and Corning drain large (>4 km^2) mines and do show declining trends in acidity production and both discharges result from hydrostatic pressure relief of a mine pool. Truetown originated as a blowout through a geologic fracture while the Corning discharge is the result of drilling into a mine to drain it.

Multiple mechanisms are in place to enable treatment and remediation of some AMD discharges from past mining areas. These include the Abandoned Mined Land Reclamation program, established under SMCRA, which can fund remediation of AMD from pre-SMCRA mines (Table 2). Also, multiple watershed improvement projects, conducted by non-government organizations and by government agencies,

Fig. 2 Acidity concentrations of waters discharging from four underground mines over two decades

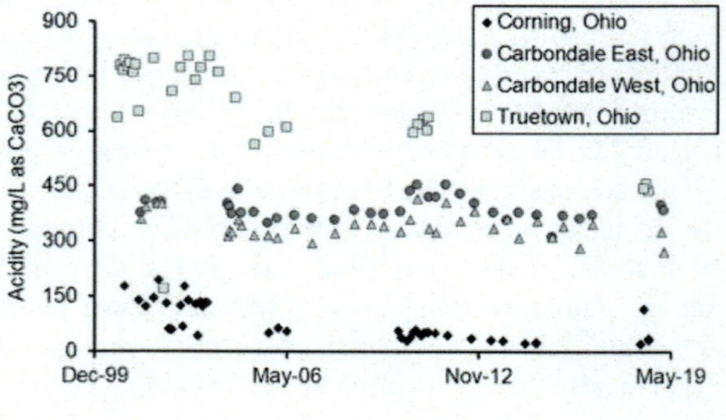

have installed water treatment and remediation measures in AMD-impacted watersheds (e.g. Simon et al. 2006; Kruse et al. 2012b, 2013; DeNicola et al. 2016). Remining of AMD source areas, while executing overburden handling practices that isolate acidic materials from hydrologic flows, has also been found to be an effective means for reducing AMD from pre-SMCRA mines in some areas (Hawkins 1995; Skousen 1997).

Despite these effects, AMD from past mining continues to degrade water resources in many areas, especially in northern Appalachia; but impact extent is far less than was documented by earlier studies.

4 Acid Mine Drainage Treatment

Both active and passive treatment systems are used to reduce AMD pollution loads, primarily targeting acidity and metals. Watershed-scale treatment systems have been installed to remediate AMD impacts to streams and rivers. They can greatly reduce acid loads and aid restoration of downstream reaches to near pre-pollution water quality as additional water dilutes the remaining pollutants and as unimpaired downstream tributaries contribute additional alkalinity (Kruse et al. 2013; Underwood et al. 2014). Some streams have improved to the point of meeting regulatory targets for macroinvertebrate and fish populations, although such improvements typically occur at some distance downstream from AMD remediation, aided by dilution with water inputs from non-impacted tributaries (Kruse et al. 2012a). More often, specific AMD sources are treated using individual active or passive systems. Thousands of AMD treatment systems have been installed throughout the Appalachian coalfield.

Treatment and reclamation of abandoned acid-generating mines in Appalachia is a multi-stakeholder process. Even when primary funding originates with the state and federal governments, local partners are often critical for the installation, maintenance, and monitoring of treatment systems. While AMD remediation varies among states, the state AML reclamation programs often partner with non-profit watershed organizations, private environmental consulting firms, universities, and industry to collect data, plan treatment, install treatment systems, and maintain and monitor them post-construction (e.g. Bowman et al. 2017, 2019). Although such partnerships have enabled widespread treatment and remediation of AMD in Appalachia, many AMD discharges remain untreated.

4.1 Active Treatment

Active treatment systems function by adding calcium or sodium-based alkaline reagents to the acidified waters as either solid-phase or liquid forms. Especially when solid-phase reagents are used, alkaline addition is often paired with mechanical mixing and oxidation. Many systems also include temporary holding ponds for the treated waters to allow settling and collection of metal precipitates. However, the ongoing cost of operations, maintenance, and sludge management can be considerable. In some cases, sludge collection and disposal are fiscally unattainable. In such cases, precipitates and sediments are not captured and collected but are released to the receiving stream where they cause continuing impairments, albeit on a smaller reach of stream. Active treatment systems are used where a passive treatment system would provide insufficient treatment or would have to be excessively large relative to available land area in order to treat the metal and acidity loadings. Active treatment systems are often deployed to treat larger point-source mine discharges or at locations where multiple diffuse sources converge (e.g. Cavazza and Beam 2010; Kruse et al. 2013). Active mining operations disturbing acidic strata must employ active treatment as a means of meeting regulatory standards. Due to their high maintenance needs, active treatment systems require ongoing access and frequent (i.e., daily or weekly) checks and thus are poorly suited to remote sites. If an active treatment system ceases operations even temporarily, impacted stream systems usually lack resilience and lasting ecological impacts may result that are not easy to reverse (Kruse et al. 2012a). A lime-dosing system treating a discharge from a long-abandoned underground coal mine is shown in Fig. 3. Many other system designs are possible. For example, common treatment methods apply alkaline reagents in liquid form to the AMD waters, with flow-control valves set to enable the reagent to enter the waters at a controlled rate (Skousen et al. 2000). Extensive research has shown that active treatment systems can reduce acidity effectively. However, AMD remediation adequate to enable significant biological recovery generally requires sufficient treatment to precipitate and retain acid-soluble metals (Bowman et al. 2017). If the treated AMD source continues to release metals at even low concentrations, such releases may impede biological recovery (Kruse et al. 2013, 2015).

4.2 Passive Treatment Systems

Passive treatment systems are installed at smaller point source discharges to treat diffuse or remote AMD inputs. Passive treatment systems rely on physical, biological, and geochemical processes to achieve AMD remediation without requiring constant or short-term resource inputs such as electric power or chemical reagents. Throughout the Appalachian coalfield, passive treatment systems are used to remediate AMD (Skousen et al. 2017). While passive treatment systems require less ongoing operation and maintenance than active systems, their installation costs tend to be greater and maintenance costs can still be significant. For some systems, frequent maintenance activities such as opening or closing of valves or clearing beaver dams or other obstructions are required. Continuing operation may require replacing alkaline or organic substrates, typically at multiple-year intervals. For AMD discharges that are clearly declining in pollutant loading, a passive treatment system with a finite operating life span may be advantageous, while maintenance must be conducted to ensure continued water-quality remediation for ongoing AMD discharges.

Fig. 3 Lime-doser system treating Hewett Fork in Raccoon Creek Watershed Ohio, as described by Kruse et al. (2012a, 2013). (Upper left) Doser silo can carry 70 tons of calcium oxide, concrete channel downstream of doser drives mixing and oxidation, doser silo is refilled periodically; (upper right) buffered iron-rich AMD mixing with untreated AMD from a second seep in concrete mixing channel downstream of doser; (Lower) water wheel inside doser is driven by AMD, turning an auger that doses calcium oxide into AMD flowing through the doser. Photos by N. Kruse Daniels

Treatment costs and effectiveness vary with multiple factors, including the land area available, other site characteristics, and the oxygen and pollutant concentrations of waters being treated. While installation costs of each system-type vary, the variation in cost is primarily due to differing performance. Some systems have extended lifespans of multiple decades, while some may fail after only a few years. Appropriate site-specific treatment-system selection and design, building upon the extensive knowledge base developed over more than three decades, can help to avoid premature failure or underperforming systems. Variation in source-water chemistry and site hydrology is an ongoing challenge for effective design and implementation of long-term AMD treatment (Zipper and Skousen 2010; Kruse et al. 2014).

Limestone leach beds (LLB) are small basins filled with high-calcium limestone fragments that receive AMD as influent waters, with basin sizes adequate to provide 30-min or longer retention times. AMD waters flow through the LLB in either a downward direction, entering as surface flow and exiting through a bottom drain, or by entering through the bottom as a water-pressure-driven upward flow. Dissolution of limestone ($CaCO_3$), enhanced by the acidic water, releases bicarbonate (HCO_3^-) for acid neutralization, as shown in Eqs. 5–7.

$$CaCO_3 + 2H^+ \rightarrow Ca^{2+} + H_2CO_3 \tag{5}$$

$$CaCO_3 + H^+ \rightarrow Ca^{2+} + HCO_3^- \tag{6}$$

$$HCO_3^- + H^+ \rightarrow H_2O + CO_{2(aq)} \tag{7}$$

If the AMD is acidic but has low metals concentrations, a downward-flow LLB may receive oxygenated AMD with little concern for precipitation of metalliferous compounds such as $Fe(OH)_3$ that can coat the limestone, rendering it less effective as an alkalinity generator, or clog pores between the rock fragments. Some metallic elements, most notably Al, tend to precipitate when neutralized even in the absence of O_2 as shown in Eq. 8:

$$Al^{3+} + 3H_2O \rightarrow Al(OH)_{3(s)} + 3H^+ \tag{8}$$

For metalliferous AMD, LLB's can be designed to avoid clogging with Fe and Al oxides and hydroxides and other oxidized metal precipitates by installing valves for periodic high-energy flushing of precipitant flocs (Cravotta 2008). Upward-flowing designs can be suitable for AMD waters with low O_2 concentrations, and with low concentrations of metals that precipitate when acid-neutralized in the absence of O_2 such as Al. In Appalachia, some LLBs have been designed for sequential failure, in which the bed is oversized such that as one part of the bed clogs, AMD flows to other unclogged areas of the bed (Bowman et al. 2017). These LLB designs rely upon the relative low cost and ubiquitous availability of limestone in some parts of the Appalachian region. LLBs increase AMD-water pH, neutralizing acid and decreasing the solubility of dissolved metals. They are best paired with sedimentation basins to capture metal precipitate flocs and sediments.

Steel slag is a low-cost waste material with a high capacity for alkalinity generation through dissolution of calcium silicates, CaO, and MgO, with the primary cost in transportation from steel production facilities. Like LLBs. **steel slag leach beds** (SSLBs) are implemented in several key ways depending upon the chemistry of the AMD (Fig. 4). In low-metal AMD, SSLBs can be used as a direct treatment with AMD flowing directly through a bed of steel slag fines, neutralizing the acidity before discharge into a receiving water body (Simmons et al. 2002). In AMD with

Fig. 4 (Left) Steel slag leach bed in East Branch of Raccoon Creek, Ohio. (Right) Periscope outflow structures to control air inflow to the underdrains of a steel slag leach bed. Photos by N. Kruse Daniels

higher metal concentrations, SSLBs are implemented to increase the alkalinity of an unimpaired water source which is then mixed with the AMD (Kruse et al. 2012b). In this design, a low-metal stream is impounded to create a source-water pond. The source water then flows vertically downwards through a bed of gravel-sized steel slag creating a low-metal, high-alkalinity water that is piped and added to an AMD-impaired stream, neutralizing the acid and reducing solubility of acid-soluble metals. Due to the high pH within SSLBs (typically 10–12), $CaCO_3$ becomes insoluble, so control of CO_2 and carbonate inflow to SSLBs is a key to their longevity. In some situations $Mg(OH)_2$ may also precipitate. This can be achieved by avoiding the use of limestone in construction of drainage channels upstream of the SSLB and the leach bed itself; and may also be achieved by including an elevation drop in the inflow pipe to exsolve CO_2, by including periscope pipes on the underdrain outflow structures to avoid air inflows into the underdrains (Fig. 4), and by piping SSLB effluent directly to the acidic receiving stream (Goetz and Riefler 2014, 2015). SSLBs typically produce an initial flush of high-alkalinity discharge (1000–2000 mg/L as $CaCO_3$) followed by longer-term concentrations on the order of 300–500 mg/L (Landers et al. 2014).

Anoxic limestone drains (ALDs) are buried anoxic limestone beds that minimize clogging through low influent metal concentrations and anaerobic conditions. ALDs can be used for low-O_2, low-metal AMD and are well suited to treating reasonably low-flow discharges from partially flooded underground mines. The sealed nature of ALDs does not allow CO_2 to escape, increasing the alkalinity generation. While the concept of an ALD is reasonably simple, they have often mistakenly been applied to inappropriate water quality, resulting in the impression that they are not a desirable treatment option. Where the residence time is insufficient or preferential flow pathways form, ALDs may render insufficient acid neutralization (Skousen and Ziemkiewicz 2005). If an ALD's influent water has high O_2, Al, and/or Fe^{3+}, results can include metal floc precipitation within the ALD, causing water-flow passages to clog and, eventually, system failure. When properly sized and used to treat waters with appropriate chemistry, ALDs can provide cost-effective AMD treatment.

Open or Oxic limestone channels (OLCs) are often used in conjunction with other treatment systems. They consist of open channels or ditches lined with limestone rocks, typically 8 cm or greater in size. Since OLCs are open to the atmosphere, metals hydrolyze and Fe forms precipitate coatings on, or "armors," the limestone surfaces. While this reduces limestone dissolution rates, armored limestone still dissolves, often generating alkalinity at up to half the rate of unarmored limestone (Ziemkiewicz et al. 1997). To minimize armoring, OLCs are established as sloping channels so that flowing waters will enable suspended particles to dislodge some of the armoring. However, residence time is critical to alkalinity generation and OLCs are sometimes installed as drainage conduits from mine reclamation sites (Fig. 5) or to convey water through a treatment site. Stand-alone OLCs reduce acidity effectively when installed with sufficient gradients and residence times.

Wetland-based systems can be used as part of a treatment train to remediate acidic, circumneutral, or alkaline mine drainages; application to AMD is for sediment retention and polishing after alkaline addition. Shallow aerobic wetlands are used for alkaline drainage or after alkaline addition to AMD and are designed to achieve sufficient residence time for metal hydrolysis and precipitation as a polishing and clarifying treatment step once Fe concentrations are low. The acidity released through metal hydrolysis is neutralized by the water's alkalinity. Such systems are easily constructed as shallow excavations, often with organic mulch or mulch-soil mixtures

Fig. 5 Open limestone channel flowing from left to right forming the drainage structure from a reclamation project at Orland Gob Pile in Hocking County, Ohio. (Photo by N. Kruse Daniels)

placed on the bottom to act as substrate and designed to achieve water depths of 10–30 cm (Skousen et al. 2017). The mine drainage flows through the excavation, which may also include living macrophytes or physical structures intended to prevent channelization of flow through just a small portion of the wetland area. Oxidation caused by the water's exposure to atmospheric O_2 and hydrolysis of dissolved metals cause them to precipitate. Metal precipitates accumulate within the wetland and, unless removed periodically, will eventually impair the system's effectiveness. A common design and construction challenge for early wetland systems was the land-area requirement which is large relative to that required for other AMD treatment-system types, especially if the AMD to be treated is highly acidic and metal laden (Shimala 2000; Skousen and Ziemkiewicz 2005; Skousen et al. 2017); thus, these applications are often limited to polishing and clarifying of neutralized waters.

Vertical flow reactors (VFRs), also referred to as vertical flow ponds, successive alkalinity producing systems (SAPS) when deployed in series, or reducing and alkalinity producing systems (RAPS), are downflow systems with a top layer of compost overlying a layer of limestone fragments (Fig. 6) (Skousen et al. 2017). Drainage pipes are embedded in the limestone and outfitted with a vertical riser to maintain a standing pool of influent water above the compost. Mine drainage waters move downward through the compost and limestone and into the drainage pipes which discharge into a settling pond. Microbial metabolism in the compost scavenges O_2, creating anoxic conditions with microbial sulfate reduction which generates alkalinity and immobilizes some of the AMD-borne metals as insoluble sulfide forms. Waters move from the compost layer downward through the limestone, generating additional alkalinity via limestone dissolution, and through the drain into the retention pond. Reducing conditions within the system limit the metal hydrolysis that can

Fig. 6 Vertical flow reactor in the foreground serving as pretreatment for source water to a steel slag leach bed treatment system in the distance. Photo by N. Kruse Daniels

lead to clogging of aerobic limestone-based systems; installation of a gravity-driven flushing system can also help to reduce clogging and prolong system life (Skousen et al. 2017). VFRs are often paired with settling ponds to re-aerate the water and drive hydrolysis reactions and settling of low-density metal precipitates. In some VFR systems where public access is likely, the compost and limestone are mixed to improve safety by providing firmer footing were a person to fall into the bed. VFRs can be deployed in series with intervening oxidation and settling ponds to meet rigorous treatment targets, given sufficient land area. Failures can occur due to undersized systems and due to accumulation of metal precipitates in the limestone layer, especially if influent waters are high in Al (Skousen et al. 2017).

Bioreactors or **sulfate-reducing bioreactors** (SRBs) are vertical flow compost-based systems with some additional biodegradable substrate (e.g., wood chips, straw, oyster shells) that form reducing conditions in which sulfate-reducing bacteria thrive, driving metal removal and alkalinity production. Microbial metabolism in the compost utilizes O_2, creating the anoxic conditions that enable sulfate-reducing bacteria to produce alkalinity and to remove metals from solution via conversion to insoluble sulfides as shown in Eqs. 9, 10:

$$SO_4^{2-} + OM \rightarrow MS + 2HCO_3^- \tag{9}$$

$$SO_4^{2-} + OM \rightarrow H_2S + 2HCO_3^- \tag{10}$$

where OM represents biodegradable organic material and MS represents insoluble metal sulfides. SRBs operate similarly to VFRs and are well suited to low-Fe systems that require removal of other metals, such as Zn, and a reduction in sulfate. Operation of SRBs is controlled through adjustment of flow rates to ensure an appropriate sulfide generation level. SRBs are often designed and implemented in tanks in series, although some have used them in larger pond-type systems (Gusek 2004; Matthies et al. 2012).

The above text provides just a cursory review of AMD passive treatment. For a more thorough review, see Skousen et al. (2017).

4.3 *Watershed Remediation*

In watersheds where AMD is produced by older and inactive mines that are not subject to current regulation, AMD treatment systems may be applied on a watershed basis. Such application would inventory all AMD sources within the watershed and prioritize treatment expenditures so as the remediate those sources creating the highest levels of ecological damage. In some watersheds, AMD produced in lower volumes and/or with lower acidity may be left untreated due to financial constraints.

In such cases, watershed remediation goals may be approached or achieved nonetheless as the lower-priority AMD waters are diluted when they enter high-volume and more-alkaline streams.

5 Effects of Remediation

The extensive impacts of AMD on streams are complex and difficult to measure fully. Even more complex is measuring recovery of those streams, as improving water chemistry is not equivalent to biological recovery. Full recovery of biological structure and stream function requires remediation of multiple factors, including water chemistry and metal deposition, and will also require alkalinity and low-conductivity water influxes from unimpaired watershed sources. Water conductivities in AMD-remediated streams generally remain higher than in regional streams unimpacted by AMD (Simmons 2010; Drerup and Vis 2017); both active and passive AMD treatment systems achieve remediation by adding soluble ions that serve as acid neutralizers but also increase water conductivity. High-conductivity mine drainage waters are characterized by reduced richness and diversity of macroinvertebrate communities and other biological effects (Merovich et al. 2021). Biological recovery is almost always slower than pH recovery and some streams that support partially recovered fish and macroinvertebrate populations still have food web limitations, particularly in basal resources (Kruse et al. 2012a, b; Kruse et al. 2013; Drerup and Vis 2017). Despite some AMD-remediated streams meeting state standards for biological communities, there are still significant differences between remediated streams and unimpaired streams (Drerup and Vis 2017).

Treatment of AMD has resulted in stream improvements across the region. For example, in Raccoon Creek Watershed, Ohio, alone, over 160 km of stream length (100 stream miles) have been remediated to meet state metrics for fish and aquatic macroinvertebrates after two decades and over $15 million of investment in treatment and reclamation (Bowman et al. 2019). More extensive stream restoration activities, as well as sufficient time to recover from the long-term impacts, may be necessary before a stream is considered no longer impacted by AMD (Merovich and Petty 2007). In the long run, depletion of the sulfidic minerals that are AMD sources will aid recovery in these AMD-impacted streams.

5.1 Chemical Response to Remediation

Stream water quality may continue to be impacted by AMD even after remediation. Treatment systems are generally successful at decreasing acid loads to receiving streams and allowing for recovery of pH to circumneutral values. Treatment of AMD may also prevent intermittent acidification in receiving streams during storm events because of excess or unreacted alkalinity deposited in stream sediments (Cravotta

2010). However, sulfate and conductivity concentrations remain elevated, allowing for potential long-term, lingering impacts to biota (Hopkins et al. 2013). Excess bicarbonate alkalinity generated during treatment may continue to provide treatment within receiving streams, but it and other dissolved constituents will also contribute to the persistent conductivity. Unless sulfate reduction is employed by the AMD-treatment system, the SO_4^{2-} produced by the AMD source passes through the treatment system and into the receiving stream. Base cations from alkaline treatments, such as Ca, Mg, and Na, may also make up a large proportion of the dissolved solids found in receiving streams.

In streams where sludge collection and disposal are not feasible or where lime dosing occurs directly in the stream, continued impact on water chemistry by acid-soluble metals is expected, albeit within a shorter stream segment than prior to treatment. This mixing zone may continue to exhibit elevated metals and total suspended solids concentrations, in addition to elevated sulfate and conductivity concentrations (Kruse et al. 2013).

Removal of Fe, Al, and Mn through precipitative processes may aid in removal of trace metals, such as Cd, Cu, Pb, and Zn, via sorption to the precipitates. The efficacy of this removal is highly dependent on pH, with 50% sorption of some metals occurring at pH levels well below those required for stream improvement (Lee et al. 2002). In systems with biological components, trace metals may be more effectively removed via bacterial sulfate reduction and subsequent sulfide precipitation but at the cost of added, potentially undesirable, constituents. Bacterial sulfate reduction produces an excess of dissolved H_2S, which will be released to receiving streams if it is not allowed to degas prior to leaving the treatment system. When organic substrates high in nutrients are used as the biodegradable substrate, excess nitrogen and phosphorus may be flushed out of the substrate at the beginning of operation.

5.2 *Biological Response to Remediation*

The primary goal of AMD treatment is recovery of biological structure and function within impacted streams. Even when all water quality improvement goals are met, biological recovery can be tenuous. If treated AMD contains even low concentrations of metals, biological recovery as measured by multimetric indices such as the Macroinvertebrate Aggregated Index for Streams (MAIS) may be stunted (Kruse et al. 2013, 2015). Residual precipitates in the stream can also continue to impact biological structure after remediation. Sensitive species such as *Ctenodaphnia magna* and mayflies have demonstrated mortality and reduced growth when exposed to AMD precipitate-covered substrates (Dsa et al. 2008). Correlations between suspended solids and total metals concentrations during storm flow demonstrate that sediments can be resuspended and continue to pose a problem after remediation (Cravotta 2010).

In a comparison of AMD-impacted, AMD-remediated, and reference streams, Bott et al. (2012) found that stream remediation using limestone-based passive methods did not result in a significant increase in macroinvertebrate densities, with

Table 3 Mean physical parameters and comparison of community structure metrics in three streams in Pennsylvania. Sites with the same letters were not significantly different ($p > 0.05$)

Parameter	References	AMD-impacted	Re-mediated
pH	5.94	3.23	6.36
Fe (mg/L)	0.15	3.10	0.06
Mn (mg/L)	0.04	12.97	0.11
Al (mg/L)	0.06	10.12	0.15
Sulfate (mg/L)	5.9	548.5	31.5
Specific conductance (μS/cm)	33.6	934.9	96.5
Total density (m^{-2})[a]	6426[b]	~1500[c]	1418[c]
EPT density (m^{-2})[a]	~2000[b]	<50[c]	~300[b]
Total richness (per 200 individuals)[a]	55[b]	23[c]	~39[bc]
EPT richness (per 200 individuals)[a]	13[b]	4[c]	~12[bc]

Adapted from Bott et al. (2012)
[a]Values designated as approximate are interpolated from Bott et al. (2012), Fig. 2

the remediated streams exhibiting >77% less total density than the reference streams. Although community structure metrics demonstrated improvement after remediation due to increased presence of sensitive taxa, macroinvertebrate communities were still significantly different from the reference stream (Table 3). Multiple stream functions, including leaf-litter decomposition and nutrient cycling, remained altered in the AMD-remediated streams relative to the references. Similar results were found by Watson et al. (2017) in West Virginia, where total and EPT richness both improved after AMD treatment but did not reach values found in reference streams; these authors also found fish assemblages at AMD treatment sites, although improved relative to AMD-impacted sites, to exhibit altered species compositions relative to reference sites.

5.3 *Effects of Treatment System Failure*

Like other engineered systems, AMD treatment systems require ongoing operations and maintenance support to maintain effectiveness. Treatment systems can fail in different ways: active treatment systems may run out of chemicals or power and passive treatment systems may have reduced hydraulic conductivity, exhausted alkalinity or organic material, or preferential flow paths. Even a short failure can have extended impacts on the biology of the receiving water body (Kruse et al. 2012a).

Kruse et al. (2012a) chronicled a two-week failure of a calcium oxide doser in Ohio. The doser was empty for two weeks during June 2010. Stream monitoring in the month following the failure found both fish and macroinvertebrates to be impaired to pre-treatment levels for approximately 10 km downstream. The impact to macroinvertebrates was similar to that of a severe drought year. Macroinvertebrate metrics recovered after one year and fish metrics recovered after two years (Kruse et al. 2012a). Several years later at the same site, the source of calcium oxide was changed and the new material bridged within the doser thereby limiting treatment (Kruse Daniels and Mackey 2016). After several months of intermittent dosing, the material was ready for replacement but approximately 15–20 tons remained in the doser. During June 2014, approximately 10 tons of calcium oxide were unloaded into the creek over an 8-h period. While the pH in the creek rose to greater than 11 within 3.5 km and to 8.5 within 10 km of the doser, biological data collected later the same summer did not suggest a lasting impact from the pulse of alkaline water (Kruse Daniels and Mackey 2016).

Over time, passive treatment systems may fail in different ways including clogging, development of preferential pathways, and exhaustion of alkaline or organic material (e.g., Kruse et al. 2012b). In Raccoon Creek Watershed of eastern Ohio, several large maintenance projects have been undertaken to overcome these issues. Both Flint Run Lake Milton and East Branch project sites have several steel slag leach beds that have needed major maintenance due to reduction in alkalinity production caused by decreased hydraulic conductivity and exhaustion of alkaline material. While there were not clear biological impacts of the failure, chemical monitoring results indicated maintenance projects were necessary to avoid biological impacts. Maintenance of Flint Run Lake Milton cost over $270,000 in 2011–2012 and East Branch cost over $240,000 in 2011–2012 (Bowman et al. 2017).

6 Summary

AMD is produced when acid-generating sulfide minerals are exposed by mining to oxygen and water in geologic materials with insufficient buffering capacity to neutralize the acidity formed. In Appalachia, AMD has high sulfate and conductivity, and elevated acidity, Fe, Al, Mn and trace metals depending on the local geology. Historically, AMD caused impairment of over 10,000 km of streams in Appalachia; but information on the current extent of AMD impacts is not available. Federal and state agencies, non-profit and private groups have undertaken treatment and reclamation of AMD across the region, which has partially restored stream chemistry by neutralizing acidity and raising pH and, in many cases, has enabled some recovery of biological communities. AMD-remediated streams, although with improved chemical and biological conditions relative to AMD-impacted streams, remain in an altered condition relative to reference streams, often with elevated water conductivity and sulfate concentrations and with altered biological communities. Given sufficient time, natural processes including sulfide mineral depletion will cause AMD impacts to subside. However, the time required extends over several decades or longer.

Multiple methods for treating AMD are available. Selection of the most effective method for a given AMD problem will depend on site-specific characteristics, including the potential for AMD-generating mineral depletion over time. Active treatment entails the addition of alkaline chemicals to neutralize the AMD acidity, and may require other ongoing inputs such as electric power. Hence, long-term funding is necessary to maintain the stream recovery, and lack of resources adequate to maintain treatment can lead to treatment-system failure and the return of stream impairment. Passive treatment systems, when selected and designed in a manner appropriate to AMD discharge chemistry and volume, can allow for effective and long-term treatment. Due to their passive nature, passive treatment systems can be resilient to change. Most passive treatment systems, however, also require long-term maintenance; such systems can be difficult to monitor in remote areas and, in the absence of AMD-source depletion, may require eventual replacement in order to maintain effective AMD remediation and ecosystem improvement.

While AMD treatment in the Appalachian region has led to recovery of many formerly acidified streams, several long-term concerns remain, including funding uncertainties for operation and maintenance of existing treatment systems, the multiple stream segments acidified by past mining that have not been remediated, the residual impacts to streams that receive AMD treatment due to precipitate deposition below treatment facilities lacking precipitant capture, and the presence of non-acid water contaminants contributing to high water conductivities.

Acknowledgements Thanks to Elyse Clark and Nora Sullivan for preparing Fig 1.

References

Appalachian regional commission (1969) Acid mine drainage in Appalachia. Appalachian Regional Commission, Washington

Bedu-Mensah H (2015) Influence of water quality and sediment transport on biological recovery downstream of lime doser systems. Ohio University, Thesis

Bott TL, Jackson JK, McTammany ME, Newbold D, Rier ST, Sweeney BW, Battle JM (2012) Abandoned coal mine drainage and its remediation: impacts on stream ecosystem structure and function. Ecol Appl 22:2144–2163

Bowman J, Sullivan N, Johnson K (2017) 2016 Stream health report. Voinovich School of Leadership and Public Affairs, Ohio University, Athens, Ohio. Retrieved from https://www.watersheddata.com/userview_file.aspx?UserFileLo=1&UserFileID=207

Bowman J, Sullivan N, Johnson K (2019) 2017–2018 Stream health report. Voinovich School of Leadership and Public Affairs, Ohio University, Athens

Burrows JE, Peters SC, Cravotta CA (2015) Temporal geochemical variations in above- and below-drainage coal mine discharge. Appl Geochem 62:84–95

Carroll KC, López DL, Stoertz MW (2003) Solute transport at low flow in an acid stream in Appalachian Ohio. Water Air Soil Poll 144:195–222

Cavazza EE, Beam RL (2010) Restoring the Bennett branch Sinnemahoning Creek, Clearfield and Elk counties, Pennsylvania. In: Proceedings of American Society of Mining and Reclamation, pp 99–122

Clark EV, Zipper CE, Soucek DJ, Daniels WL (2021) Contaminants in Appalachian water resources generated by non-acid-forming coal-mining materials. Chapter 9 in: Zipper CE, Skousen JG (eds) Appalachia's coal-mined landscapes. Springer

Cravotta CA (2008) Laboratory and field evaluation of a flushable oxic limestone drain for treatment of net-acidic drainage from a flooded anthracite mine, Pennsylvania, USA. Appl Geochem 23:3404–3422

Cravotta CA (2010) Abandoned mine drainage in the Swatara Creek basin, southern anthracite coalfield, Pennsylvania, USA: 2. Performance of treatment systems. Mine Water Environ 29:200–216

Cravotta CA, Brady KBC (2015) Priority pollutants and associated constituents in untreated and treated discharges from coal mining or processing facilities in Pennsylvania, USA. Appl Geochem 62:108–130

Demchak J, Skousen J, McDonald LM (2004) Longevity of acid discharges from underground mines located above the regional water table. J Environ Qual 33:656–668

DeNicola DM, Stapleton MG (2016) Using macroinvertebrates to assess ecological integrity of streams remediated for acid mine drainage. Restor Ecol 24:656–667

Drerup SA, Vis ML (2017) Extracellular enzyme activity suggests phosphorus limitation of biofilm productivity in acid mine drainage remediated streams. Hydrobiologia 784:225–236

Dsa JV, Johnson KS, Lopez D, Kanuckel C, Tumlinson J (2008) Residual toxicity of acid mine drainage-contaminated sediment to stream macroinvertebrates: Relative contribution of acidity vs. metals. Water Air Soil Poll 194:185–197

Eriksson KA, Daniels WL (2021) Environmental implications of regional geology and coal mining in the Appalachians. Chapter 2 in: Zipper CE, Skousen JG (eds) Appalachia's coal-mined landscapes. Springer

Evangelou VP (1995) Pyrite oxidation and its control. CRC Press, Boca Raton

Freund JG, Petty JT (2007) Response of fish and macroinvertebrate bioassessment indices to water chemistry in a mined Appalachian watershed. Environ Manage 39:707–720

Goetz ER, Riefler RG (2014) Performance of steel slag leach beds in acid mine drainage treatment. Chem Eng J 240:579–588

Goetz ER, Riefler RG (2015) Geochemistry of CO_2 in steel slag leach beds. Mine Water Environ 34:42–49

Gusek J (2004) Scaling up design challenges for large scale sulfate reducing bioreactors. In: Proceedings of American Society of Mining and Reclamation, pp 752–765

Hansen E, Hereford A, McIlmoil R (2010) Orange water, green jobs. Solutions 1(4):53–61

Hawkins JW (1995) Characterization and effectiveness of remining abandoned coal mines in Pennsylvania. U.S. Bureau of Mines Report of Investigations 9562

Herlihy AT, Kaufmann PR, Mitch ME, Brown DD (1990) Regional estimates of acid mine drainage impact on streams in the Mid-Atlantic and Southeastern United States. Water Air Soil Poll 50:91–107

Hopkins RL, Altier BM, Haselman D, Merry AD, White JJ (2013) Exploring legacy effects of surface coal mining on stream chemistry. Hydrobiologia 713:87–95

Kirby CS, Cravotta CA (2005) Net alkalinity and net acidity 1: Theoretical considerations. Appl Geochem 20:1920–1940

Kruse NA, Bowman JR, Mackey AL, McCament B, Johnson KS (2012a) The lasting impacts of offline periods in lime dosed streams: A case study in Raccoon Creek, Ohio. Mine Water Environ 31:266–272

Kruse NA, DeRose L, Korenowsky R, Bowman JR, Lopez D, Johnson K, Rankin E (2013) The role of remediation, natural alkalinity sources and physical stream parameters in stream recovery. J Environ Manage 128:1000–1011

Kruse NA, Mackey AL, Bowman JR, Brewster K, Riefler R (2012b) Alkalinity production as an indicator of failure in steel slag leach beds treating acid mine drainage. Environ Earth Sci 67:1389–1395

Kruse NA, Stoertz MW, Green DH, Bowman JR, Lopez DL (2014) Acidity loading behavior in coal-mined watersheds. Mine Water Environ 33:177–186

Kruse Daniels N, Mackey A (2016) The effects of a plug of alkaline water in an acid stressed watershed. In: Proceedings of International Mine Water Association, Freiberg Germany, pp 425–429

Landers S, Maj S, Mackey A, Kruse N (2014) Steel slag leach bed longevity analysis. In: Proceedings of American Society of Mining and Reclamation, Oklahoma City

Lee G, Bigham JM, Faure G (2002) Removal of trace metals by coprecipitation with Fe, Al, and Mn from natural waters contaminated with acid mine drainage in the Ducktown Mining District, Tennessee. Appl Geochem 17:569–581

Matthies R, Aplin AC, Boyce AJ, Jarvis AP (2012) Geochemical and stable isotopic constraints on the generation and passive treatment of acidic, Fe–SO_4 rich waters. Sci Total Environ 420:238–249

Meek FA (1996) Evaluation of acid prevention techniques used in surface mining. Chapter 11 In: Acid mine drainage control and treatment. National Mine Land Reclamation Center, West Virginia University, Morgantown

Merovich GT, Hitt NP, Merriam ER, Jones JW (2021) Response of aquatic life to coal mining in Appalachia. Chapter 10 in: Zipper CE, Skousen JG (eds) Appalachia's coal-mined landscapes. Springer

Merovich GT, Petty JT (2007) Interactive effects of multiple stressors and restoration priorities in a mined Appalachian watershed. Hydrobiologia 575:13–31

Nordstrom DK (2001) Mine waters: acidic to circumneutral. Elements 7:393–398

Shimala J (2000) Hydrogeochemical characterization of the Carbondale Wetland, Athens County, Ohio: Evaluation of acid mine drainage remediation alternatives. Ohio University, Thesis

Simmons JA (2010) Phosphorus removal by sediment in streams contaminated with acid mine drainage. Water Air Soil Poll 209:123–132

Simmons JA, Lawrence ER, Jones TG (2005) Treated and untreated acid mine drainage effects on stream periphyton biomass, leaf decomposition, and macroinvertebrate diversity. J Fresh Ecol 20:413–424

Simmons J, Ziemkiewicz P, Black DC (2002) Use of steel slag leach beds for the treatment of acid mine drainage. Mine Water Environ 21(2):91–99

Simon M, Cherry D, Currie R, Zipper CE (2006) The ecotoxicological recovery of Ely Creek and tributaries (Lee County, VA) after remediation of acid mine drainage. Environ Mon Assess 123:109–124

Singer PC, Stumm W (1970) Acidic mine drainage: the rate-determining step. Science 167:1121–1123

Skousen, J (1997) Water quality changes and costs of remining in Pennsylvania and West Virginia. In: Proceedings of American Society of Mining and Reclamation, pp 64–76

Skousen J (2017) A methodology for geologic testing for land disturbance: acid-base accounting for surface mines. Geoderma 308:302–311

Skousen JG, McDonald LM, Ziemkiewicz P (2019) Acid mine drainage formation, control and treatment: Approaches and strategies. Extr Ind Soc 6:241–249

Skousen J, Rose A, Geidel G, Foreman J, Evans R, Hellier W (1999) Handbook of technologies for avoidance and remediation of acid mine drainage. West Virginia University, Morgantown

Skousen J, Sexstone A, Ziemkiewicz P (2000) Acid mine drainage treatment and control. Chapter 6, in: Barnhisel R et al. (eds) Reclamation of drastically disturbed lands. American Society of Agronomy

Skousen J, Simmons J, McDonald LM, Ziemkiewicz P (2002) Acid–base accounting to predict post-mining drainage quality on surface mines. J Environ Qual 31:2034–2044

Skousen J, Ziemkiewicz P (2005) Performance of 116 passive treatment systems for acid mine drainage. In: Proceedings of American Society of Mining and Reclamation, pp 1100–1133

Skousen JG, Zipper CE (2021) Coal mining and reclamation in Appalachia. Chapter 3 in: Zipper CE, Skousen JG (eds) Appalachia's coal-mined landscapes. Springer

Skousen J, Zipper CE, Rose A, Ziemkiewicz PF, Nairn R, McDonald LM, Kleinmann RL (2017) Review of passive systems for acid mine drainage treatment. Mine Water Environ 36:133–153

Underwood BE, Kruse NA, Bowman JR (2014) Long-term chemical and biological improvement in an acid mine drainage-impacted watershed. Environ Monit Assess 186:7539–7553

U.S. Environmental Protection Agency (USEPA) (1995) Streams with fisheries Impacted by acid mine drainage in MD, OH, PA, VA and WV. Region 3, Wheeling WV

U.S. Office of Surface Mining Reclamation and Enforcement (OSMRE) (2019) Abandoned mine land inventory system (eAMLIS). https://amlis.osmre.gov/Default.aspx

Warner DL (1970) Water pollution from mining activities in the United States. Prepared for the Federal Water Quality Administration

Watson AS, Merovich GT, Petty JT, Gutta JB (2017) Evaluating expected outcomes of acid remediation in an intensively mined Appalachian watershed. Environ Monit Assess 189:339

Wood SC, Younger PL, Robins NS (1999) Long-term changes in the quality of polluted minewater discharges from abandoned underground coal workings in Scotland. Q J Eng Geol 32:69–79

Younger PL, Banwart SA, Hedin RS (2002) Mine water hydrology, pollution, remediation. Kluwer Academic Publ, Boston

Ziemkiewicz PF, Skousen JG, Brant DL, Sterner PL, Lovett RJ (1997) Acid mine drainage treatment with armored limestone in open limestone channels. J Environ Qual 26:1017–1024

Zipper CE (2000) Coal mine reclamation, acid mine drainage and the Clean Water Act. p. 169–191, in: Barnhisel R et al (eds) Reclamation of drastically disturbed lands. American Society of Agronomy

Zipper CE, Skousen JG (2010) Influent water quality affects performance of passive treatment systems for acid mine drainage. Mine Water Environ 29:135–143

Contaminants in Appalachian Water Resources Generated by Non-acid-forming Coal-Mining Materials

Elyse V. Clark, Carl E. Zipper, David J. Soucek, and W. Lee Daniels

Abstract This chapter summarizes environmental processes causing elevated total dissolved solids (TDS) and associated trace elements in streams influenced by mining in non-acid-forming geologic strata, discusses the spatial distribution of mining impacts on water quality, and describes the contaminants' biological effects. Appalachian coal mining affects water quality by enabling accelerated mineral weathering and consequent release of dissolved elements; such effects are evident during and after mining. In non-acid-forming geologic systems where calcium-bearing minerals neutralize the acidity produced by pyrite oxidation, mineral weathering results in circumneutral pH or alkaline waters with elevated concentrations of calcium, magnesium, sulfate, and bicarbonate, which are the dominant major ions in Appalachia's surface-mining-influenced streams. The TDS concentration in streams is proportional to specific conductance (SC), which can be measured in the field. Many mining-influenced streams have SCs >500 $\mu S\ cm^{-1}$ and often exceed 2000 $\mu S\ cm^{-1}$, while streams in unmined forested areas generally have SCs <100 $\mu S\ cm^{-1}$, which is characteristic of natural background conditions. Aquatic macroinvertebrates of pollution-sensitive taxa are adversely affected in streams with elevated SCs. Although many metals may be present at levels below aquatic toxicity thresholds, other elements, especially selenium, impact aquatic organisms through bioaccumulation and consequent toxic effects.

Keywords Water quality · Freshwater salinization · Headwater stream · Valley fill

E. V. Clark (✉)
Fitchburg State University, Fitchburg, MA, USA
e-mail: eclark12@fitchburgstate.edu

C. E. Zipper · W. L. Daniels
Virginia Tech, Blacksburg, Virginia, USA

D. J. Soucek
Illinois Natural History Survey, Champaign, Illinois, USA

C. E. Zipper and J. Skousen (eds.), *Appalachia's Coal-Mined Landscapes*,
https://doi.org/10.1007/978-3-030-57780-3_9

1 Introduction

Appalachian coal mining causes release of contaminants including major ions (e.g. sulfate, SO_4^{2-}; bicarbonate, HCO_3^-; calcium, Ca^{2+}; and magnesium, Mg^{2+}) and minor and trace elements (e.g., aluminum, Al; arsenic, As; iron, Fe; manganese, Mn; nickel, Ni; selenium, Se) from geologic materials to regional water resources. Those dissolved constituents may cause physiological effects on the aquatic biota of mining-influenced streams. Large volumes of rock are disturbed by coal mining, resulting in the exposure of previously buried materials to air and water, accelerating physical and chemical weathering of the materials, and release of weathered products to water. Precipitation infiltrates and drains freely through surface mine spoils and underground mining voids, causing ions and elements to dissolve in the water and increasing the dissolved ion concentrations; groundwater may also enter mining-disturbance areas and be affected similarly. These waters drain freely from surface coal mines during and after mining and are actively pumped from underground mine voids during mining, resulting in altered chemical composition of the streams receiving the drainage from mine sites. In this context, water contaminants are dissolved ions or elements generated by mine-rock mineral weathering and released at concentrations exceeding natural background. Stream water with elevated levels of mining-origin contaminants can be detrimental to aquatic organisms; such effects depend on the nature and concentration of the water contaminants, the sensitivity of organisms to those contaminants, and the contaminant's potential for bioconcentration in higher trophic levels of aquatic food chains.

Here we discuss how water-borne contaminants in mining-influenced streams are generated and the factors influencing their generation, with a primary focus on surface mining. We also provide a regional overview of the quality of mining-influenced Appalachian water resources and discuss effects of mining-generated contaminants on the organisms residing in mining-influenced streams. In areas where acid-producing materials are associated with coal, the drainage tends to be acidic and the dominant dissolved elements in the water are Fe, Al, Mn, SO_4^{2-}, and other acid-soluble elements that are dissolved and mobilized by acidic mine waters (Brady et al. 1998); these effects are addressed by Kruse Daniels et al. (2021). This chapter focuses on moderate to high pH (5.5 to 8.2) discharges from current and past coal mining in predominantly non-acid-generating strata.

2 Movement of Water Over and Through Mine Sites

The surface mining process involves removing layers of rock (termed "overburden") from above a coal seam by fracturing the rock with explosives, moving the fractured rock with equipment to expose and extract the coal, and using the fractured rock to reconstruct and rebuild the land surface. During land reconstruction, the waste rock (termed "mine spoil") is strategically placed around the mine site depending on landscape reconstruction plans; such placement affects rock potentials to generate water contaminants. During permitting, mining companies pre-test each rock layer for its potential to produce acidity and related contaminants prior to mining, then plan to place waste rock with lower acid-generation potentials in areas likely to be in contact with water. Recently, some mining firms have begun rock testing and planning to limit post-mining release of other contaminants (e.g., Se). Contaminant generation on mine sites is directly influenced by the interaction of environmental water with mine rocks.

Landscape reconstruction often occurs by placing mine spoils in excavated pits created during coal extraction; the mine spoils are typically placed to resemble the landscape's approximate original contours (Skousen and Zipper 2021). In some cases, however, alternative landforms are constructed. Rock fracturing causes an expansion in the volume of mining-disturbed materials, and these excess spoils are placed in alternative landforms termed valley fills (Miller and Zegre 2014). Valley fills are V-shaped landforms constructed in valleys adjacent to and topographically below the mine site (Skousen and Zipper 2021). Construction is initiated by building a sediment containment pond down-gradient of the valley fill drainage water outlet, then excavating soil and vegetation from the valley area that will be filled. The mine spoils in excess of what is needed to construct post-mining landforms are placed in the valley, usually by pushing or loose-dumping. Such processes cause large boulders to gravitationally segregate along the valley floor, producing a porous structure that allows water drainage into the containment pond. Valley fills may also be constructed by placing a rock drain in the valley floor then covering the drain with mine spoils constructed into a series of horizontal layers (Hester et al. 2019). After construction, the valley fill surface is graded with bulldozers and revegetated with grasses and occasionally with shrubs and trees. Sediment containment ponds are also constructed to receive drainage from surface mining areas that lack valley fills and for water pumped from underground mines. The water discharged from sediment containment ponds and/or drainage ditches is monitored for regulatory compliance; if water quality is in regulatory compliance after mining is complete and the site is revegetated, the ponds are generally removed.

The transformation of the natural Appalachian terrain to constructed post-mining landforms influences water quality, and valley fills are especially influential because they channel water from the mine site through fractured mine spoils (Cormier et al. 2013a). The water discharged from a mine site and associated valley fill, if present, is a combination of infiltrated water, overland flow (runoff), and subsurface flow through the mine spoil materials (Fig. 1); the pathways differ in their potential to generate water contaminants and release them from the mine site (Hester et al. 2019).

2.1 *Overland Flow*

Overland flow is the process by which water runs over the land surface, often discharging into a stream, and can occur when precipitation cannot infiltrate fully into

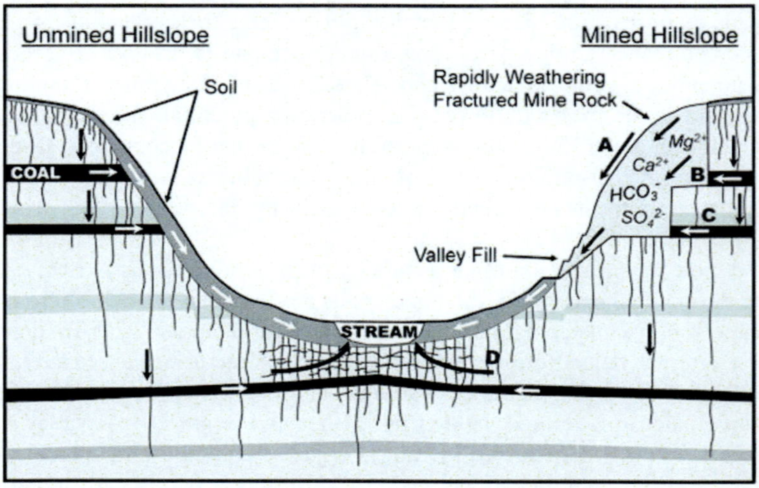

Fig. 1 Conceptual model of regional hydrology for an unmined Appalachian hillslope (left) and mined hillslope (right). Arrows indicate predominate flow paths. Fractures in the rock are illustrated. Flowpaths on and within the mined hillslope include **A** overland flow, **B** flow through the mine spoils that discharges from a valley fill, **C** underground mine and/or coal seam discharges into the mine spoils, and **D** local and regional groundwater discharges to a stream. The dominant ions generated during spoil weathering are shown (HCO_3^-, SO_4^{2-}, Ca^{2+} and Mg^{2+}) in addition to the ions typically associated with underground mining at depths below regional drainage (Na^+ and Cl^-). Figure based on Evans et al. (2015)

the soil because the surface has insufficient porosity to absorb the rainfall (infiltration-excess), or when the precipitation cannot infiltrate into the land surface due to all surficial pore spaces being saturated with water (saturation-excess). Prior studies of Appalachian coal mine sites have indicated that infiltration-excess overland flow is often the dominant runoff mechanism, especially in young mine soils (Ritter and Gardner 1993; Guebert and Gardner 2001). For example, Ritter and Gardner (1993) reported that infiltration-excess overland flow was the dominant runoff process if a mine soil has an infiltration rate less than 3 cm hr^{-1} and saturation-excess overland flow becomes dominant as a soil develops. A study of 97 storm events in five mining-influenced streams in Virginia found that infiltration-excess overland flow was evident in large storm events (Clark et al. 2016). Overland flow minimizes dissolved contaminant generation on surface coal mines because of low water–rock contact time; however, overland flow can cause water resource degradation via soil erosion, especially on mine sites with poorly-established vegetation (e.g. Bonta 2000; Simmons et al. 2008). Sediment control structures are strategically placed below active mines to capture runoff and eroded sediment; these structures are cleaned out periodically to ensure efficient containment of sediment but are removed when mining is completed.

2.2 *Infiltration*

The fraction of precipitation that infiltrates the mined-land surface varies due to factors including vegetative cover, mine soil development, soil physiochemical properties, and compaction (Jorgensen and Gardner 1987). Unless they have been highly compacted, mine soils tend to increase in their ability to infiltrate precipitation as soil develops in response to vegetation establishment. Infiltration rates in Appalachian coal mining materials may be 1–2 cm hr^{-1} in the initial years after mining, but may increase to 2.5–8 cm hr^{-1} in the years following reclamation if mining equipment has not severely compacted the soils (Ritter and Gardner 1993; Guebert and Gardner 2001; Negley and Eshleman 2006; Simmons et al. 2008). Infiltration rates can vary depending on the type of vegetation present. A study compared two 14-year-old mine soils developed on slopes ranging from 45 to 50% and reported infiltration rates of 16.5 cm hr^{-1} on the soil reclaimed with trees and 10.4 cm hr^{-1} on the soil reclaimed with grasses (Clark and Zipper 2016). These higher rates of infiltration were attributed to the influence of roots on soil development (Rodrigue and Burger 2004) as well as the presence of macropores and large bridging voids between mine spoil rocks in the subsurface (Guebert and Gardner 2001; Clark and Zipper 2016).

Higher infiltration rates indicate that more water will be in contact with mine spoils which can increase contaminant generation depending on the type of spoil material and the amount of time infiltrated water contacts the spoil materials. When groundwater from unmined areas adjacent to a mine drains into and flows through the mine spoils, that water also acquires contaminants.

2.3 *Subsurface Flow*

For water that does infiltrate into a mine soil, studies show that the subsurface of a mine site may resemble a "pseudokarst" environment and that downward movement into the subsurface occurs as infiltrated water travels through voids between large rocks, discontinuous channels, and macropores in the spoils (Caruccio and Geidel 1985; Greer et al. 2017). A study of Pennsylvanian mine spoils found hydraulic conductivities ranging from 4.45×10^{-9} m s^{-1} to upwards of 7.58×10^{-2} m s^{-1} indicating the presence of significant flow paths at depth in mine spoils (Hawkins 1998). Electrical resistivity imaging of a Virginia valley fill indicated upper layers of small rock fragments (10^{-5} to 10^{-2} m diameter), pockets of larger boulders (10^{-2} to > 1.0 m diameter) close to but below the surface, and large boulders with void spaces at the bottom of the materials, which corresponded to areas of deeper preferential flow paths within the valley fill (Greer et al. 2017). Subsurface flow waters in valley fills often contain both infiltrated water from precipitation and groundwater from the mine site. Subsurface water interacting with mine spoils in valley fills and mine backfills generates contaminants that are discharged from the spoils to enter streams draining mine sites.

2.4 *Discharge*

Mined areas have been found to discharge both higher baseflows (Nippgen et al. 2017) and higher stormflows (Bonta et al. 1997; Negley and Eshleman 2006) than unmined watersheds in similar terrain. In some cases, higher baseflows and stormflows occur together in the same watersheds (Messinger and Paybins 2003), perhaps in response to decreased evapotranspiration caused by forest loss on the land surface (Evans et al. 2015), but neither effect is observed consistently in the literature (Wiley et al. 2001). Other researchers have found mined landscapes to hold large volumes of water within the pores and voids of mine spoils, while discharge varies in response to factors such as water infiltration rates and volumes, void space characteristics, and flow paths out of the mined materials (Wunsch et al. 1996). Studies show that many mined landforms, including those with valley fills, perennially discharge water, indicating that water is stored, released, and discharged more evenly over the year (Nippgen et al. 2017). Others show distinctly higher flows in winter and lower flows in summer, on average, than unmined landforms, although some discharges are intermittent (Dickens et al. 1989; Merricks et al. 2007; Clark et al. 2016). The discharge from valley fills and other mined landforms often respond to storm events in a manner that is termed "flashy," with high discharges shortly after rain events and a quick return to baseflow conditions (Bonta et al. 1997; Murphy et al. 2014). Discharges from mined areas appear to be comprised of groundwater and subsurface flow during baseflow conditions, and a combination of runoff, shallow subsurface flow, and groundwater during stormflow conditions (Miller and Zegre 2014). Since discharges are the combination of multiple source waters of differing qualities, contaminant concentrations in mining-influenced streams can vary highly as well (Clark et al. 2016). Contaminant generation in mining-influenced streams tends to vary seasonally, with higher levels generally occurring during summer baseflows and lower levels during winter and stormflows (Clark et al. 2016; Timpano et al. 2018b).

3 Contaminant Generation in Non-acid-forming Mine Spoils

Since water–rock contact is an important influence on the quality of water draining from Appalachian coal surface mines, regulations require mining firms to pre-test materials for their contaminant-generation potentials prior to mining. Historically, mine spoils of differing mineralogy, lithology, and sulfur contents were combined and used as the substrate for reclamation, which resulted in acidic mine drainage in areas with acid-generating spoils (Skousen et al. 2019; Kruse Daniels et al. 2021). In recent times, spoil mineralogy and sulfur content analyses can identify acid-forming rock layers (Skousen 2017), enabling mining companies to isolate those layers to minimize acid generation via water–rock interaction.

3.1 *Mine Spoil Mineralogy*

Contaminant generation is directly influenced by weathering of the minerals within mine spoils. Research has shown that the major mineral groups that tend to occur in Appalachian coal-bearing strata include silicates, carbonates, sulfides, and clay minerals (Howard et al. 1988; Dulong et al. 1997; Miller et al. 2012; Clark et al. 2018a). The silicate minerals tend to include quartz (SiO_2), feldspars ($(Na,Ca,)AlSi_3O_8$), muscovite ($KAl_2(AlSi_3O_8)(OH)_2$), biotite ($K(Mg,Fe)_3AlSi_3O_{10}(F,OH)_2$), and chlorite ($(Fe,Mg,Al)_6(Si,Al)_4O_{10}(OH)_8$). Carbonate minerals in mine spoils typically include calcite ($CaCO_3$), dolomite ($CaMg(CO_3)_2$), and siderite ($FeCO_3$). Secondary minerals and other less-abundant minerals include the sulfide iron pyrite (FeS_2), clay minerals including kaolinite and illite, and the Fe-bearing oxide goethite. (See Eriksson and Daniels (2021) for further detail).

3.2 *Physical and Chemical Weathering*

Surface mining accelerates physical and chemical weathering of mine spoils by exposing previously-buried strata to the ambient environment. Physical weathering occurs due to shattering of rocks with explosives, but also due to the spoil dumping and re-grading during the land reconstruction phase. Contaminant generation in mined landscapes is due to exposure of fractured rock to oxygen (O_2) and water (H_2O), the main drivers of chemical weathering.

Certain minerals found in Appalachian mine spoils are highly reactive and prone to contaminant generation through oxidation, neutralization, and hydrolysis reactions. Acid mine drainage occurs due to the interaction of H_2O and O_2 with sulfur-bearing minerals, especially pyrite (Kruse Daniels et al. 2021). Although recent mining practices isolate spoils with high pyrite content, many mine spoils with low potentials to generate acidity still contain trace amounts of reactive pyrite (Clark et al. 2018a). Pyrite oxidation is highly acid-forming, releasing ${SO_4}^{2-}$ and H^+ ions (Singer and Stumm 1970):

$$FeS_2 + 3.75\,O_2 + 3.5\,H_2O = Fe(OH)_3 + 2\,SO_4^{2-} + 4\,H^+ \tag{1}$$

Many mine spoils also contain trace amounts of carbonates (Clark et al. 2018a), which can neutralize the pyrite-generated and other acidity,

$$CaCO_3 + H^+ = Ca^{2+} + HCO_3^- \tag{2}$$

releasing HCO_3^- and cations including Ca^{2+}, Mg^{2+}, Mn^{2+} and Fe^{2+} from carbonate minerals (Rose and Cravotta 1998; Clark et al. 2018a). Acid neutralization also occurs due to feldspar hydrolysis, exemplified by the following reaction of potassium feldspar:

$$2KAlSi_3O_8 + 2H^+ + 10H_2O = 2K^+ + 4Si(OH)_4 + Al_2Si_2O_5(OH)_4 \quad (3)$$

which releases K^+ ions while forming the clay mineral kaolinite. The weathering of Ca-rich and Na-rich feldspars may also neutralize acidity, including the acidity generated by pyrite oxidation, and release soluble Na^+ and Ca^{2+}. Additional minerals commonly found in Appalachian mine spoils, including muscovite, biotite, chlorite, goethite, and other trace minerals, can have elements or impurities such as Al, As, copper (Cu), Fe, Mn, Ni, lead (Pb), Se and zinc (Zn) that substitute into the mineral structure and are subsequently released during weathering (Tuttle et al. 2009; Blowes et al. 2014). In general, however, the influence of slow-reacting minerals on surface water quality and contribution to total dissolved solids in non-acid-forming environments is far less than in areas with rocks containing highly reactive minerals such as sulfides.

3.3 *Factors Affecting Contaminant Generation*

The physical and chemical weathering of Appalachian mine spoils is responsible for the generation of water-borne contaminants including the major ions Ca^{2+}, Mg^{2+}, HCO_3^-, SO_4^{2-}, potassium (K^+), sodium (Na^+), and chloride (Cl^-) and minor and trace elements including, but not limited to, Al, Mn, Fe, As, Cu, Ni, Pb, Se, and Zn (Cravotta and Brady 2015). Collectively, these contaminants are released and dissolved in water, increasing the concentration of total dissolved solids (TDS) in a stream. Increased TDS concentrations in streams are also termed salinity, and elevated-TDS streams can be described as salinized (e.g., Timpano et al. 2015). Freshwater salinization caused by human activities is a worldwide phenomenon and concern due to its effects on aquatic biota and human water-use potentials (Cañedo-Argüelles et al. 2013). Mining-induced salinization of freshwaters also occurs in other world regions (e.g., Schreck 1995; Goetsch and Palmer 1997; Hancock et al. 2005), although the specific ions released may vary from those released by Appalachian mining due to the nature of reactive minerals present in the geologic materials that are disturbed.

In central Appalachian mining-influenced streams, the TDS are typically dominated by Ca^{2+}, Mg^{2+}, SO_4^{2-} and HCO_3^- with smaller proportions of other major, minor, and trace ions (Pond et al. 2008, 2014; Timpano et al. 2015, 2018a, b). The proxy variables electrical conductivity (EC) and specific conductance (SC, which is EC adjusted to 25 °C), which can be measured in situ in streams, are typically measured and interpreted to indicate TDS concentration. In undisturbed Appalachian forested reference streams, SC values are generally <200 $\mu S\ cm^{-1}$ and often <100 $\mu S\ cm^{-1}$, whereas mining-influenced stream SC values are typically >500 $\mu S\ cm^{-1}$ with reported levels >2000 $\mu S\ cm^{-1}$ in certain studies (Merricks et al. 2007; Pond et al. 2008, 2014; Sena et al 2014; Evans et al. 2014; Clark et al. 2016).

Contaminants of geologic origin are mobilized from mineral sources when rock layers are fractured by mining and placed near the surface when exposed to O_2 and H_2O; such contaminant generation continues after mining has ceased. The degree of TDS generation is influenced by the weathering status of the strata, its mineralogical composition, and its exposure to water. Pre-weathered sandstones, generally brown in color, typically have the lowest potentials for TDS release, whereas unweathered sandstones, generally gray in color, and shales typically generate higher levels of TDS (Sena et al. 2014; Daniels et al. 2016; Clark et al. 2017). Pre-weathered geologic materials extend from beneath the soil to depths of 10 to 20 m where they have been chemically weathered over time by infiltrating waters and oxygen, whereas unweathered sandstones originate from deeper in the geologic column and have not experienced a high degree of chemical weathering. A field study of loose-dumped mine spoils in Kentucky found that pre-weathered spoils had an average discharge EC of 829 $\mu S\ cm^{-1}$ after two years of placement relative to the measured discharge EC of 1032 $\mu S\ cm^{-1}$ in unweathered spoils (Agouridis et al. 2012). Laboratory-based column leaching studies have similarly concluded that pre-weathered sandstones often generate less TDS. A study of 39 central Appalachian mine spoil samples showed that at the onset of leaching, pre-weathered sandstones produced leachate waters with an average SC of 618 $\mu S\ cm^{-1}$, whereas unweathered sandstones had an average SC of 1181 $\mu S\ cm^{-1}$; SC levels in both spoil types declined with continued leaching (Clark et al. 2017). Spoil materials originating from closer to a coal seam and shale materials with high organic coal-like content will generate even higher levels of TDS (Daniels et al. 2016; Clark et al. 2017). Spoil materials also vary greatly in their potential to generate trace elements including Se, which is of high ecotoxicological concern in Appalachia (Merovich et al. 2021). Like TDS, Se is typically released at higher concentrations from unweathered spoils and tends to occur at especially high levels in carboniferous rocks adjacent to coal seams (Vesper et al. 2008).

The above text describes chemical characteristics of waters discharged from surface mines and near-surface underground mines, which typically disturb rock within ~100 m of the ground surface and at elevations above stream initiation. Underground mine discharges, however, can affect surface water quality and flows (Lambert et al. 2004; Cravotta et al. 2014). Underground mines located relatively close to the surface affect rock strata similar to those affected by surface mines, and discharge waters with the common central Appalachian mining signature (i.e. elevated Ca^{2+}, Mg^{2+}, SO_4^{2-} and HCO_3^-). Deeper underground mines, such as those located below regional drainage and hundreds of meters below the surface, operate in geologic strata that retain brines dominated by Na^+ and Cl^-. Strata within a zone of intermediate depth tend to have drainage waters characteristic of both the surface chemical signature and the deeper chemical signature (Callaghan et al. 2000; Siegel et al. 2014), thus waters discharged by sub-drainage underground mines can be elevated in Na^+ and occasionally Cl^-, which differs from the surface mine signature. Discharges from sub-drainage underground mines typically cease once pumps intended to prevent the mine works from becoming inundated conclude operation and mining is completed, whereas above-drainage underground mine cavities continue to channelize and discharge groundwater after mining has ceased.

4 Contaminant Release to Water Resources

Mine spoils differ in their potentials to generate and release TDS to surface water. Potentials for TDS release to surface waters by mine sites are also influenced by the water–rock interaction time and by water-flow pathways. Since the hydrology of mine sites is complex and characterized by multiple flow pathways (Fig. 1), baseflow concentrations of water generally reflect the groundwater SC, since groundwater has the highest water–rock contact time. The "pseudokarst" flow regimes and porous nature of mine spoil fills enable storage of large water volumes and prolonged discharges of high-TDS water as baseflow over time (Miller and Zegre 2014). Storm events tend to dilute discharge SC due to rapid influx and runoff of low-TDS precipitation (Evans et al. 2015; Clark et al. 2016).

4.1 Mining-Influenced and Reference Stream Water Quality

Specific conductance levels in Appalachian mining-influenced streams are usually much greater than those in forested reference streams with no watershed mining. Research indicates that, in the absence of further mining or similar disturbance, TDS concentrations of water discharged from reclaimed mine sites will decline with time (Evans et al. 2014), but more research is necessary to learn about those effects. A study of 15 central Appalachian valley fills found SC levels remaining at 690 $\mu S\ cm^{-1}$, on average, 11–33 years after spoil placement and revegetation (Pond et al. 2014). A study of four West Virginia valley fills, three to seven years old with discharge SCs ranging from 1310 to 3050 $\mu S\ cm^{-1}$ found no association between SC and valley fill age (Merricks et al. 2007). A study of five 2.5- to 20-year-old valley fills in Virginia found mean baseflow SCs varying from 690 to 1660 $\mu S\ cm^{-1}$ and no effect of age on SC (Clark et al. 2016).

Laboratory studies and other field studies of ion release from non-acid-forming mine spoils indicate that SC release declines over time with continued leaching (e.g. Daniels et al. 2016; Clark et al. 2017; Fig. 2a), suggesting that the lack of temporal effects observed by the field studies described above are likely due to monitoring times insufficient to capture SC declines. Declining ion release with continued leaching occurs as the primary mechanism of soluble ion release, pyrite oxidation, decreases in part due to pyrite depletion, resulting in declines of SO_4^{2-} release as leaching continues (Fig. 2a). Subsequent leaching becomes dominated by alkalinity-generating reactions including carbonate dissolution and feldspar hydrolysis (Clark et al. 2018b) that may cause HCO_3^- concentrations to increase as leaching

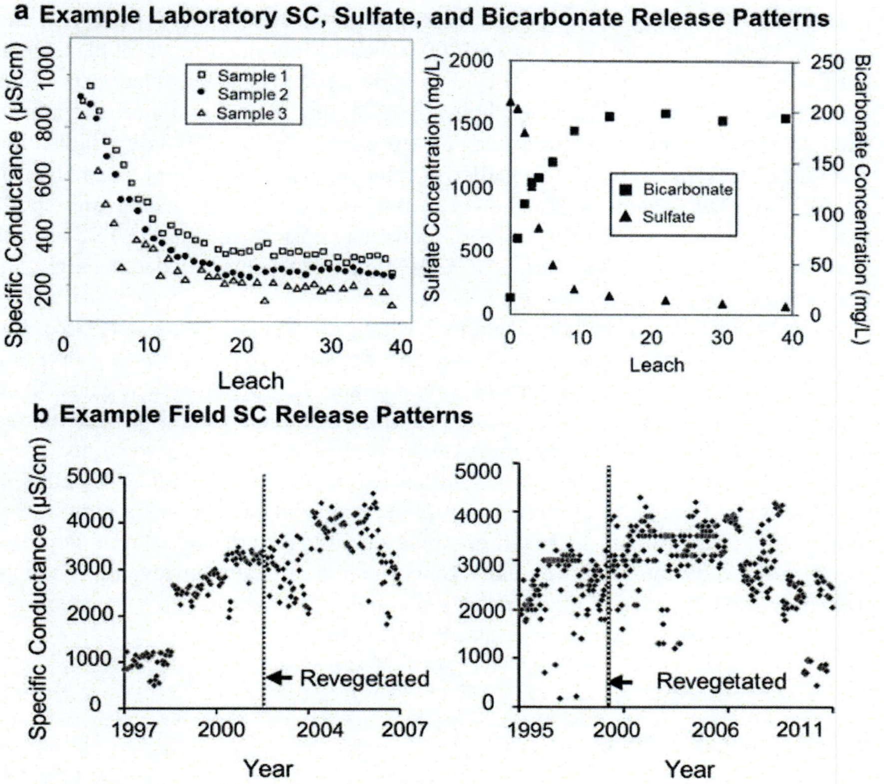

Fig. 2 Example patterns of specific conductance (SC), sulfate, and bicarbonate release from **a** laboratory studies of mine spoil leaching patterns showing SC release from three spoil samples analyzed in Clark et al. (2016), and one spoil sample analyzed in Clark et al. (2018a, b) showing sulfate and bicarbonate leaching patterns and **b** SC of waters discharged by two Virginia valley fills studied in Evans et al. (2014) during construction and following revegetation

progresses (Fig. 2a). Most of the ions present in mine spoil leachate, with the exception of HCO_3^- and Fe, decline in concentration with continued leaching (Clark et al. 2018b). A study of 137 Virginia valley fills found that 15–25 years on average were required for stream SC to decline to $< 500\ \mu S\ cm^{-1}$ (Evans et al. 2014; Fig. 2b). These data were generated from Virginia valley fills, many of which were smaller than those constructed in areas with more extensive mountaintop mining as studied by Merricks et al. (2007) and Pond et al. (2014). Even for the small Virginia valley fills, Evans et al. (2014) observed that several years after valley fill completion were required before SCs began declining, likely due to the large volumes of rock responding to limited amounts of precipitation and consequent leaching. Cianciolo et al. (2020a) studied temporal trends of SC in 18 mining-influenced streams over nine years and found five with declining SCs, all at slow rates ranging from <1 to ~ 4% per year; but also found SO_4^{2-}/HCO_3^- ratios, an indicator of mine spoils' geochemical weathering

(Clark et al. 2018b; Fig. 2a), to be declining in 10 of the 18 streams. A field lysimeter study in Kentucky found EC values >1000 μS cm^{-1} initially after spoil placement, but those levels declined and stabilized at ~500 μS cm^{-1} within nine years (Sena et al. 2014). Laboratory column leaching studies of the same spoil materials from the Kentucky field lysimeter study showed correspondence to the field studies, indicating that SC releases from mine spoil weathering typically decline over time (Daniels et al. 2016). This effect is not universal, however, as a minority of Appalachian mine spoils that contain trace amounts of reactive minerals may initially increase in TDS release [2 of 39 samples studied by Clark et al. (2017)], then subsequently decline after 2–3 pore volumes in laboratory studies.

Basic geochemical principles, column leaching studies (Daniels et al. 2016), and field studies such as Evans et al. (2014) and Sena et al. (2014) suggest eventual temporal decline of mining-generated contaminants in Appalachian streams. Such studies also documented declining concentrations of trace elements with continued leaching and found that Se and most other trace element releases declined more rapidly than those of major ions (Ziemkiewicz et al. 2011; Clark et al. 2018b). Given sufficient time after mining is complete, geochemical principles suggest that mining-influenced stream SCs will continue declining, perhaps to levels approximating natural background eventually. Scientific data to date, however, provide no field-based confirmation of that expectation, nor do they indicate if the time required will be on the scale of human generations or longer.

5 Biological Effects

5.1 *Major Ions*

Multiple taxa of aquatic biota have been studied in salinized Appalachian streams (Merovich et al. 2021). Benthic macroinvertebrate communities are altered in Appalachian mining-influenced streams with elevated SCs; researchers have also found assemblage structures of fishes and salamanders (Merovich et al. 2021) and microbial communities (Vander Vorste et al. 2019) to be altered in salinized Appalachian streams.

Much of the research in this area has focused on benthic macroinvertebrate communities, which are dominated by aquatic insects. One reason for this focus is the common use of benthic macroinvertebrates for biomonitoring and bioassessment by state and federal agencies engaged in Clean Water Act enforcement. Also, fish communities tend to be low in diversity or absent from headwaters, which contributes, in part, to the high diversity and abundance of aquatic insects in these habitats (Meyer et al. 2007); headwaters are often studied to determine mining impacts because they are affected directly. In addition, invertebrates tend to be more sensitive to elevated major ion concentrations than fish in laboratory toxicity tests, although exceptions exist (Wang et al. 2016). Therefore, where they exist, aquatic life benchmarks and water quality standards for major ions or their surrogates (e.g., SC) are driven by effects on invertebrates.

Although it is clear that biota are altered in mining-influenced streams with elevated SC, mechanisms governing those effects remain under study. While frequently used to report major ion effect levels (e.g. Pond et al. 2008, 2014; US EPA 2011; Timpano et al. 2015), SC is not a stressor nor a cause of toxicity itself, but it is an indicator variable that can be used to predict toxicity if the ionic composition of the water is roughly known. Measures of SC alone do not differentiate between various combinations of ions; solutions of NaCl and KCl, for example, with similar SC values will have very different toxicities to a freshwater mussel (Wang et al. 2018). Furthermore in tests with a single species (e.g. an amphipod) and a single salt (sodium sulfate), Soucek (2007) reported acute toxicity thresholds based on SCs that ranged from 1000 to >8000 $\mu S\ cm^{-1}$ depending on Cl^- concentrations and water hardness.

Osmoregulation is the active maintenance of electrolyte (dissolved salts) concentrations in body fluids relative to the external medium (water), and mining-influenced streams have elevated concentrations of these electrolytes. Therefore, it makes sense to predict that disruption of osmoregulation is the driver of effects on aquatic organisms in these headwater streams. However, it remains unclear whether toxicity of major ions to freshwater invertebrates, and particularly to insects like mayflies, is due to (1) a non-specific osmotic effect, (2) the action of specific ions or combinations of ions (e.g., Mg^{2+} or $SO_4{}^{2-}$ toxicity), or (3) more indirectly, disruption of other physiological processes when more energy is devoted to osmoregulation (Bott et al. 2012; Kefford 2019). Buchwalter et al. (2019) presented an argument for the latter case being important for mayflies, but mechanisms across various taxonomic groups may differ, as do sensitivities.

Numerous studies have documented the fact that major ion toxicity to laboratory test organisms like crustaceans, insects, and fish is dependent on the ionic composition of a water or effluent. The most frequently documented ion interaction is that of the influence of hardness, or more specifically (but not necessarily), Ca^{2+} concentration on major ion toxicity. The toxicities of both Na^+ and Mg^{2+} salts have been observed to decrease with increasing background water hardness/Ca (Mount et al. 1997, 2016; Soucek and Kennedy 2005; Davies and Hall 2007; Soucek 2007; Elphick et al. 2011a, b; Soucek et al. 2011, 2018). It is also known that depending on the type of organism tested, solutions tend to be more toxic when dominated by particular major ions. For example, a commonly tested cladoceran, *Ceriodaphnia dubia,* is more sensitive to solutions dominated by K^+, Mg^{2+}, and HCO_3^- than those dominated by Na^+, Cl^-, and $SO_4{}^{2-}$ (Mount et al. 1997, 2016).

Little is known about how the various ions associated with Appalachian coal mining-influenced streams interact to cause and/or ameliorate toxicity, although in full-life tests with a mayfly, Kunz et al. (2013) observed greater chronic sensitivity to solutions dominated by Ca^{2+}, Mg^{2+}, $SO_4{}^{2-}$, and HCO_3^- compared to solutions dominated by Na^+ and $SO_4{}^{2-}$. The ameliorative effect of Ca/hardness probably does not apply to mining-influenced headwater streams because Ca^{2+} and Mg^{2+} are the predominant cations associated with elevated salinity in this case. However, other interactions may be important if, as has been suggested by recent studies, ion activity is a predominant driver of toxicity (Erickson et al. 2017). Ion activity is essentially the effective concentration of an ion after accounting for its interactions with other ions in more concentrated solutions. For example, Ca^{2+} has a strong influence on the activity of SO_4^{2-}. More research is needed to further determine specific causes of toxicity by high SC Appalachian mine waters, but in the meantime, effects have been reported in a variety of manners.

Over the last decade and a half, many field studies have documented impacts on benthic macroinvertebrate communities, particularly mayflies, associated with elevated major ion concentrations where pH remained circumneutral or slightly alkaline. Effects were frequently reported in terms of SC, although in some cases effects were reported in terms of SO_4^{2-} concentration. One such study observed a lack of mayflies in streams with SC >3000 $\mu S\ cm^{-1}$ dominated by Na^+ and SO_4^{2-} downstream of an active coal-processing effluent in Ohio (Kennedy et al. 2003). In central Appalachia, Merricks et al. (2007) documented a lack of mayflies in benthic macroinvertebrate communities inhabiting streams below valley fill settling ponds. These West Virginia (WV) streams had SC values ranging from 1200 to 3000 $\mu S\ cm^{-1}$. Also in WV, Pond et al. (2008) compared 10 unmined streams with 27 streams that had valley fills located within their catchments, and reported that all sites with a mean SC value of 500 $\mu S\ cm^{-1}$ or higher were considered impaired using a multimetric benthic macroinvertebrate community index. Furthermore, whereas mayflies accounted for 20% to nearly 60% of the community in unmined streams, mayflies in mining-influenced streams with SCs >500 $\mu S\ cm^{-1}$ constituted less than 30% of the community and those present were typically Baetidae, which is a ubiquitous and relatively contaminant-tolerant family. The impacts noted by Pond et al. (2008) were correlated with SC as well as with a number of other physical and chemical characteristics, but the authors concluded that the strongest signal was from the elevated major ion concentrations and SC. Pond (2010) did further work to document a change point in mayfly percent abundance relative to reference streams at an SC of 175 $\mu S\ cm^{-1}$.

Following this work in WV, US EPA (2011; see also Cormier et al. 2013c) performed an analysis of chemical and biological data from more than 2500 sites in WV. For each invertebrate genus included in the database, they calculated the stream SC value above which there was only a 5% probability of detecting that genus in a sample (a 95% extirpation concentration or XC_{95}). They then compiled the XC_{95} values for all genera and created a sensitivity distribution for which they calculated a 5th percentile hazardous concentration. This value was calculated to be 300 $\mu S\ cm^{-1}$, which became the SC benchmark used by federal and state agencies for headwater streams in the applicable ecoregions (central Appalachian streams dominated by elevated Ca^{2+}, Mg^{2+}, SO_4^{2-} and HCO_3^-). Streams impacted by coal mining in these ecoregions with an SC <300 $\mu S\ cm^{-1}$ are predicted to avoid local extirpation of 95% of native macroinvertebrate genera.

The US EPA (2011) study included sites that were selected probabilistically and therefore likely had other stressors in some cases, but the authors conducted a careful analysis of confounding factors and concluded that the primary cause of taxa loss was elevated major ions, measured as SC. Timpano et al. (2015, 2018a) took a more careful approach to site selection with the goal of finding sites at which elevated major ions were the only substantial stressor; they observed impairment according to a family-level multi-metric benthic macroinvertebrate community index at SC levels between 560 and 903 $\mu S\ cm^{-1}$ or 214 to 441 $mg\ L^{-1}$ in terms of $SO_4{}^{2-}$, and further observed a departure from reference condition for various mayfly metrics when SC reached 243 $\mu S\ cm^{-1}$ on average. It appears to be a consistent finding that in central Appalachian headwater streams dominated by Ca^{2+}, Mg^{2+}, $SO_4{}^{2-}$, and HCO_3^-, when SC is elevated to around 240–500 $\mu S\ cm^{-1}$, benthic community structure changes, particularly due to loss of mayflies.

The apparent sensitivity of mayflies to stream salinization in central Appalachia has been corroborated by laboratory studies, albeit mostly using different ion combinations. For example, in tests with Na_2SO_4, a median lethal concentration (LC50) of 1338 mg $SO_4{}^{2-}$ L^{-1} was reported for the mayfly *Neocloeon triangulifer* in water with a hardness of 90 $mg\ L^{-1}$ (Soucek et al. 2018); this was lower than values for the seven other species for which data exist under similar background conditions: an amphipod, a fingernail clam, two cladocerans, a midge, a mussel, and a fish (Mount et al. 1997, 2016; Soucek and Kennedy 2005; Davies and Hall 2007; Soucek 2007; Wang et al. 2016). Reported LC50s for these latter taxa ranged from 1874 to 14,134 mg $SO_4{}^{2-}$ L^{-1}. These sulfate concentrations, which caused mortality over the course of two- to four-day exposures, are much higher than those typically observed in central Appalachian streams, but longer-term tests have produced effect levels that are more relevant to actual field measured concentrations. For example, in a full life exposure of *N. triangulifer* to Na_2SO_4, Soucek and Dickinson (2015) observed significant changes relative to control responses at concentrations ranging from 145 to 528 mg $SO_4{}^{2-}$ L^{-1} and calculated a SC effect level of 725 $\mu S\ cm^{-1}$. Kennedy et al. (2004) reported SC LC50s of 737 to 773 $\mu S\ cm^{-1}$ for the mayfly *Isonychia bicolor* exposed to a simulated mine effluent dominated by Na^+ and $SO_4{}^{2-}$. In a study simulating Appalachian mining-influenced ionic compositions, Kunz et al. (2013) observed significant toxicity to *N. triangulifer* at sulfate concentrations ranging from 386 to ~770 mg $SO_4{}^{2-}$ L^{-1}, or in terms of conductivity, ~800 and 1300 $\mu S\ cm^{-1}$, respectively, and Buchwalter et al. (2019) observed effects with sodium sulfate to this species over a similar SC range. Regardless of ionic composition, *Neocloeon* (which was formerly called *Centroptilum*) appears to be sensitive to SC values in the range of 700 to 1300 $\mu S\ cm^{-1}$ in laboratory studies; this compares well with the XC_{95} for *Centroptilum* from US EPA (2011) of 1092 $\mu S\ cm^{-1}$.

In summary, both laboratory and field studies are in agreement that mayflies tend to be especially sensitive to elevated concentrations of major ions of a variety of ionic compositions and are at particular risk in central Appalachian headwater streams impacted by coal mining. Field studies also make it clear that not only mayflies but other macroinvertebrate taxa are also responding to the elevated major ion concentrations and SCs that characterize mining-influenced Appalachian streams, although mechanisms governing those effects have received less study.

5.2 Trace Elements

The weathering of geologic materials disturbed by mining releases trace elements as well as major ions. Trace elements have received less attention in mining-influenced streams, however, because dissolved concentrations in water tend to be low, often below levels defined as protective by the US EPA (2019). Studies of biological effects in these systems typically focus on elevated major ions as the primary stressor. For example, Cormier et al. (2013b, c) listed Se, Mn, Fe, and Al concentrations in water as potential confounding factors for their SC benchmark development but concluded that major ions were likely the primary cause of observed benthic macroinvertebrate community declines. Pond et al. (2008) compared 10 unmined streams and 27 streams influenced by valley fills, analyzing water samples for dissolved concentrations of several trace elements and finding that Mn, Ni, and Se were elevated in the mining-influenced streams. Pond et al. (2014), again comparing WV reference ($n = 7$) and valley fill impacted streams ($n = 15$), reported Se as the only significantly elevated trace element in the mining-influenced streams. Similarly, Merricks et al. (2007) observed elevated concentrations of Se, Zn, and Mn in streams below valley fill drainages. More recently, Whitmore (2016) measured dissolved trace elements in water samples from nine streams in VA and WV; Cu, Ni, Se, and strontium (Sr) concentrations were elevated in the mining-influenced streams. A laboratory investigation found measurable release of As, Cd, Cu, and Pb, in addition to Mn, Ni, and Se from fresh Appalachian mine spoils (Clark et al. 2018b). These studies demonstrate that multiple trace elements are released from weathering mine spoils. Additional scientific literature stresses the importance of dietary exposure and bioaccumulation of some trace elements, especially Se.

Selenium is an essential micronutrient required for proper function of some proteins, but it can be toxic to aquatic biota at concentrations elevated marginally over essential levels (Janz et al. 2010). Selenium bioaccumulation in aquatic food chains is initiated by uptake from water by bacteria, fungi, and plants, followed by dietary transfer to higher trophic levels (Presser and Luoma 2010). This creates hazards for higher trophic levels in aquatic food chains [e.g., fish and birds, because of toxic effects including reproductive abnormalities observed at relatively low tissue concentrations (e.g., Lemly 1996)].

Selenium bioaccumulation is a common occurrence in mining-influenced streams where Se concentrations in water are commonly above natural background (e.g., Pond et al. 2008, 2014; Timpano et al. 2018a, b; Whitmore et al. 2018). Whitmore et al. (2018) observed enrichment factors (i.e. the ratio of Se concentrations in particulate media (detritus, leaf litter, periphyton) to the Se concentration in water) ranging from the 100s to nearly 2000; trophic transfer factors from particulate media to primary consumer insects ranging from ~3 to ~10; and trophic transfer factors from primary consumer to predatory insects ranging from 1.3 to 2.6. The increasing concentrations of Se through progressive trophic levels indicate biomagnification consistent with studies in other regions (Presser and Luoma 2010). Differences between stream water and insect Se concentrations ranged from 3000x to >10,000x (Whitmore et al. 2018), confirming the fact that seemingly low water Se concentrations do not indicate lack of risk to higher trophic levels. Selenium is also transferred from aquatic insects to insectivorous fish at trophic transfer ratios > 1 (Presser and Luoma 2010). Selenium-related fish deformities were observed by Arnold et al. (2014) in a mining-influenced stream where macroinvertebrate Se concentrations (10.1 mg kg^{-1}) were within the range reported by Whitmore et al. (2018) and where water-borne Se concentrations (4.7 $\mu g\ L^{-1}$) were lower than elevated Se concentrations reported by other studies including Pond et al. (2008, 2014).

Recent studies indicate that macroinvertebrates may also be at risk of bioaccumulation effects themselves, in addition to serving as a medium for transfer of Se to higher trophic levels including fish (Lemly 1993, 1996). Conley et al (2009) conducted laboratory experiments to investigate Se toxicity to a genus of mayfly, representing a taxonomic group of particular pertinence to Appalachia, and found reduced fecundity with Se tissue burdens between 30 and 100 $\mu g\ g^{-1}$ dry weight. Other laboratory studies have found sub-lethal effects such as reduced growth by macroinvertebrates with Se body burdens within the range of 1–33 $\mu g\ g^{-1}$ dry weight (DeBruyn and Chapman 2007). Macroinvertebrates collected from Appalachian streams with elevated water Se have exhibited body burden concentrations in the range of <10 to >30 $\mu g\ g^{-1}$ dry weight (Presser 2013; Arnold et al. 2014; Whitmore et al. 2018; Cianciolo et al. 2020b), suggesting possible sub-lethal effects of reduced growth and, perhaps in some cases, reduced fecundity.

Several studies, including Whitmore et al. (2018), have found elevated Se concentrations occurring in high SC streams, consistent with the Clark et al. (2018b) finding that fresh mine spoils tend to release both major ions and Se at elevated levels, especially early in the weathering sequence. This suggests that Appalachian mined landscapes are releasing waters that exert multiple stresses on macroinvertebrates and other aquatic animals.

Other mining-origin trace elements with bioaccumulation potentials are also present in these systems although generally at levels below ecotoxicological thresholds such as those defined by US EPA (2019); their roles as potential contributors to multiple stressor effects have not been widely investigated. Although it has lower bioaccumulation potentials, Fe is an element of concern in this region with both potential toxic effects and indirect effects from precipitation of ferric hydroxides that can smother food and habitat for invertebrates (Linton et al. 2007). In the case of Mn, Merricks et al. (2007) observed sediment concentrations that were 10 times greater downstream of valley fills relative to reference streams. In a laboratory study, Dittman and Buchwalter (2010) observed modest Mn bioaccumulation (bioaccumulation factors up to ~2) in a variety of aquatic insect species; much of the Mn was adsorbed to the exoskeletons of the insects. Studies in other regions have documented bioaccumulation of multiple elements that are released from weathering mine spoils, including As, Cd, Cr, Cu, and Zn as well as Se (DeForest et al. 2007). As is the case with Se, dietary exposure to these other trace elements is thought to have a greater influence on uptake in tissues than aqueous exposure. Studies in other regions have also demonstrated that the presence of multiple stressors can exert measurable effects on aquatic biota, even when all individual stressors are present at levels below ecotoxicological thresholds (e.g. Walter et al. 2002; Gautier et al. 2014), but published studies of such effects are lacking for Appalachia.

6 Regional Water Quality

6.1 *Specific Conductance and Major Ions*

Publicly-available water quality data for the Appalachian coalfields can be used to assess the spatial distribution of coal mining impacts on water resources. Data were compiled for a total of 4714 water quality monitoring stations of state agencies in Kentucky ($n = 258$), Ohio ($n = 326$), Pennsylvania ($n = 117$), Tennessee ($n = 263$), Virginia ($n = 96$), and West Virginia ($n = 3249$). All stations had <25 mi^2 (65 km^2) watersheds, and all data observations from each monitoring location between January 1, 2007 and December 31, 2016 were averaged. Methods are described in Supplementary Information (prior to references).

The SC of streamwater samples from 4714 locations with watersheds <25 mi^2 in the Appalachian coalfields ranged from 10 to 15,535 μS cm^{-1}, with an average of 512 μS cm^{-1} and median of 302 μS cm^{-1} (Table 1). Of the monitoring location analyzed, 50% had SCs >300 μS cm^{-1}, 32% had SCs >500 μS cm^{-1}, and 13% had SCs >1000 μS cm^{-1}. The maximum SC values, specifically 13,615 and 15,535 μS cm^{-1} for Kentucky and West Virginia, respectively, were associated with high Na^+ and Cl^- concentrations, suggesting influence by non-surface-mining sources such as underground mines, natural gas extraction, or road de-icing salts. Sulfate, as the main anion component of TDS, had the highest concentrations in eastern Ohio, southwestern Pennsylvania, northern and southern West Virginia, and

Table 1 Specific conductance (μS cm^{-1}) data for selected water monitoring locations in the Appalachian coalfields. These results should not be used for water quality comparisons among states given the differing methods of selecting monitoring locations employed by the individual states

State	*n* stations*	Mean basin area (mi^2)	Percentile						
			1st	10th	25th	Median—50th	75th	90th	99th
KY	*280*	5.3	15	40	84	212	489	1131	4282
OH	*365*	11.0	166	235	328	492	702	1366	2697
PA	*118*	7.2	20	27	36	65	176	314	2946
TN	*266*	5.4	14	32	59	153	300	650	1700
VA	*99*	11.1	12	37	159	343	742	987	2260
WV	*3539*	4.2	24	79	157	310	648	1210	3060

* *Source* = publicly-available water quality data from January 1, 2007 to December 31, 2016. See Supplementary Information for further information about data sources and analysis methods

eastern Kentucky (Fig. 3); the median SO_4^{2-} concentration for all stations was 57,955 µg L^{-1}. Median Ca, Mg, and K concentrations were 30,341, 9400, and 2150 µg L^{-1}, respectively, for all stations. Median Na and Cl concentrations were 9248 and 6893 µg L^{-1}, respectively, for all stations.

The spatial distribution of SC and SO_4^{2-} values indicates regional areas of high SC corresponding to heavily-mined areas of the Appalachians and areas of low SCs corresponding to lower mining activities. Although spatial distribution maps of Ca, Mg, K, Na, and Cl are not shown, their patterns of distribution are similar to those of SO_4^{2-}. Since many of the <25 mi^2 watersheds had SC values >1000 µS cm^{-1}, it is expected that impacts on aquatic biota are occurring in these areas because the SC is elevated above the 300 µS cm^{-1} benchmark for headwater streams as described above (US EPA 2011).

6.2 *Minor and Trace Elements*

A survey of regional minor and trace element concentrations dissolved in <25 mi^2 (65 km^2) watershed streams in the Appalachian coalfields indicated the spatial distribution of elements including Al, As, Cd, Cu, Fe, Mn, Ni, Pb, Se, and Zn. Median Al, Fe, and Mn concentrations for all stations were 226, 414, and 90 µg L^{-1}, respectively (Table 2). Dissolved As, Cd, and Pb data for 490 water quality monitoring stations in the Appalachian coalfields indicated that most measurements were below detection, thus data are not presented. The median Zn concentration in the <25 mi^2 coalfield streams was 5 µg L^{-1}; the median Cu concentration was 1.6 µg L^{-1}. A total of 2040 stations determined the spatial distribution of Se excluding stations in Pennsylvania. The median Se concentration was 1.0 µg L^{-1}, with a 99th percentile concentration of 10.5 µg L^{-1}; 480 stations had dissolved Se concentrations > 3.1 µg L^{-1}, which is the US EPA (2016) water quality criterion for Se in streams where fish tissue concentrations are not available.

7 Summary and Conclusions

Coal surface mining in Appalachia affects water quality. Geologic materials disturbed by mining weather through chemical reactions in the presence of O_2 and H_2O and generate soluble ions that are released to streams. While these effects have been most well-studied for surface mines, underground mining also influences water quality. Materials of differing mineralogical compositions and lithologies influence the quality of mine water discharges and of the streams receiving drainage. The flow pathway that water takes (e.g. runoff, soil water, groundwater) similarly influences eventual discharge water quality. Water quality effects include elevated total dissolved solids dominated by the cations Ca^{2+} and Mg^{2+}, the anions SO_4^{2-} and HCO_3^- and other major ions and multiple trace elements, including the known toxicant, Se,

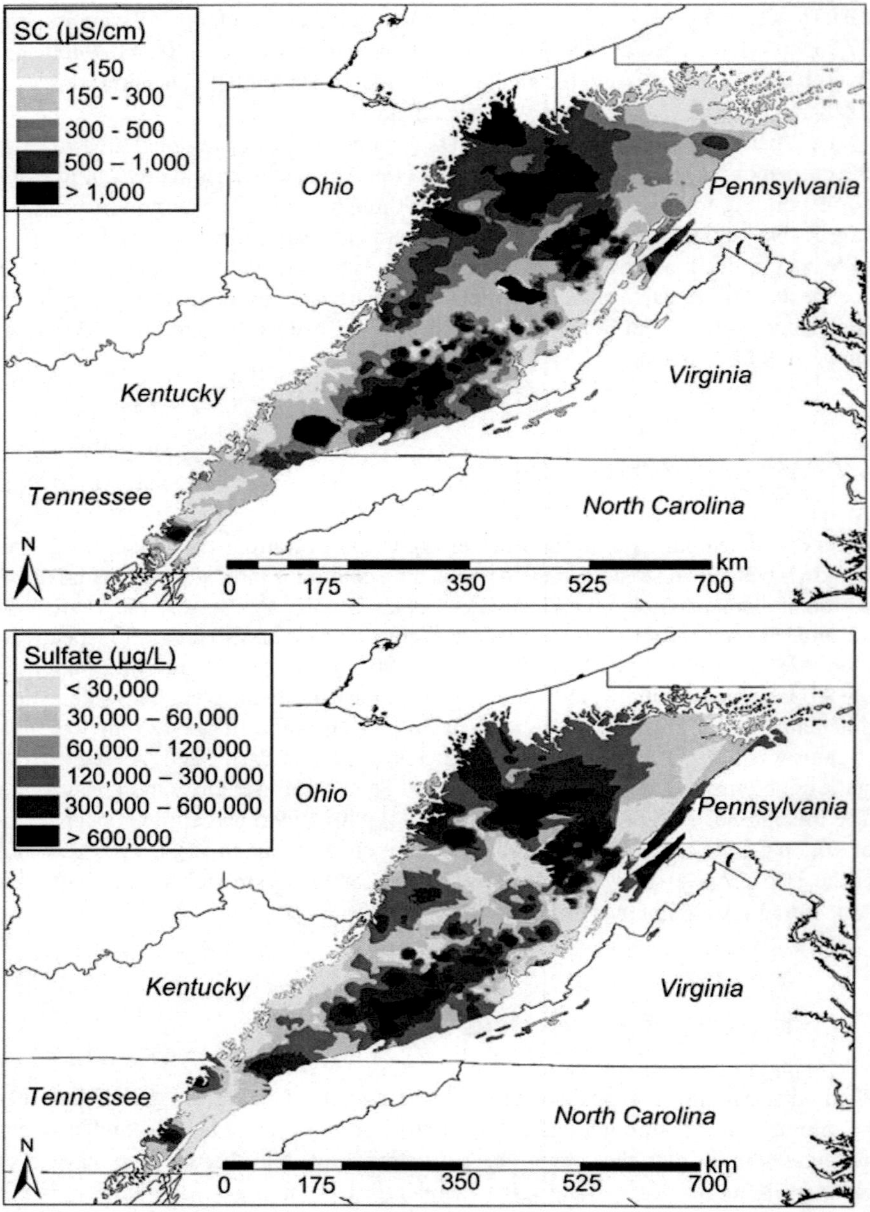

Fig. 3 Water quality conditions interpolated from 4714 water quality monitoring stations in streams within the Appalachian coalfields with <25 mi^2 watersheds, specific conductance values (μS cm^{-1}, above) and sulfate concentrations (μg L^{-1}, below)

Table 2 Dissolved minor and trace element concentration (μg L^{-1}) percentiles, comparison to reference stream data, and US EPA (2019) recommended criterion continuous concentrations (CCC)

Element[a]	*n* Stations	Percentile				Reference Stream Mean[b]	CCC (μg L^{-1})
		50th	75th	90th	99th		
Al	*3029*	226	574	1811	15,364	92.5	– [c]
Cu	*1217*	1.6	3.0	5.0	16.8	2.9	9.0[d]
Fe	*3465*	414	906	2052	14,730	176	1000[e]
Mn	*2750*	90	300	1102	5260	34.1	–
Ni	*570*	1.6	3.4	8.4	99.6	<10	52[e]
Se	*2040*	1.0	1.8	5.0	10.5	<1.5	3.1[e]
Zn	*1301*	5.0	9.5	18.1	218	10.2	120[e]

[a]The elements As, Cd, and Pb are not included due to most values below detection
[b]Data from Pond et al. (2008)
[c]US EPA Aquatic Life Criteria for Chronic Exposure, Hardness and pH-dependent, must be calculated
[d]US EPA Aquatic Life Criteria for Chronic Exposure, assuming 100 mg/$CaCO_3$ L hardness
[e]US EPA Aquatic Life Criteria for Chronic Exposure

occurring at lesser concentrations. These effects occur in association with non-acid-forming mine drainage, as well as in association with acidic drainage (Kruse Daniels et al. 2021).

Elevated TDS concentrations and SCs in mining-influenced streams are associated with changes in macrobenthic invertebrate communities, especially when SC levels are >300 μS cm^{-1}. Studies to date indicate several mayfly taxa are highly sensitive to Appalachian coal mining effects on downstream water quality, but other taxa also demonstrate sensitivity. Although multiple trace elements have been studied, most do not exceed regulatory toxicity thresholds. Field and laboratory evidence suggests Se can bioaccumulate at levels indicating potential toxicity to macroinvertebrates and to higher trophic levels in Appalachian mining-influenced streams. The scientific evidence indicates a likelihood that, in the absence of further disturbance, water quality impacts by current and past mining areas are likely to decline with time, but that time scales on the order of decades or longer will likely be required for substantial mitigation of the effects.

Supplementary Information

Data Sources for Tables 1 and 2, Figure 3:
West Virginia: West Virginia Department of Environmental Protection.
Virginia: Virginia Department of Environmental Quality.
Pennsylvania: Storet (see https://www.waterqualitydata.us/), 21PA_WQX, Pennsylvania Department of Environmental Protection

Tennessee: https://www.tn.gov/environment/program-areas/wr-water-resources/water-quality/water-resources-data-map-viewers.html
Kentucky: Kentucky Department of Environmental Protection 21KY_WQX
Ohio: Storet (see https://www.waterqualitydata.us/)

Data Processing:

1. All stations' data were converted to similar units to have a uniform data set.
2. Water quality monitoring stations in non-coal-producing counties were removed.
3. All data points before 1 January 2007 and after 31 December 2016 were removed.
4. Stations with only one water quality monitoring sample or data point were removed.
5. Water quality data from multiple sampling events at a single station were averaged for the entire time frame of 1 January 2007 to 31 December 2016.
6. For stations that did not have watershed areas, watershed area was determined using the XY coordinates and the EPA Model.
7. The complete database was imported into ArcMap for GIS analysis. Water quality concentrations were plotted using their XY coordinates and the distribution map of SC in the coalfield was produced using the Kriging Interpolation tool in the Spatial Analyst toolbox.

References

Agouridis CT, Angel PN, Taylor TJ, Barton CD, Warner RC et al (2012) Water quality characteristics of discharge from reforested loose-dumped mine spoil in eastern Kentucky. J Environ Qual 41:454–468

Arnold MC, Lindberg TT, Liu YT, Porter KA, Hsu-Kim H et al (2014) Bioaccumulation and speciation of selenium in fish and insects collected from a mountaintop removal coal mining-impacted stream in West Virginia. Ecotoxicology 23:929–938

Blowes DW, Ptacek CJ, Jambor JL, Weisener CG, Paktunc D et al (2014) The Geochemistry of Acid Mine Drainage. In: Holland HD, Turekian KK (eds), Treatise on Geochemistry, 2nd ed. Elsevier, p 131–190

Bonta JV (2000) Impact of coal surface mining and reclamation on suspended sediment in three Ohio watersheds. J Am Water Resourc Assoc 36:869–887

Bonta JV, Amerman CR, Harlokowicz TJ, Dick WA (1997) Impact of coal surface mining on three Ohio watersheds: surface-water hydrology. J Am Water Resourc Assoc 33:907–917

Bott TL, Jackson JK, McTammany M, Newbold JD, Rier S et al (2012) Abandoned coal mine drainage and its remediation: impacts on stream ecosystem structure and function. Ecol Appl 22:2144–2163

Brady KBC, Smith MW, Schueck J (eds) (1998) Coal Mine Drainage Prediction and Pollution Prevention in Pennsylvania. Pa Dep Environ Prot, Harrisburg, p 375

Buchwalter D, Scheibener S, Chou H, Soucek DJ, Elphick J (2019) Are sulfate effects in the mayfly *Neocloeon triangulifer* driven by the cost of ion regulation? Phil Trans R Soc B 374:20180013

Callaghan T, Brady K, Chisholm W, Sames G (2000) Hydrology of the Appalachian Bituminous Coal Basin. In Kleinmann R (ed) Prediction of Water Quality at Surface Coal Mines. Acid Drainage Technology Initiative. West Virginia University, Morgantown, pp 36–72

Cañedo-Argüelles M, Kefford BJ, Piscart C, Prat N, Schafer RB, Schulz C (2013) Salinisation of rivers: an urgent ecological issue. Environ Pollut 173:157–167

Caruccio FT, Geidel G (1984) Induced Alkaline Recharge Zones to Mitigate Acidic Aeeps. In: Graves DH, De Vore RW (eds) In Proc Sympos Surface Min Hydrol Sedimentol Reclam. Lexington KY. pp 27–36

Cianciolo TR, McLaughlin DL, Zipper CE, Timpano AJ, Soucek DJ, Schoenholtz SH (2020a) Impacts to water quality and biota persist in mining-influenced Appalachian streams. Sci Tot Environ 717:137216

Cianciolo TR, McLaughlin DL, Zipper CE, Timpano AJ, Soucek DJ et al. (2020b) Selenium bioaccumulation across trophic levels and along a longitudinal gradient in headwater streams. Environ Toxicol Chem 39:692–704

Clark EV, Daniels WL, Zipper CE, Eriksson K (2018a) Mineralogical influences on water quality from weathering of surface coal mine spoils. Appl Geochem 91:97–106

Clark EV, Greer BM, Zipper CE, Hester ET (2016) Specific conductance-stage relationships in Appalachian valley fill streams. Environ Earth Sci 75:1222

Clark EV, Zipper CE (2016) Vegetation influences near-surface hydrological characteristics on a surface coal mine in eastern USA. Catena 139:241–249. https://doi.org/10.1016/j.catena.2016.01.004

Clark EV, Zipper CE, Daniels WL, Keefe MJ (2018b) Appalachian coal mine spoil elemental release patterns and depletion. Appl Geochem 98:109–120

Clark EV, Zipper CE, Daniels WL, Orndorff ZW, Keefe MJ (2017) Modeling patterns of total dissolved solids release from central Appalachia, USA mine spoils. J Environ Qual 46:55–63

Conley JM, Funk DH, Buchwalter DB (2009) Selenium bioaccumulation and maternal transfer in the mayfly *Centroptilum triangulifer* in a life-cycle, periphyton-biofilm trophic assay. Environ Sci Technol 43:7952–7957

Cormier SM, Wilkes SP, Zheng L (2013a) Relationship of land use and elevated ionic strength in Appalachian watersheds. Env Tox Chem 32:296–303. 10.1002/etc.2055

Cormier SM, Suter GW, Zheng L (2013b) Assessing causation of the extirpation of stream macroinvertebrates by a mixture of ions. Environ Toxicol Chem 32:277–287

Cormier SM, Suter GW, Zheng L (2013c) Derivation of a benchmark for freshwater ionic strength. Environ Toxicol Chem 32:263–271

Cravotta CA, Brady KBC (2015) Priority pollutants and associated constituents in untreated and treated discharges from coal mining or processing facilities in Pennsylvania, USA. Appl Geochem 62:108–130

Cravotta CA, Goode DJ, Bartles MD, Risser DW, Galeone DG (2014) Surface water and groundwater interactions in an extensively mined watershed, upper Schuylkill River, Pennsylvania, USA. Hydrol Process 28:3574–3601

Daniels WL, Zipper CE, Orndorff CW, Skousen JG, Barton CD et al (2016) Predicting total dissolved solids release from central Appalachian coal mine spoils. Environ Pollut 216:371–379

Davies TD, Hall KJ (2007) Importance of calcium in modifying the acute toxicity of sodium sulphate to *Hyalella azteca* and *Daphnia magna*. Environ Toxicol Chem 26:1243–1247

DeBruyn AMH, Chapman PM (2007) Selenium toxicity to invertebrates: will proposed thresholds for toxicity to fish and birds also protect their prey? Environ Sci Technol 41:1766–1770

DeForest DK, Brix KV, Adams WJ (2007) Assessing metal bioaccumulation in aquatic environments: The inverse relationship between bioaccumulation factors, trophic transfer factors and exposure concentration. Aquat Toxicol 84:236–246

Dickens PS, Minear RA, Tschantz BA (1989) Hydrologic alteration of mountain watersheds from surface mining. Water Environ Fed 61:1249–1260

Dittman EK, Buchwalter DB (2010) Manganese bioconcentration in aquatic insects: Mn oxide coatings, molting loss, and Mn(II) thiol scavenging. Environ Sci Technol 44:9182–9188

Dulong FT, Fedorko N, Renton JJ, Cecil CB (1997) Chemical and Mineralogical Analyses of Coal-bearing Strata in the Appalachian Basin. US Geol Surv Open-File Rep 02–489

Elphick JR, Bergh KD, Bailey HC (2011a) Chronic toxicity of chloride to freshwater species: Effects of hardness and implications for water quality guidelines. Environ Toxicol Chem 30:239–246

Elphick JR, Davies M, Gilron G, Canaria EC, Lo B, Bailey HC (2011b) An aquatic toxicological evaluation of sulfate: The case for considering hardness as a modifying factor in setting water quality guidelines. Environ Toxicol Chem 30:247–253

Erickson RJ, Mount DR, Highland TL, Hockett JR, Hoff DJ et al (2017) The acute toxicity of major ion salts to *Ceriodaphnia dubia.* II. Empirical relationships in binary salt mixtures. Environ Toxicol Chem 36:1525–1537

Eriksson KA, Daniels WL (2021) Environmental implications of regional geology and coal mining in the Appalachians. Chapter 2 In: Zipper CE, Skousen JG (eds) Appalachia's Coal-Mined Landscapes. Springer

Evans DM, Zipper CE, Donovan PF, Daniels WL (2014) Long-term trends of specific conductance in waters discharged by coal mine valley fills in Central Appalachia, USA. J Am Water Resour Assoc 50:1449–1460

Evans DM, Zipper CE, Hester ET, Schoenholtz S (2015) Hydrologic effects of surface coal mining in Appalachia (USA). J Am Water Resourc Assoc 51:1436–1452

Gautier PT, Norwood WP, Prepas EE, Pyle GG (2014) Metal-PAH mixtures in the aquatic environment: a review of co-toxic mechanisms leading to more-than-additive outcomes. Aquat Toxicol 154:253–269

Goetsch PA, Palmer CG (1997) Salinity tolerances of selected macroinvertebrates of the Sabie River, Kruger National Park, South Africa. Arch Environ Contam Toxicol 32:32–41

Greer BM, Burbey TJ, Zipper CE, Hester ET (2017) Electrical resistivity imaging of hydrologic flow through surface coal mine valley fills with comparison to other landforms. Hydrol Process 31:2244–2260

Guebert MD, Gardner TW (2001) Macropore flow on reclaimed surface mines: infiltration and hillslope hydrology. Geomorphology 39:151–169

Hancock GR, Wright A, De Silva H (2005) Long-term final void salinity prediction for a post-mining landscape in the Hunter Valley, New SouthWales, Australia. Hydrol Process 19:387–401

Hawkins JW (1998) Hydrologic Characteristics of Surface-mine Spoil. In: Brady KBC, Smith MW, Schueck J (eds) Coal Mine Drainage Prediction and Pollution Prevention in Pennsylvania. Pa Dep Environ Prot, 3.1–3.11

Hester ET, Little KL, Buckwalter JD, Zipper CE, Burbey TJ (2019) Variability of subsurface structure and infiltration hydrology among surface coal mine valley fills. Sci Tot Envi 651:2648–2661

Howard JL, Amos DF, Daniels WL (1988) Phosphorus and potassium relationships in southwestern Virginia coal-mine spoils. J Environ Qual 17:695–700

Janz DM, DeForest DK, Brooks ML, Chapman PM, Gilron G et al (2010) Selenium Toxicity to Aquatic Organisms. In: Chapman PM, Adams WJ, Brooks ML, Delos CG, Luoma SN et al (eds) Ecological Assessment of Selenium in the Aquatic Environment. CRC Press, Boca Raton FL, pp 141–231

Jorgensen D, Gardner TW (1987) Infiltration capacity of disturbed soils: temporal change and lithologic control. Water Resour Bull 23:1161–1172

Kefford BJ (2019) Why are mayflies (*Ephemeroptera*) lost following small increases in salinity? Three conceptual osmophysiological hypotheses. Phil Trans R Soc B 374:2018001

Kennedy AJ, Cherry DS, Currie RJ (2003) Field and laboratory assessment of a coal processing effluent in the Leading Creek watershed, Meigs County, Ohio. Arch Environ Contam Toxicol 44:324–331

Kennedy AJ, Cherry DS, Currie RJ (2004) Evaluation of ecologically relevant bioassays for a lotic system impacted by a coal-mine effluent, using *Isonychia.* Environ Monit Assess 95:37–55

Kruse Daniels N, LaBar J, McDonald L (2021) Acid mine drainage: Sources, legacy, and treatment. Chapter 8 In: Zipper CE, Skousen JG (eds) Appalachia's Coal-Mined Landscapes. Springer

Kunz JL, Conley JM, Buchwalter DB, Norberg-King TJ, Kemble NE et al (2013) Use of reconstituted waters to evaluate effects of elevated major ions associated with mountaintop coal mining on freshwater invertebrates. Environ Toxicol Chem 32:2826–2835

Lambert DC, McDonough KM, Dzombak DA (2004) Long-term changes in quality of discharge water from abandoned underground coal mines in Uniontown Syncline, Fayette County, PA, USA. Water Res 38:277–288

Lemly AD (1993) Teratogenic effects of selenium in natural populations of freshwater fish. Ecotoxicol Environ Safety 26:181–204

Lemly AD (1996) Assessing the toxic threat of selenium to fish and aquatic birds. Environ Monit Assess 43:19–35

Linton TK, Pacheco MA, McIntyre DO, Clement WH, Goodrich-Mahoney J (2007) Development of bioassessment-based benchmarks for iron. Environ Toxicol Chem 26(6):1291–1298

Merovich GT Jr, Hitt NP, Merriam ER, Jones JW (2021) Response of aquatic biota to Appalachia's coal-mined landscapes. Chapter 10 In: Zipper CE, Skousen JG (eds) Appalachia's Coal-Mined Landscapes. Springer

Merricks TC, Cherry DS, Zipper CE, Currie R, Valenti T (2007) Coal-mine hollow fill and settling pond influences on headwater streams in southern West Virginia, USA. Environ Monit Assess 129:359–378

Messinger T, Paybins KS (2003) Relations Between Precipitation and Daily and Monthly Mean Flows in Gaged, Unmined and Valley-filled Watersheds, Ballard Fork, West Virginia, 1999–2001. US Geol Surv Water Resourc Investigations Rep 03–4113

Meyer JL, Strayer DL, Wallace JB, Eggert SL, Helfman GS, Leonard NE (2007) The contribution of headwater streams to biodiversity in river networks. J Am Water Resour as 43:86–103

Miller AJ, Zegre NP (2014) Mountaintop removal mining and catchment hydrology. Water 6:472–479

Miller J, Barton C, Agouridis C, Fogel A, Dowdy T (2012) Evaluating soil genesis and reforestation success on a surface coal mine in Appalachia. Soil Sci Soc Am J 76:950–960

Mount DR, Erickson RJ, Highland TL, Hockett JR, Hoff DJ et al (2016) The acute toxicity of major ion salts to *Ceriodaphnia dubia*: I. Influence of background water chemistry. Environ Toxicol Chem 35:3039–3057

Mount DR, Gulley DD, Hockett JR, Garrison TD, Evans JM (1997) Statistical models to predict the toxicity of major ions to *Ceriodaphnia dubia*, *Daphnia magna*, and *Pimephales promelas* (fathead minnows). Environ Toxicol Chem 16:2009–2019

Murphy JC, Hornberger JM, Liddle RG (2014) Concentration-discharge relationships in the coal mined region of the New River basin and Indian Fork sub-basin, Tennessee, USA. Hydrol Process 28:718–728

Negley T, Eshleman K (2006) Comparison of stormflow responses of surface-mined and forested watersheds in the Appalachian mountains, USA. Hydrol Process 20:3467–3483

Nippgen F, Ross MRV, Bernhardt ES, McGlynn BL (2017) Creating a more perennial problem? Mountaintop removal coal mining enhances and sustains saline baseflows of Appalachian watersheds. Environ Sci Technol 51:8324–8334

Pond GJ (2010) Patterns of *Ephemeroptera* taxa loss in Appalachian headwater streams (Kentucky, USA). Hydrobiologia 641:185–201

Pond GJ, Passmore ME, Borsuk FA, Reynolds L, Rose CJ (2008) Downstream effects of mountaintop coal mining: comparing biological conditions using family- and genus-level macroinvertebrate bioassessment tools. J N Am Benthol Soc 27:717–737

Pond GJ, Passmore ME, Pointon ND, Felbinger JK, Walker CA et al (2014) Long-term impacts on macroinvertebrates downstream of reclaimed mountaintop mining valley fills in central Appalachia. Environ Manag 54:919–933

Presser TS (2013) Selenium in Ecosystems Within the Mountaintop Coal Mining and Valley-fill Region of Southern West Virginia: Assessment and Ecosystem-scale Modeling. US Geol Surv Prof Pap 1803

Presser TS, Luoma SN (2010) A methodology for ecosystem-scale modeling of selenium. Integr Environ Assess Manag 6:685–710
Ritter JB, Gardner TW (1993) Hydrologic evolution of drainage basins disturbed by surface mining, Central Pennsylvania. Geol Soc Am Bull 105:101–115
Rodrigue JA, Burger JA (2004) Forest soil productivity of mined land in the midwestern and eastern coalfield regions. Soil Sci Soc Am J 68:833–844
Rose AW, Cravotta CA (1998) Geochemistry of coal mine drainage. In: Smith MW, Schueck J, Brady KBC (eds) Coal Mine Drainage Prediction and Pollution Prevention in Pennsylvania. Pa Dep Environ Prot, 22p
Schreck P (1995) Environmental impact of uncontrolled waste disposal in mining and industrial areas in central Germany. Environ Geol 35:66–72
Sena K, Barton C, Angel P, Agouridis C, Warner R (2014) Influence of spoil type on chemistry and hydrology of interflow on a surface coal mine in the eastern US coalfield. Water Air Soil Pollut 225:2171
Siegel DI, Smith B, Perry E, Bothun R, Hollingsworth M (2014) Pre-drilling water-quality data of groundwater prior to shale gas drilling in the Appalachian Basin: analysis of the Chesapeake Energy Corporation dataset. Appl Geochem 63:37–57
Simmons J, Currie W, Eshleman K, Kuers K, Monteleone S et al (2008) Forest to reclaimed land use change leads to altered ecosystem structure and function. Ecol Appl 18:104–118
Singer PC, Stumm W (1970) Acidic mine drainage: the rate-determining step. Science 167(3921):1121–1123
Skousen J (2017) A methodology for geologic testing for land disturbance: acid-base accounting for surface mines. Geoderma 308:302–311
Skousen J, Ziemkiewicz P, McDonald L (2019) Acid mine drainage formation, control, and treatment: approaches and strategies. Extract Ind Soc 6:241–249. https://doi.org/10.1016/j.exis.2018.09008
Skousen J, Zipper CE (2021) Coal mining and reclamation in Appalachia. Chapter 3 In: Zipper CE, Skousen JG (eds) Appalachia's Coal-Mined Landscapes. Springer
Soucek DJ (2007) Comparison of hardness- and chloride-regulated acute effects of sodium sulfate on two freshwater crustaceans. Environ Toxicol Chem 26:773–779
Soucek DJ, Dickinson A (2015) Full-life chronic toxicity of sodium salts to the mayfly *Neocloeon triangulifer* in tests with laboratory cultured food. Environ Toxicol Chem 34:2126–2137
Soucek DJ, Kennedy AJ (2005) Effects of hardness, chloride, and acclimation on the acute toxicity of sulfate to freshwater invertebrates. Environ Toxicol Chem 24:1204–1210
Soucek DJ, Linton TK, Tarr CD, Dickinson A, Wickramanayake N et al (2011) Influence of water hardness and sulfate on the acute toxicity of chloride to sensitive freshwater invertebrates. Environ Toxicol Chem 30:930–938
Soucek DJ, Mount DR, Dickinson A, Hockett JR (2018) Influence of dilution water ionic composition on acute major ion toxicity to the mayfly *Neocloeon triangulifer*. Environ Toxicol Chem 37:1330–1339
Timpano AJ, Schoenholtz SH, Soucek DJ, Zipper CE (2015) Salinity as a limiting factor for biological condition in mining-influenced Central Appalachian headwater streams. J Am Water Resour Assoc 51:240–250
Timpano AJ, Schoenholtz SH, Soucek DJ, Zipper CE (2018a) Benthic macroinvertebrate community response to salinization in headwater streams in Appalachia USA over multiple years. Ecol Indic 91:645–656
Timpano AJ, Zipper CE, Soucek DJ, Schoenholtz SH (2018b) Seasonal pattern of anthropogenic salinization in temperate forested headwater streams. Water Res 133:8–18
Tuttle ML, Breit GN, Goldhaber MB (2009) Weathering of the New Albany Shale, Kentucky: II. Redistribution of minor and trace elements. Appl Geochem 24:1565–1578
Vander Vorste R, Timpano AJ, Cappellin C, Badgley BD, Zipper CE, Schoenholtz SH (2019) Microbial and macroinvertebrate communities, but not leaf decomposition, change along a mining-induced salinity gradient. Freshwater Biol 4:671–684

Vesper DJ, Roy M, Rhoades CJ (2008) Selenium distribution and mode of occurrence in the Kanawha Formation, southern West Virginia, USA. Int J Coal Geol 73:237–249

United States Environmental Protection Agency (US EPA) (2011) A Field-based Aquatic Life Benchmark for Conductivity in Central Appalachian Streams. Office of Research and Development, Washington DC. EPA/600/R-10/023

United States Environmental Protection Agency (US EPA) (2016) Aquatic life ambient water quality criterion for selenium—Freshwater 2016. EPA 822-R-16–006. Washington DC

United States Environmental Protection Agency (US EPA) (2019) Water Quality Criteria. https://www.epa.gov/wqc

Walter H, Consolaro F, Gramatica P, Scholze M, Altenburger R (2002) Mixture toxicity of priority pollutants and no observed effect concentrations. Ecotoxicology 11:299–301

Wang N, Dorman RA, Ingersoll CG, Hardesty DK, Brumbaugh WG et al (2016) Acute and chronic toxicity of sodium sulfate to four freshwater organisms in water-only exposures. Environ Toxicol Chem 35:115–127

Wang N, Kunz JL, Dorman RA, Ingersoll CG, Steevens JA et al (2018) Evaluation of chronic toxicity of sodium chloride or potassium chloride to a unionid mussel (*Lampsilis siliquoidea*) in water exposures using standard and refined toxicity testing methods. Environ Toxicol Chem 37:3050–3062

Wiley JB, Evaldi RD, Eychaner JH, Chambers DB (2001) Reconnaissance of stream geomorphology, low streamflow, and stream temperature in the Mountaintop Coal-Mining Region, Southern West Virginia, 1999–2000. Water-Resources Investigations Report 01-4092. US Geological Survey, Charleston, West Virginia

Whitmore KM (2016) Selenium dynamics in headwater streams of the central Appalachian coalfields: An investigation of enrichment and bioaccumulation. MS thesis, Virginia Tech, Blacksburg, VA, USA

Whitmore KM, Schoenholtz SH, Soucek DJ, Hopkins WA, Zipper CE (2018) Selenium dynamics in headwater streams of the Central Appalachian coalfield. Environ Toxicol Chem 37:2714–2726

Wunsch DR, Dinger JS, Taylor PB, Carey DI, Graham CDR (1996) Hydrogeology, Hydrochemistry, and Spoil Settlement at a Large Mine-spoil Area in Eastern Kentucky: Starfire Tract. KY Geol Surv Rep Investigations 50

Ziemkiewicz PF, O'Neal M, Lovett RJ (2011) Selenium leaching kinetics and in situ control. Mine Water Environ 30:141–150

Response of Aquatic Life to Coal Mining in Appalachia

George T. Merovich Jr., Nathaniel P. Hitt, Eric R. Merriam, and Jess W. Jones

Abstract Rivers and streams of Appalachia harbor some of the most biologically diverse, endemically rich, freshwater faunas in the world. Many aquatic species, however, are threatened by a wide array of anthropogenic stressors, including past and present mining. This chapter reviews impacts of mining on aquatic life in Appalachia, including interactive effects on food web dynamics, effects on ecosystem functional processes, and effects that are manifested in aquatic communities at watershed scales. Responses to mining depend on whether mine effluent is acidic in nature or highly alkaline; this dichotomy structures much of the discussion. Mining directly impacts aquatic biota because effluents can be toxic due to elevated major ions, trace metals, notably selenium, and/or acidity. Drainage from mined areas also affects complicated changes to biota indirectly by altering food webs, biotic interactions, and movement processes in stream networks. Finally, we review the response and recovery of aquatic biota to remediation efforts that treat mine effluent, and we discuss the use of aquatic organisms in assessing and managing impacts of mining activities. Throughout, we present uncertainties and gaps in our knowledge about impacts of mining activity on aquatic life in Appalachia, which represent important directions for future research.

Keywords Mountaintop removal · Mine drainage · Fishes and salamanders · Mussels and crayfish · Stream ecosystem function · Recovery of freshwater biota

---

G. T. Merovich Jr. (✉)
Juniata College, Huntingdon, Pennsylvania, USA
e-mail: merovich@juniata.edu

N. P. Hitt
U.S. Geological Survey, Leetown Science Center, Kearneysville, West Virginia, USA

E. R. Merriam
U.S. Army Corps of Engineers, Pittsburgh, Pennsylvania, USA

J. W. Jones
U.S. Fish and Wildlife Service, Blacksburg, Virginia, USA

C. E. Zipper and J. Skousen (eds.), *Appalachia's Coal-Mined Landscapes*,
https://doi.org/10.1007/978-3-030-57780-3_10

1 Introduction

Appalachia is one of the most biologically diverse regions of the world, with freshwater faunal richness higher than any other temperate region (Jelks et al. 2008). For example, Appalachia represents one of two known hotspots for crayfish diversity in the world (Crandall and Buhay 2008) and also has the highest rate of endemism for fishes in North America (Warren et al. 2000). Salamander and amphibians are exceptionally diverse (Kozak and Wiens 2010), and Appalachia contains the majority of the world's mussel diversity (Mynsberge et al. 2009). However, freshwater ecosystems are also among the most threatened ecosystems in the world (Abell et al. 2008). This is particularly true in Appalachia, where unique natural resources are threatened by landscape and water quality changes brought about by numerous activities including coal mining (Bernhardt and Palmer 2011).

Coal mining, through various effect pathways, has considerable negative impacts on the chemical and physical conditions of receiving streams (Kruse Daniels et al. 2021; Clark et al. 2021). Organisms, communities, and trophic interactions within impacted watersheds change significantly as a consequence of coal mining. Mining-influenced streams affect terrestrial organisms that use water directly as a resource and affect dispersal dynamics by modifying animal movement and behavior along watercourses that are travel corridors (Lituma et al. 2021). This chapter describes the response of aquatic organisms to mining—life forms that use aquatic environments directly as habitat for at least part of their life cycle. Appalachian surface coal mining, and in some cases underground mining, influences water and aquatic biota during active operations, and for years and decades following mine closure (Kruse Daniels et al. 2021; Clark et al. 2021). The discussion focuses entirely on lotic systems (streams and rivers), the primary aquatic ecosystems impacted by coal mining throughout Appalachia.

We begin by describing the effects of coal mining and mined lands on primary producers (mostly algal) and microbial heterotrophs (bacterial and fungal components) because these life forms often are linked closely within biofilm (periphyton) on rocks on stream bottoms. We then address secondary production (animals) and examine impacts on macroinvertebrates (i.e., insects and non-insect invertebrates such as crayfishes and mussels), salamanders, and fish. We look at the direct effects of mining on each group of organisms, including chemical toxicity of mining effluent and physical habitat alteration due to mining. Chemical toxicity is discussed along three pathways that include (1) elevated trace metal concentrations (typically Fe, Al, and Mn) that have increased solubility in acidic water, (2) elevated major ion concentrations including dissolved cations (e.g., Ca, Mg, Na, K) and anions (e.g., HCO_3^-, chloride, SO_4^{2-}) that increase salinity, and (3) elevated trace element concentrations that behave independently of pH, most importantly Se. Major ions and trace elements often are measured collectively as total dissolved solids (TDS) or as specific conductance (Clark et al. 2021). Physical habitat alteration occurs primarily as a result of two processes that include (1) sediment deposition resulting from erosional processes occurring with mining activity on the landscape, and (2) precipitation of metal hydroxide flocculants (floc) as a result of decreased solubility of acid-soluble metals (particularly Fe and Al) when acidic drainage is neutralized by dilution or interaction with alkaline waters, such as a receiving stream. Precipitates may also form when metals in highly saline mine waters emerge from valley fill underdrains or groundwater seeps and experience changes in solubility due to changes in atmospheric exposure. Stream bottoms become physically covered by deposited sediments, and, in the case of flocs, are often armored or cemented in an immovable, impenetrable barrier to benthic organisms.

Direct impacts of mining on aquatic life are decreased biodiversity and simplified community structure when sensitive species are lost. These changes link coal mining activity to changes in trophic interactions, ecosystem function, and animal dispersal processes indirectly. Consequently, mining and mined lands may have effects that extend beyond impacted streams and affect ecological conditions at watershed scales. Understanding these processes has important implications for prioritizing remediation efforts.

The specific nature of direct chemical toxicity and physical habitat alteration as a result of mining activity often aligns with the types of mining practices employed. Surface mining (Skousen and Zipper 2021), including mountaintop removal/valley fill mining (MTR/VF), constitutes the dominant source of land use change in Appalachia (Pericak et al. 2018) and typically produces alkaline mine drainage high in major ions and trace elements. Moreover, crushed rock from removal techniques is deposited in adjacent valleys, burying streams in valley fills (VF). Underground mining may also produce waters elevated in TDS, although in some cases with differing chemical character. Surface and deep mining techniques both may produce acid mine drainage (AMD; Kruse Daniels et al. 2021) with high concentrations of various toxic trace metals, occurring most commonly in northern Appalachia. Acidic drainages from active mines were more common before federal regulations required preventive practices (i.e., Surface Mining Control and Reclamation Act of 1977), but many acidic discharges from past mining remain evident as mining's legacy.

Response and recovery of aquatic life to remediation of mining impacts is complex and requires better scientific understanding. We devote a portion of the chapter to management considerations that discuss these complexities and consider other land use stressors that may interact with mining impacts. Management considerations include a review and description of biotic indices and methods that have been developed using aquatic life to detect and quantify mining influence in terms of toxicity or ecosystem-level biological integrity. We discuss these concepts in light of opportunities for research that will improve understanding of biotic response to mining for informed management and conservation of aquatic resources in this landscape.

2 Biofilm

Biofilm, or periphyton, is an organic microlayer of algal, bacterial, and fungal cells attached to epilithic (rock) and other surfaces in aquatic ecosystems (Allan 1995). Thus, biofilms contain both autotrophs and heterotrophs. In shaded streams lacking macrophytes (aquatic plants) as in most Appalachian headwaters, algae in biofilm known as diatoms are the predominant primary producers, but bacteria (cyanobacteria) are also important, as are other macro- and filamentous green algae in waters with elevated nutrients and adequate light. Autotrophs in biofilm fix inorganic carbon into organic matter, a resource that readily supplies energy to higher trophic levels. Although quantities of organic material produced in biofilm are generally far lower than those entering streams from riparian vegetation (such as fallen leaves), the organic material in biofilm is often of higher nutritional quality and thus plays an important role in aquatic food webs (Cashman et al. 2013). Autotrophs provide an essential source of particulate and dissolved organic carbon for heterotrophs (bacteria and fungi), which are essential to nutrient cycling in benthic and hyporheic zones (bottom sediments) of rivers (Smucker and Vis 2011).

2.1 *Chemical Toxicity*

Direct impacts of mining to periphyton and macroalgae stem from chemical toxicity of mining effluents. Under acidic conditions, acid-soluble trace metals are elevated in concentration to harmful levels. These metals can block protein function or modify protein and cell membrane structure of algal cells. Metals may also displace or compete with similarly structured essential metals needed for various metabolic reactions, and consequently disrupt cell functions (Das et al. 2009).

In non-acidic mine effluents, major dissolved ions that elevate salinity can be harmful, but mechanisms and patterns of toxicity are less well understood, perhaps due to particular solute compositions studied. In general, high salinity can cause osmo-regulatory imbalances and ionic stress in autotrophs (Bray et al. 2008). For example, hyper-salinization of waters with sodium chloride (NaCl) in the laboratory depresses photosynthesis in cyanobacteria by inactivating electron transport in photosynthesis and cellular respiration (Allakhverdiev et al. 2000).

Laboratory studies with NaCl indicate that salinity levels observed in typical Appalachian streams likely would not induce osmotic and ionic stress on periphyton. For example, Cook and Francoeur (2013) observed moderately reduced photosynthetic efficiencies by lotic periphyton in waters with short-term NaCl exposures to conductivities as low as 4428 μS/cm, relative to natural water with conductivity of 1609 μS/cm. Costello et al. (2018) found that moderately elevated NaCl concentrations (up to approximately 10 times the ambient reference of 80 mg/L chloride) had no substantive effect on periphyton photosynthetic rates; but further increases of

NaCl produced moderate photosynthetic-rate increases (approximately 50% at 100 times ambient level).

The bicarbonate ion, however, is a major contributor to salinity in mining effluent in Appalachia, and in some cases the dominant contributor in waters discharged from older mined lands. This anion is important because it can serve as an inorganic carbon source for algae. For example, Silva et al. (2000) found that chlorophyll-*a* content of algae on artificial substrates was enhanced at moderate salinities of a complex saline solution (260–1000 mg/L, with bicarbonate as the dominant anion) but depressed at higher salinities (~2260 mg/L, with chloride as dominant). Thus, the supply of bicarbonate in mining effluents may increase algal production. More study is needed to establish the link between salinity and primary production in in situ mining effluent and to understand its implication for consumers. Nevertheless, Nielsen et al. (2003) reported that most freshwater algae do not tolerate salinity over 10,000 mg/L. Furthermore, since chemical toxicity is concentration- and ion-dependent (DeNicola and Stapleton 2002), management of mining impacts will require assessment of specific ions rather than conductance or salinity as a surrogate for solute content (Simmons 2012).

2.2 *Physical Impacts*

Physical effects of mining on biofilm and primary production may include direct impacts from precipitation of metal hydroxides (Smucker and Vis 2011). Precipitates armor benthic substrates and smother the biofilm community preventing growth (Hogsden et al. 2013; Smucker et al. 2014). Precipitates may be chemically benign to aquatic biota compared to dissolved metals in the water column (DeNicola and Stapleton 2002). Precipitates also decrease dissolved nutrient availability through nutrient adsorption, particularly phosphorus that binds to Fe and Al hydroxides. Nutrient loss in this manner limits periphyton production and adds an additional stressor to microbial growth and thus consumer trophic levels (DeNicola and Lellock 2015). For example, Smucker and Vis (2011) found increases in activity of algal phosphatase (an enzyme used to acquire organically bound sources of phosphate) with AMD impact, indicating phosphorus limitation in AMD streams.

Altered sediment and flow regimes within mining-impacted streams are direct, physical stressors to biofilm. Excessive supply of fine sediments released from surface mining landscapes buries stream bottoms (Griffith et al. 2012), creating unstable surfaces and clogging interstitial habitats (Mueller et al. 2013). There is currently limited understanding of how, and the extent to which, sustained flows from MTR/VF-influenced watersheds (Nippgen et al. 2017) interact with chemical changes and altered instream characteristics such as sediment size distribution (Griffith et al. 2012) to influence biofilm communities, primary production, and nutrient cycling.

2.3 Community-Level Effects

Typical responses of biofilm communities to mining-induced acidification include reduced species richness, diversity, and biomass (Bray et al. 2008), and a shift in biofilm community composition (Luís et al. 2009). Heavily impacted streams are dominated by assemblages of tolerant, acidophilic taxa. Food web complexity declines (Smucker et al. 2014) and biofilm communities also become less dynamic spatially and temporally as AMD increases, likely because severe conditions limit niche opportunities (Smucker et al. 2014).

Overall, primary production has been shown to decline in mining-acidified streams (Das et al. 2009). In acidic waters, $pH < 3.5$ was found to be a threshold for algal richness loss (Smucker et al. 2014). In severe acidic drainage where precipitates are absent, growth and biomass of a few acid-tolerant taxa have been observed to be high, approaching levels of un-impacted streams. Tolerant algae and fungi can form metal complexes in extracellular spaces that prevent cellular entry of metals, thus rendering them resistant to metal toxicity (Das et al. 2009). Consequently, growth, abundance, and biomass in algal assemblages have been observed to respond to AMD in a U-shaped pattern, attaining high levels in severe AMD (where toxic acid-soluble metals remain in solution) and in un-impacted reference streams, and falling to low levels in moderately acidified streams where flocculates form, smother algae, and hinder growth (Smucker et al. 2014). Dissolved metals may supply some micronutrients needed for growth of acid-tolerant algal species in severe AMD (Das et al. 2009). The absence of primary consumers such as grazing insects and/or the unpalatability of acid-tolerant algae may also partly explain the positive growth response with severe AMD (Smucker et al. 2014). For example, Drerup and Vis (2016) found AMD alters phospholipid fatty acid profiles in biofilm to profiles considered poorer quality food for consumers that cannot themselves synthesize essential fatty acids.

As biofilms grow in streams receiving mine effluent, they may become a concentrated source of toxic metals due to microbial uptake of elements including Se, Cd, Mn, Zn, and Ni from the water column (Bray et al. 2008; Das et al. 2009; Bier et al. 2015). Selenium, for example, is an essential micronutrient for aquatic algae; hence, those algae remove Se from the water column and bioconcentrate Se (Janz et al. 2010). Selenium in biofilms of mining-influenced streams, for example, can range as high as 10 $\mu g\ g^{-1}$ dry weight, approximately 1000 times higher relative to streams draining unmined watersheds (Arnold et al. 2017; Whitmore et al. 2018). Primary consumers can transfer trace elements such as Se from biofilm to higher trophic levels where they bioaccumulate (Das et al. 2009). Such processes have been noted for Se in mining-influenced streams in Appalachia (Presser 2013; Arnold et al. 2017; Whitmore et al. 2018) and for other trace elements that occur commonly in mining effluent elsewhere (e.g., DeForest et al. 2007). Conley et al. (2009), for instance, analyzed periphyton cultures spiked with Se and found that biofilms readily adsorb Se, providing a pathway for uptake in consumers. Using the mayfly *Neocloeon triangulifer*, they reported reduced postpartum fecundity and maternal transfer of dietary Se to eggs. Thus, contaminated biofilms might impart a bottom-up control on mayflies, other scraper taxa, and fishes (Sect. 7), but this requires further testing.

3 Macroinvertebrates

Aquatic macroinvertebrates play a critical role in stream food webs and nutrient cycling, serving as the main processors of organic material in rivers. Macroinvertebrates are essential primary consumers in most freshwater ecosystems, including Appalachian headwaters, where they convert autotrophic biomass into organic matter forms that can be consumed by higher-trophic-level fauna. Macroinvertebrate taxa are assigned to feeding guilds called functional feeding groups (FFGs) based on traits related to mouthpart morphology (and consequently their food types). Key FFGs include shredders (coarse particulate leaf detritus), scrapers (periphyton), filtering and gathering collectors (fine particulate detritus), and predators (prey). The River Continuum Concept (Vannote et al. 1980) predicts a set of longitudinal physical, chemical, and biological changes from headwaters to large-river reaches, including changes in the relative abundance of FFGs. Coarse particulate organic matter (i.e., woody debris and leaf litter) represents the base of the food web in small headwater streams, and macroinvertebrate shredders process up to 73% of the riparian leaf litter entering these systems (Covich et al. 1999). Further downstream, primarily grazers consume algae and periphyton, and collectors consume fine particulate organic matter from the water column and/or benthic sediments. Individuals convert this material into biomass (i.e., secondary production), supplying food to predatory invertebrates and vertebrates (e.g., fish, amphibians, and birds) occupying higher trophic levels.

3.1 Chemical Toxicity

Much of the empirical evidence regarding causal mechanisms for loss of macroinvertebrate taxa throughout the MTR/VF coal mining region points to increases in major ions resulting from mining-spoil leachates and related sources. Appalachian macroinvertebrates evolved in chemically dilute waters, and their exposure to increased salinity may cause physiological stress and/or mortality due to difficulty maintaining internal ion concentrations (Cormier et al. 2013a; Kefford 2019). Elevated HCO_3^- and SO_4^{2-} ions in mined-land discharge may interfere with uptake of chloride and other necessary ions through gill epithelia (Cormier et al. 2013a; Kefford 2019). However, given the importance of specific ionic mixtures and the large number of potential physiological pathways to lethal and sub-lethal responses, further research is needed to better understand mechanisms of toxicity, which may vary depending on specific ions present.

Many early studies investigating the toxicity of mining-related ions were conducted with standard test organisms (e.g., *Ceriodaphnia dubia*, *Daphnia magna*, and *Hyalella azteca*). These studies documented and began to characterize complexities [e.g., effects of differing ion mixtures (Kennedy et al. 2005; Soucek 2007b)] in the lethal and sub-lethal effects [e.g., declines in feeding rates and oxygen consumption (Soucek 2007a)] of variable mining-related ionic mixtures on a variety of macroinvertebrate species. However, standard test species are generally not relevant to Appalachian streams and thus have limited utility when attempting to establish management criteria meant to protect benthic communities.

An increasing number of studies characterize the toxicity of mine effluent to regionally relevant taxa—predominantly Ephemeroptera (mayfly) species (i.e., *Isonychia bicolor*, *N. triangulifer*) given their well-documented sensitivity to ionic stress—and compare these results to those obtained for standard organisms. Mayfly taxa consistently show more sensitivity to elevated ionic concentrations as compared to the various test species. For example, the lowest observable effects concentration to elevated water conductivity due to a high-sodium mining effluent was more than twice as high for *C. dubia* (3730 μS/cm) than for the mayfly *I. bicolor* (1562 μS/cm) (Kennedy et al. 2004). In a similar study, mayflies (*I. bicolor*) were more sensitive to conductivity from a high-sodium coal processing effluent (lowest observable effect concentrations ranging from 1508 to 4101 μS/cm) as compared to *C. dubia* (2132–4240 μS/cm) (Echols et al. 2010). Kunz et al. (2013) documented variable toxicity of reconstituted salt mixtures from three MTR/VF-impacted streams to the mayfly *N. triangulifer* that was highly dependent on ionic composition, corroborating previous work characterizing contaminant-specific responses in test organisms (Kennedy et al. 2005; Soucek 2007b).

A number of field studies demonstrate the loss of sensitive taxa in response to elevated ionic concentrations and associated specific conductances in mining effluents, with mayflies consistently exhibiting the greatest sensitivity (Pond et al. 2008; Timpano et al. 2018). Cormier et al. (2013a) estimated extirpation concentrations of specific conductance for 163 genera and used these values to develop an aquatic life benchmark (i.e., 300 μS/cm) expected to prevent local extirpation of 95% of taxa in waters with ionic signatures dominated by HCO_3^- and SO_4^{2-}. This benchmark is consistent with results of other field studies documenting cumulative declines in overall community diversity and richness, as well as Ephemeroptera taxa and associated bioassessment metrics (Sect. 9), at or below conductivities of 300 μS/cm (Pond 2010; Timpano et al. 2018). Genus-level extirpation criterion developed by Cormier et al. (2013a) for *Neocloeon* at 1092 μS/cm (then classified as *Centroptilum*; Jacobus and Wiersema 2014) and *Isonychia* at 1180 μS/cm are also generally consistent with laboratory toxicity tests conducted on *N. triangulifer* between 800 and 1300 μS/cm (Kunz et al. 2013) and *I. bicolor* at 1562 μS/cm (Kennedy et al. 2004). However, these values are higher than the overall aquatic life benchmark of 300 μS/cm. These results suggest *N. triangulifer* and *I. bicolor* may not be representative of overall community tolerance, highlighting the value of integrating field and laboratory studies to understand toxicity and community response (Buchwalter et al. 2017).

Community response in AMD-impacted streams is predominantly driven by large shifts in pH and increases in concentrations of dissolved trace metals that impair regulation of ions and metabolically active metals (Rainbow 2002). Toxic effects of AMD have been observed in streams with episodic (MacCausland and McTammany 2007) and chronic (Battaglia et al. 2005) chemical degradation—both of which are extremely common throughout mined watersheds in Appalachia (Petty and Barker 2004).

3.2 *Physical Impacts*

Water quality—both acid and alkaline mine drainage—is generally considered the dominant driver of altered macroinvertebrate communities within mining-impacted systems (Pond et al. 2008; Merriam et al. 2011). However, physical habitat degradation also impacts invertebrate communities throughout the Appalachian coalfields. Mountaintop mining is associated with increased fine sediments—a pattern attributed to both valley fills and altered hydrologic regimes and associated increases in stream bank erosion (see Griffith et al. 2012). Increased fine sediments and embeddedness downstream of MTR/VF operations [reviewed by Griffith et al. (2012)] correlate negatively with invertebrate community condition (Pond 2004). Fine sediments can alter benthic macroinvertebrate communities via a number of pathways, including physical (e.g., abrasion and burial of individuals) and chemical impacts (e.g., oxygen limitation and sediment toxicants), altered food (e.g., algal and periphyton growth) and nutrient availability (e.g., decreased exchange between surface and hyporheic benthic habitats), and habitat loss (e.g., filling of interstitial spaces) (Jones 2015). In AMD-impacted streams with pH >3.5 where metals are insoluble, metal hydroxides form and can cover gills, smothering respiratory surfaces (Soucek et al. 2000). Precipitation of metal hydroxides alters physical habitat quality in streams impacted by AMD, further impacting macroinvertebrate community structure (Hogsden and Harding 2012). Metal hydroxides coat the stream bottom, armoring sediments together, which reduces potential for animal colonization and exchange of nutrients between surface waters and hyporheic (sub-stream bed) zones.

4 Mussels

Freshwater mussels are benthic macroinvertebrates but are considered separately because of their unique life history and special conservation needs. Mussels are bivalve mollusks. They burrow and secure themselves in gravel, sand, and silt substrates, where they carry out important roles such as nutrient cycling, water filtration, and modification of stream sediments that may benefit other organisms (Vaughn 2018). They are long-lived, reaching maximum ages of 20–50 years or more depending on the species. Mussels require clean, unpolluted water and stable substrates (i.e., not comprised of loose or unconsolidated materials) to feed, grow, and reproduce (Jones 2015). Mussels feed by filtering suspended organic particles, <20 μm in size, from the water column, but can also filter deposited particles through the shell-gap when burrowed in the stream bottom (Strayer et al. 2004). The organic particles mussels consume are coated by biofilm, which is an additional important food source. Mussels reproduce in a specialized way that requires fish to serve as hosts to their larvae. Female mussels release larval offspring, called glochidia, into the water. Glochidia attach and encyst as an ecto-parasite on gills and fins of a suitable fish host. The fish host therefore provides larvae with long-distance dispersal capabilities and allows them to metamorphose to the juvenile stage. Juveniles then excyst and fall away from their fish host at a small size, typically <100–300 μm, and settle into the stream bottom to grow into adults (Jones 2015).

4.1 Distribution and Conservation Status

Appalachia contains some of the most diverse and endemic mussel resources in the United States, especially in the upper Ohio, Cumberland, Tennessee, Tombigbee, and Alabama river basins—river systems that drain parts of Pennsylvania, Ohio, Kentucky, West Virginia, Virginia, Tennessee and Alabama. Many streams in these watersheds occur in areas affected by coal mining, and in some cases, contain globally significant mussel faunas that are in decline or are otherwise imperiled (Jones 2015).

Currently, mussels are the most imperiled group of animals in the United States, with nearly 75% of the 298 species occurring in North America listed at the federal or state levels as extinct, endangered, threatened, or of special concern (Williams et al. 2017). Various anthropogenic impacts have played a role in their decline, including hydrological alterations to streams from dams, sedimentation from agricultural and urban landscapes, dredging of stream bottoms, water pollution from industrial and agricultural sources, toxic spills, and chemical and physical alterations to streams from mining (Neves et al. 1997). Thus, an understanding of the impacts of mining on mussels is essential for effective conservation.

While mussel biodiversity is high in Appalachia, it is patchy at the stream scale. Generally, 1st through 3rd order headwater streams contain fewer species than larger 4th through 6th order rivers, but headwater streams can harbor rare and endemic mussels. Streams that drain portions of the Appalachian Plateau in the upper Ohio, Cumberland and Tennessee River basins in the states of Kentucky, Tennessee, Virginia, and West Virginia contain many areas of conservation concern for mussels. For example, 49 extant mussel species occur in the Clinch River and its tributary the Powell River of southwestern Virginia and northeastern Tennessee, where 24 species are listed as federally endangered, the highest number of endangered mussels of any river system in the US (Ahlstedt et al. 2016; Jones et al. 2018). The Big South Fork Cumberland River and its tributary the Little South Fork of northeastern Tennessee and southeastern Kentucky is another hotspot for rare and endangered mussels, where

26 species occur, six of which are listed as federally endangered (Warren and Haag 2005).

Mining has been identified as the primary suspected cause of mussel declines in both the Clinch/Powell and Big South Fork/Little South Fork Cumberland River systems, affecting populations not only in small headwater streams but also in medium-sized to large-river systems where mussel biodiversity is greatest (Zipper et al. 2014; Rogers et al. 2018). Streams further south in the Appalachian Mountain chain in northern Alabama, such as the Mulberry and Locust Forks of the Black Warrior River (Tombigbee basin) contain mussel faunas in watersheds affected by coal mining and have declined greatly since the 1970s after initiation of mining (McGregor et al. 2013). Freshwater mussels also inhabit northern Appalachian streams. For example, more than 50 mussel species inhabit the upper Ohio River basin in western Pennsylvania and in West Virginia where coal mining has been prevalent for decades (Ortmann 1909).

4.2 Chemical and Physical Effects

The complex life history traits of mussels, such as dependency on fish to metamorphose their larvae, a small sensitive juvenile stage, filter-feeding, and long-lived bottom-dwelling adults, make them especially vulnerable to chemical and physical changes in rivers induced by mining activity (Strayer et al. 2004; Jones 2015). Most apparent is how mining may affect mussels through fishes, because reduced fish diversity and abundance will weaken the capacity of mussels to reproduce and disperse (Sect. 7). Also important is that large-scale landscape alterations associated with surface mining result in increased transport and deposition of fine sediments in streams. This material buries natural stream bottoms where mussels live.

Recent field and laboratory studies have shown that increased concentrations of mining-related contaminants further impact mussel viability (Jones 2015). Field studies conducted in the Clinch River within Tennessee and Virginia (Rogers et al. 2018) and the Powell River (Phipps 2019) documented increased levels of major ions and multiple trace metals in surface and interstitial waters in areas where mussel populations have declined. In the laboratory, major ions, trace elements (i.e., Al, Cu, Fe, Mn Se, Zn), and other metals (Griffith et al. 2012; Cormier et al. 2013b) are directly toxic to juvenile mussels (Kunz et al. 2013). High concentrations are suspected to disrupt osmoregulation and feeding (Gillis 2011; Wang et al. 2018). Moreover, Rogers et al. (2018) reported that elevated major ions and trace metals in a mining-affected section of the Clinch River was correlated with decreased survival and growth of caged *Villosa iris* despite low concentrations of contaminants. However, laboratory studies have shown contrasting results. For example, Ciparis et al. (2015) saw no effects of major ion mixtures on *Villosa iris* at concentrations similar to mining-impacted portions of the Clinch and Powell Rivers. Further research that combines lab- and field-based investigations into complex contaminant mixtures are needed to identify reasons for survival and growth declines in mussels (Rogers et al. 2018). These findings may shed light on the reasons for reduced recruitment and shifts to older age classes that population assessments have indicated.

4.3 Community-Level Effects

Although the search for specific contaminants responsible for mussel population declines continues (Zipper et al. 2014), monitoring data indicate that mining-impacted streams are associated with simplified mussel assemblage composition. This finding is driven by reduced richness of rare and sensitive taxa and by dominance of a few tolerant ones.

Poor quality food resources may explain this shift. Mining can decrease the richness and abundance of species comprising the biofilm food resource for mussels. Furthermore, trophic transfer of bioaccumulated mining contaminants from biofilms to mussels may be responsible (Sect. 2). Thus, toxicants in mine-influenced water may decrease the food quality and increase exposure of mussels to food contaminants, which may ultimately impair the physiological condition of adults and juveniles, thereby impairing reproduction and recruitment. More research is needed to elucidate the impacts coal mining has on mussels that translates to mussel communities as simplified versions of what occurs without mining impacts. Routine monitoring of mussel assemblages in watersheds of conservation concern where mining, or recovery from mining impact, is occurring will aid mussel management and conservation efforts. Key elements of a monitoring program for mussels should include quantitative assessment of populations at regular time intervals, and assessment of contaminant levels in surface and interstitial waters of the stream substrates where mussels live (Phipps 2019).

5 Crayfishes

Like freshwater mussels, crayfish diversity and endemism in Appalachia is exceptional. Extensive adaptive radiation has led to high regional diversity, but stream-level diversity is normally low, often with only one or two species. In streams, crayfish perform numerous ecosystem functions. They are omnivores, feeding on nearly any type of energy source available including coarse particulate organic matter (e.g., leaves), dead animals, macrophytes, and fish eggs. They are predators on benthic insect larvae and fish, and may compete for food resources with these organisms as well, while providing food sources for predatory fishes. They also can significantly modify inorganic materials through physical (e.g., burrowing) action (Creed and Reed 2004).

Crayfishes are highly imperiled globally and in the US. Small natural ranges play a major role in the imperilment of the group. Endemic species with limited natural ranges are particularly vulnerable to extirpation from catastrophic disturbance events, including catchment-wide habitat degradation typically found in surface mined landscapes (Welsh and Loughman 2015). Recent research on crayfish conservation ecology, distribution, systematics, and taxonomy has recognized a number of new species and imminent threats to crayfish diversity in mining regions of Appalachia (Loughman et al. 2017a). For example, the recently described Tug Valley crayfish (*Cambarus hatfieldi*) and Conhaway crayfish (*C. appalachiensis*) occur in restricted drainages within the MTR/VF region that makes them vulnerable to mining impacts (Loughman et al. 2013, 2017b). Similarly, the Guyandotte River crayfish (*C. veteranus*) and the Big Sandy crayfish (*C. callainus*), both recently identified and endemic to a restricted range in Appalachia, have been classified as federally endangered and threatened, respectively, due to declining populations in areas impacted by surface mining. Recently, areas of critical habitat needed to protect these two species have been identified (USFWS 2019a; USFWS 2019b).

Key threats to crayfishes stemming from coal mining include elimination of headwaters via stream burial, scouring and sedimentation in downstream habitats from flooding, and water chemistry effects. Mining is likely responsible for the decline and extirpation of various crayfishes such as the Guyandotte River crayfish in southern West Virginia and adjacent areas (Jones et al. 2010; Loughman and Welsh 2010). Welch and Loughman (2015) modeled species occurrence with abiotic factors in southern West Virginia and found that species varied considerably in their relationship to physical habitat and salinity gradients. Species in the genus *Cambarus* tended to be more sensitive than *Faxonius* (formerly *Orconectes*) to salinity. Similarly, Peake et al. (2004) reported distinctive environmental drivers of crayfish distribution in the Kentucky coalfields. *Faxonius* was more associated with mining and residential land uses, while *Cambarus* preferred natural catchments. In a survey of the Guyandotte River crayfish, Loughman et al. (2017a) modeled environmental factors associated with presence/absence and found physical habitat and conductivity were the strongest predictors. They hypothesized that population declines via mortality or reduced fitness is linked to high SO_4^{2-} or other major ions from coal mining. In contrast, the presence of the Big Sandy crayfish was found to be influenced by sedimentation and not water quality. Whitmore et al. (2018) reported that crayfish accumulate Se to a lesser degree than other macroinvertebrates, but the implications of this accumulation in terms of toxicity and trophic transfer along the food chain remain unknown.

In streams influenced by AMD, crayfish abundances are typically reduced by acidic conditions, metal hydroxide precipitates, and armoring of substrates that prevent burrowing and reduce food availability (Dillon and Lee 2002). Response to acidity varies by genus. Berril et al. (1985) found *Cambarus* less sensitive than *Faxonius*. Elevated Al concentrations did not affect survival. Crayfishes can persist in mild levels of AMD, but timing of life stages with severity of AMD is important. For example, juveniles and adults in ecdysis (molting) are sensitive to elevated acidity and conductivity (Hartman et al. 2010). Species that do persist in AMD-impacted streams may flourish because of the absence of fish predators, despite the absence of benthic refugia due to substrate armoring (Loughman and Welsh 2010). Shifts in resource availability may also influence crayfish accumulation of toxins. Aluma et al. (2017)

found that *F. sanbornii* accumulated mercury and methyl mercury more slowly in AMD impaired streams than in streams with lesser AMD influence, possibly due to greater feeding reliance on detrital material rather than tissue of animals, including those that may have bioaccumulated Hg.

6 Salamanders

Salamanders are important vertebrates in stream ecosystems where mining activities have environmental impacts. In fishless headwaters, salamanders are the dominant vertebrate species with biomass as significant as benthic macroinvertebrates. Davic and Welsh (2004) presented data showing total salamander biomass >9 g/m^2 in a North Carolina headwater stream. Though highly variable, production of benthic invertebrates in eastern US headwater streams averages about the same per year (Wallace and Eggert 2009). The ecological roles of salamanders as predators, nutrient sources to aquatic systems, and linkages between terrestrial and aquatic ecosystems are still poorly understood (Davic and Welsh 2004). Like crayfishes, adaptive radiation of salamanders within the Appalachians has enriched the region with some of the greatest species diversity and endemism in the world (Mitchell et al. 1999). They represent the greatest predator density in forest ecosystems and are particularly abundant in riparian areas. As with other amphibians, they are sensitive bio-indicators of ecosystem health. Some salamanders spend their entire life in water, while others are only temporary inhabitants. The sensitive, porous skin of salamanders functions in respiratory gas exchange and thus is a key feature that makes them vulnerable to environmental contaminants.

Amphibians are declining globally, and salamanders in particular are highly vulnerable to environmental impacts brought on by mining that could further threaten global amphibian diversity (Keitzer and Goforth 2013). Impacts to salamanders are similar to the impacts experienced by other biota that largely make their living in the benthos—chemical toxicity of metals, major ions, and trace elements, physical armoring of substrate by metal hydroxides, and altered flow and sediment dynamics that bury habitat. In West Virginia, abundances of five salamander species were reduced in streams containing waters from MTR/VF (Wood and Williams 2013). Declines were attributed to high levels of silt, dissolved metals, and trace ions that can affect salamanders and their macroinvertebrate prey. Similarly, reduced richness was reported for salamanders in MTR/VF streams in Kentucky due to elevated SO_4^{2-} and major ions Ca, Mg, K, and Na (Muncy et al. 2014). Wood and Williams (2013) further speculated that the potential is high for bioaccumulation of Se in salamanders because of elevated Se in VF-influenced streams. Recent research suggests that salamanders under these conditions bioaccumulate Se in excess of protective levels (Cianciolo et al. 2020). Petty et al. (2013) observed the replacement of forest salamanders (e.g., Northern Dusky Salamander, *Desmognathus fuscus*) typical of headwater streams, with species typical of open areas (e.g., Northern Spring Peeper, *Pseudacris crucifer*) in southern West Virginia streams that had been reconstructed on mined areas. In AMD-impacted streams, salamander abundance was low compared to circumneutral streams (Schorr et al. 2013). High acidity and metal concentrations, which impair osmoregulation and respiration ability, were hypothesized as factors responsible for the decline.

A resurgence in research activity on conservation of salamanders relative to mining impacts, like what has occurred recently for crayfishes, would improve scientific knowledge of their status. Half of the salamander species that occupy much of Appalachia's mining districts have some kind of conservation concern (Mitchell et al. 1999). Increased monitoring, and perhaps enhanced protections, are likely warranted for even dominant species because the salamander guild within local areas is often occupied by just one species that functions as the keystone predator (Davic and Welsh 2004).

7 Fishes

Appalachia is a global hotspot for freshwater fish species diversity (Abell et al. 2008) supporting more than 660 native fish taxa (Warren et al. 2000), many of which are imperiled due to anthropogenic stressors including coal mining (Jelks et al. 2008). Underground mining primarily affects downstream fish populations through point-sources of AMD. Effects of surface mining are characterized by spatially complex hydrological, chemical, and physical alterations to stream habitat (Fig. 1).

7.1 Chemical Toxicity

Acid mine drainage is associated with decreased stream fish diversity and biomass (Cannon and Kimmel 1992). Direct physiological effects involve ionic stress due to sodium ion efflux from acid-mediated ion transport across branchial epithelia of fish gills (Claiborne et al. 2002). Grippo and Dunson (1996) experimentally demonstrated this physiological response using reconstituted mine effluent water. Low pH may also impair fish egg membrane development, resulting in decreased hatch rates (Schofield and Driscoll 1987). Additionally, low pH is associated with Al toxicity due to its increased solubility and bioavailability (Wang et al. 2015). In a field study, Carline et al. (1992) demonstrated spikes in Al toxicity for stream fishes from stormflow-induced pH changes, resulting in decreased fish abundance and increased fish movement to avoid exposure. Regional-scale analyses have also shown that Appalachian streams with low acid-neutralizing capacity tend to support fewer fish species (Herlihy et al. 1993). Native fish in poorly buffered streams are subject to increased risks from acidification, e.g., Brook Trout *Salvelinus fontinalis* (McClurg et al. 2007). Limestone treatment of AMD has successfully restored fish species richness and biomass in some cases but not all (cf. Downey et al. 1994; Eggleton et al. 1996).

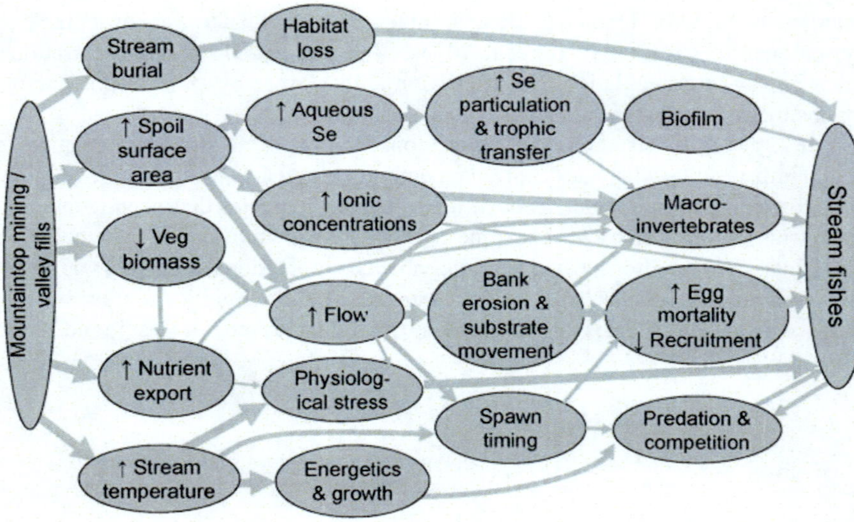

Fig. 1 Conceptual model of mechanisms by which mountaintop mining and valley fills (MTR/VF) affect stream fish assemblages in Appalachia. Circles represent ecosystem components and arrows represent effect pathways. Arrow width indicates the hypothesized relative importance (thicker = greater importance) of effect pathways described in the text. For example, MTR/VF increases spoil surface area and downstream flow, which in turn has a moderate impacts on benthic macroinvertebrate and prey availability to affect change in fish assemblages

Surface mining in Appalachia is associated with decreased diversity and abundance of downstream fish populations (Hitt and Chambers 2014) including federally threatened species, e.g., Blackside Dace, *Chrosomus cumberlandensis*, Kentucky Arrow Darter, *Etheostoma spilotum* (Black et al. 2013; Hitt et al. 2016). MTR/VFs increase total surface area for weathering of parent materials (Ross et al. 2016) and this affects downstream water quality and quantity (Fig. 1).

Elevated conductivity downstream from MTR/VF operations is well-documented (USEPA 2011) and some fish populations show threshold responses to increased salinity, measured as conductivity (Kimmel and Argent 2010; Black et al. 2013; Hitt et al. 2016). The conductivity signal from surface mining is primarily due to elevated SO_4^{2-} concentrations along with associated cations, generally Mg and Ca (Griffith 2014), but SO_4^{2-} transport sites across gill membranes have not been identified (Griffith 2017), so the physiological effects of SO_4^{2-} exposure remain unclear. Although mining-induced ionic toxicity has been demonstrated for aquatic macroinvertebrates (Pond 2010; Mount et al. 2016), mechanistic pathways for fishes are less well understood (Griffith 2017). For example, experimental exposures of juvenile Blackside Dace to reconstituted mine water did not induce mortality, although some sublethal histological effects were observed (J. Kunz, USGS Columbia Environmental Research Center, personal communication). Moreover, potentially confounding effects associated with other land uses and potential

Fig. 2 Creek chub (*Semotilus atromaculatus*) exhibiting spinal deformity characteristic of selenium toxicity. This fish was observed downstream from mountaintop mining operations in the Guyandotte River basin, West Virginia. Length units are inches (top) and centimeters (bottom). Photo by N.P. Hitt, U.S. Geological Survey

stressors can obscure the underlying mechanisms affecting stream biota downstream from MTR/VF operations (Merriam et al. 2013). Electrofishing detection efficiency decreases in high-conductivity streams (Hill and Willis 1994), but this cannot fully account for fish community responses to mining because observed response thresholds (Hitt and Chambers 2014; Hitt et al. 2016) are within the conductivity range for effective electrofishing (Dean et al. 2019).

Selenium toxicity has been linked to fish population declines in lakes receiving Se from coal-ash residue (Brandt et al. 2017) and in stream ecosystems downstream from MTR/VF operations (Lindberg et al. 2011; Arnold et al. 2014). Selenium naturally occurs in rock formations associated with some coal seams in Appalachia (Clark et al. 2018) and although a micronutrient, at higher doses it can cause teratogenic effects in fish and wildlife due to the substitution of Se for S in tertiary protein structures (Hodson et al. 2010). In fish, such teratogenic effects are often exhibited as deformities of the spine (Fig. 2), craniofacial bones, and fin structures (Lemly 1997) as well as edema in early life stages (Gillespie and Baumann 1986). Such deformities are observed more commonly in larval fish than adults due to Se incorporation in eggs during vitellogenesis and subsequently increased larval morality rates (e.g., Hamilton et al. 2000). Food sources also influence Se exposure and toxicity to Appalachian stream fish communities (Hodson et al. 2010) (e.g., Sect. 2).

Water-quality standards for Se have been developed by several Appalachian states, but such aqueous-based standards may have limited utility because dietary Se concentrations are more directly linked to fish toxicity than aqueous concentrations (Stewart et al. 2010). Alternatively, Se standards based on fish tissue are limited by site-specific Se dynamics (Presser 2013), temporal variation in Se body burdens (Arnold et al. 2015), and sample sizes necessary to detect Se thresholds in fish tissue with sufficient statistical power (Hitt and Smith 2015). Selenium toxicity often cannot be separated from additional confounding effects of MTR/VF on fish communities (Lindberg et al. 2011; Hitt and Chambers 2014) but instead represents one of several potentially interacting pathways (Fig. 1). Federal water-quality standards for Se include both water-concentration and fish-tissue components (USEPA 2016).

7.2 Physical Impacts

Direct physical effects of AMD on fishes include changes to stream physical habitat due to precipitation of iron hydroxide and other metals (Rose and Ghazi 1998) that armor substrates and reduce interstitial space. In turn, substrate embeddedness can affect fish feeding behavior (McGinley et al. 2013) and decrease reproductive success for gravel-spawning fishes (Sutherland et al. 2002). For example, in a long-term empirical study, Hitt and Roberts (2012) showed that fish community responses to sedimentation favored fishes that use spawning mounds that are constructed by fish and therefore less influenced by ambient sedimentation. Moreover, fish community recovery has been observed in response to removing fine substrates and decreasing substrate embeddedness (Shirey et al. 2016).

Mountaintop mining techniques specifically affect several aspects of stream fish habitat, yielding multiple potentially interactive effect pathways for fish populations (Fig. 1). The most obvious direct effect of MTR involves the burial of perennial streams with overburden extracted during the mining process (i.e., VFs). Over 2000 km of streams in Appalachia have been buried in this way (Bernhardt and Palmer 2011). McGarvey and Johnston (2013) estimated that direct habitat loss from MTR/VF significantly reduces Brook Trout and Smallmouth Bass *Micropterus dolomieu* distribution and abundance in West Virginia, thereby diminishing regional economic benefits associated with recreational fishing. However, stream burial probably comprises a relatively small component of MTR's overall effects on fishes because few species typically inhabit the uppermost headwater streams subject to burial, whereas fish species richness and abundance increases in a downstream direction (Hitt and Roberts 2012). Instead, hydrological and chemical effects of MTR/VF may extend downstream for tens of kilometers (USEPA 2011) into areas characterized as a global hotspot for freshwater fish biodiversity (Abell et al. 2008).

Mountaintop removal mining often increases base flow and peak flows due to increased run-off and decreased evapotranspiration in post-mining sites and valley fills (Wiley et al. 2001). In turn, increased stream flows can cause bank erosion and sediment transport into downstream areas (Paybins et al. 2000). Hitt and Chambers (2014) observed fewer lithophilic spawning fish downstream of MTR/VF sites than in control sites, suggesting spawning-limitation due to decreased substrate size. High flows can also reduce fish abundance by scouring spawning redds or early larval-stage fishes (Kanno et al. 2016).

Mountaintop removal mining also affects downstream water temperature regimes. Wiley et al. (2001) documented seasonal effects whereby spring, fall, and winter water temperatures in mining-influenced streams were warmer than reference conditions, and summer temperatures were cooler than reference conditions. They also documented a dampening of daily and annual variation in water temperature relative to reference conditions. Water temperature affects many aspects of fish physiology, growth, behavior, and competitive interactions (Lynch et al. 2016), and MTR/VF may have several effects in this regard. Physiological stress associated with high summer temperatures (e.g., Snyder et al. 2015) or low winter temperatures (e.g.,

Donaldson et al. 2008) may be mitigated to some extent by VF outflows. However, dampening thermal variation may obscure thermal cues for spawning initiation in spring months (King et al. 2015). In addition, thermal alterations can affect benthic macroinvertebrate growth and community structure (Ward and Stanford 1982) which in turn affect prey availability and quality for stream fishes. Warmer spring and fall conditions in MTR-influenced streams may further cause energetic stress due to the increased demand for food resources in high-temperature environments (Railsback and Rose 1999).

7.3 Food Web Trophic Effects

Several lines of evidence suggest that energetic limitations may explain observed fish community responses to MTR/VF (Hitt et al. 2016). Although total benthic macroinvertebrate density may not change after exposure to MTR/VF impacts (Hartman et al. 2005; Johnson et al. 2013), predictable compositional changes, including loss of mayflies (Sect. 3) (Hartman et al. 2005; Pond 2010), may affect the energetics of invertivorous fishes due to their selectivity of benthic macroinvertebrate prey (Worischka et al. 2015). This hypothesis is supported by the observation that obligate invertivorous fishes (e.g., darters) declined more than species with other feeding strategies in high-conductivity streams (Hitt and Chambers 2014). Moreover, some fishes that transition to piscivory in adult life stages appear to be able to persist downstream from MTR/VF operations (i.e., Green Sunfish, *Lepomis cyanellus*; Creek Chub, *Semotilus atromaculatus*) in contrast to obligate invertivores that were greatly reduced in number or extirpated (Hitt and Chambers 2014). Hitt and Chambers (2014) also observed self-thinning in fish populations downstream from MTR/VF (i.e., increasing mean fish weight with decreasing abundances), further indicating the potential importance of energetic limitation for stream fishes downstream from coal mining operations. To the extent that macroinvertebrate community alterations result from persistent water-quality effects of geologic disturbance (Clark et al. 2021), these fish community effects can be expected to continue for years and decades beyond the cessation of active mining.

8 Ecosystem Function

Ecosystem function refers to processes that occur in nature (e.g., nutrient cycling, biomass production) as a result of biological activity. Arguably, functional capacity may be more important to understand and quantify than measuring taxonomic composition, because river function reflects its ability to carry out ecological processes that maintain ecosystem integrity and provide ecosystem services. Functions also may be redundant across a variety of taxonomic groups. Functional attributes of ecosystems, however, are more complex to study, and so these studies are less common than those documenting compositional changes in aquatic communities.

8.1 Organic Matter Processing

In Appalachian streams impacted by mining, organic matter processing and macroinvertebrate functional feeding group composition have been a focus of study to understand how mining influences stream function (Rosenberg et al. 2008). Artificially constructed packs of naturally fallen, dried leaves typically are employed to estimate leaf litter decomposition as a component of organic matter processing. Results are often varied. Many invertebrate species that process leaf litter within Appalachian headwaters [e.g., Plecoptera (stoneflies) and Trichoptera (caddisflies)] are extirpated by mining impacts (Hartman et al. 2005; Pond 2012), resulting in decreased rates of organic matter processing within streams impacted by both acid (Simmons et al. 2008; Hogsden and Harding 2012) and alkaline (Fritz et al. 2010; Petty et al. 2013) mine drainage. In comparisons of unmined versus VF streams (reclaimed 11–33 years), Pond et al. (2014) found two of five FFGs (scrapers and shredders) were significantly altered downstream of VFs. In VF streams, scraper taxa (namely mayflies) were reduced or absent, while the relative abundance of shredders showed a concomitant increase, though dominated by metal and ion-tolerant stoneflies (*Amphinemura* and *Leuctra*). In contrast, Hartman et al. (2005) reported declines in both % scrapers and % shredders below several VFs in West Virginia. Fritz et al. (2010) found decreases in organic matter breakdown rates in perennial and intermittent streams in MTR/VFs compared to streams in unmined watersheds, and these rates correlated with shredder diversity. Drover et al. (2019) found that FFG density varied seasonally in mining-influenced streams with high conductivity/salinity. In streams receiving AMD, Niyogi et al. (2013) found that microbial processing of organic matter, and heterotrophy in general, is reduced in moderate AMD when metal flocculates occur, but returns to reference-like conditions in streams with pH <3. This pattern is similar to the variation in algal growth across an AMD gradient (see Sect. 2).

Several studies have documented no change in organic matter processing rates due to headwater mining operations, a result attributed to increases in salinity-tolerant macroinvertebrates that are shredders in MTR/VF landscapes (Vander Vorste et al. 2019). Vander Vorste et al. (2019) found bacterial and fungal composition varied along a mining-induced salinity gradient, but no salinity-induced trends in organic matter decomposition were observed. Functional redundancy with species turnover may not always be expected and may depend on stressor type and magnitude (Vander Vorste et al. 2019). While biofilm and macroinvertebrate community compositions are simplified with mining impacts, how shifts influence ecological roles in nutrient and food web dynamics in river ecosystems receiving mine drainage requires more study to substantiate findings (Smucker and Vis 2011; Bier et al. 2015).

8.2 Secondary Production

Mining activity also affects secondary production, the process of accumulating biomass in consumers (i.e., growth). The overwhelming loss of sensitive taxa has been shown to decrease secondary production in mined systems (Voss and Bernhardt 2017). This is true for waters receiving severe AMD, which are often devoid of macroinvertebrates and fishes. Production of tolerant taxa within alkaline mine drainage settings can offset lost productivity of sensitive taxa (Johnson et al. 2013). Stream ecosystems supply energy via secondary production to terrestrial ecosystems. For example, terrestrial foragers (e.g., birds, etc.) use aquatic resources both instream (i.e., macroinvertebrate prey) and/or in riparian zones (i.e., flying invertebrate prey). The extent to which mining alters this energy flux needs further study; reduced quality of subsidies from stream ecosystems may negatively impact riparian foragers (see Frantz et al. 2018). On the other hand, aquatic subsidies in mined watersheds may deliver contaminants to riparian ecosystems. Naslund et al. (2020) found Se concentrations in riparian spiders that exceeded the daily dietary level for birds in MTR coal mining landscapes of West Virginia. These trends were related to elevated Se in biofilm, thus providing evidence that mining impacts on aquatic ecosystems expose terrestrial food webs to Se toxicity.

Additionally, more research is needed to improve understanding of the extent to which mining alters important abiotic processes and linked biotic functions. For example, the degree to which sustained stream flows from MTR/VFs may play a role in secondary production (Sect. 7, for example), physical break down of leaf material in leaf packs, and increasing nutrient spiral length (both uptake length and turnover time) is unknown (Ensign and Doyle 2006). As organisms assimilate nutrients from flowing water and river sediments, nutrients are locally retained temporarily until remineralized. Thus, nutrients appear to spiral in a downstream direction as they move between these biotic and abiotic compartments as water flows. This phenomenon is termed nutrient spiraling (Newbold et al. 1981). Processes that negatively impact nutrient retention and assimilation will likely lengthen nutrient spirals and consequently will exacerbate nutrient loading and eutrophication to downstream rivers. Understanding mechanisms affecting these critical organic matter processes carried out by biofilm and macroinvertebrates will be important to managing downstream nutrient dynamics.

9 Movement and Dispersal

Animals move to access various resources and conditions needed for life, because these requirements often vary spatially and temporally. Movement processes may include emigration, immigration, dispersal, and migration on larger spatial and temporal scales, or limited movements on smaller time scales for diurnal activities.

Movement processes of aquatic biota in streams are largely confined to the water environment in one dimension, i.e., animals access requirements by moving upstream and downstream. Thus, streams are corridors for movement. Even the flying adult stage of aquatic insects typically travel within stream corridors for mating and oviposition; dispersal via overland flight is relatively rare (Griffith et al. 1998). Stream network structure, i.e., the pattern of stream connectivity (typically dendritic) and interactions with stream size, plays a role in controlling these movement processes. However, mining effluent can impart chemical conditions that impair hydrologic connectivity for organisms dispersing among river segments in the network, and thus impeding movement (Freund and Petty 2007).

Since the majority of Appalachian surface mining disturbance (and VFs) occurs at or near stream origins, loss and isolation of stream channels from downstream segments can have consequences on macroinvertebrate colonization dynamics (Campbell Grant 2011; Merriam and Petty 2016). Macroinvertebrates move via drift, crawling, swimming, and flight (in adult insects). After adult oviposition in upstream areas, downstream drift is the primary mode of larval dispersal (Brittain and Eikeland 1988). In central Appalachian headwater streams, Pond et al. (2014) described the daily, baseflow drift of thousands of colonizing individuals from dozens of taxa to downstream habitats. Watershed area and season were important factors determining drift richness, volume, and rate. Within a mined watershed, non-polluted headwaters are needed to counteract downstream extinction rates by providing constant supplies of drifting colonizers. However, when present, these drifters could also mask the detection of stressors to a resident downstream assemblage, because the fate of sensitive taxa is unknown (Pond et al. 2014). This conclusion is supported by increased diversity of sensitive taxa in poor water-quality segments when upstream intervening tributaries are present (Merovich and Petty 2010; Pond et al. 2014). However, Pond et al. (2014) hypothesized that a source-and-sink scenario like this cannot prevent local extirpations in chronically polluted receiving streams. Parkyn and Smith (2011) conceptualized a lack of time for macroinvertebrate recovery when both upstream source areas and adjacent landscape conditions were highly disturbed; this can often be the case in Appalachian surface mined areas.

Mining impacts in Appalachia affect the requirements for fish to move as well, and therefore understanding the response of fishes to environmental conditions requires a regional perspective from which to evaluate local patterns. Stream fish assemblages are organized not only by local environmental factors, but also by regional movements to access habitat for life history expression, such as feeding and spawning locations (Schlosser 1991), to escape locally unsuitable habitat and seek refugia (e.g., Carline et al. 1992), and to maintain links among discrete populations (Schlosser and Angermeier 1995). For example, stream connectivity to riverine source populations can increase local fish species richness and abundance (Hitt and Angermeier 2008a) and buffer local population declines in response to declining environmental quality (Hitt and Angermeier 2008b). Moreover, habitat loss at the regional scale may incur local extirpations if local populations are supported by a regional pool of immigrants that becomes eliminated (Leibold et al. 2004). For instance, Freund and Petty (2007) suggested that low fish diversity in stream networks influenced by AMD may indicate regional responses rather than localized responses, whereas benthic macroinvertebrate communities by contrast appeared to be controlled by localized conditions. An understanding of such regional influences is necessary to prioritize restoration and conservation strategies (Merovich et al. 2013) and to account for effects of mining in the context of climate change (McDonnell et al. 2015).

10 Management Considerations

10.1 Response and Recovery of Aquatic Life to Remediation of Mining Impacts

Considerable effort has been invested by watershed groups, state water agencies, and other non-governmental organizations to mitigate environmental impacts of mining activities in Appalachia. A large portion of these efforts target chemical treatment of mine drainage effluent (Kruse Daniels et al. 2021). Because aquatic life is impacted by mining in predictable ways (e.g., reduced richness, simplified community structure), it is of interest to understand to what degree aquatic assemblages improve or recover with efforts that attempt to improve or remediate physico-chemical conditions.

Figure 3 presents an idealize model of these expectations. The model traces the fate of biological condition over time as it experiences anthropogenic impact from mining and subsequent recovery as remediation improves in-stream abiotic conditions. Biological communities provide ecosystem services (Sect. 8); impairment of conditions diminishes these services. As remediation improves physico-chemical conditions, biotic conditions improve so that concomitant services improve but lag from historical levels. The lag represents the legacy of mining effects, as even well-executed remediation does not restore pre-mining environmental conditions; and ecological lift refers to the degree of recovery attributed to remediation efforts (Fig. 3). The condition of aquatic life directly relates to the condition of the aquatic environment, therefore it is important to know how biota respond to remediation efforts and recover from legacy impacts. Although much is known about treating acidic abandoned mine drainage to improve water quality, post-treatment monitoring of biotic conditions is important also and additional monitoring is needed (Trout Unlimited 2011).

Studies that publish on AMD remediation generally find large improvements in water chemistry, such as increased pH and net alkaline conditions. Dissolved metals typically found in AMD including Fe and Al are reduced but can remain elevated compared to background levels, and in many cases still exceed water quality criteria (Watson et al. 2017). Sulfate and conductivity typically remain high in treated waters as well (Merovich et al. 2007; Kruse Daniels et al. 2021). When active surface mines are closed and the land is reclaimed, the water-quality effects of mining (e.g., elevated conductivity) commonly persist for decades, although often with patterns of slow decline (Clark et al. 2021).

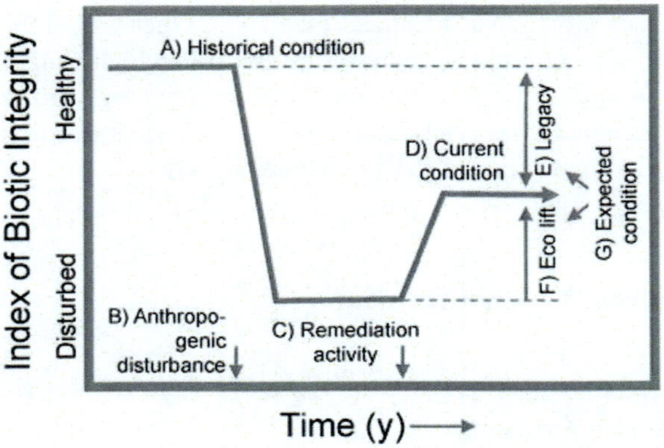

Fig. 3 Hypothetical pattern of ecosystem service response (measured as an index of biotic integrity, IBI, for example) to anthropogenic disturbance and subsequent remediation. At point **A**, Historical ecosystem and biotic conditions were undisturbed and existed in fully functioning states. Disturbance **B** decimates conditions, which persists for some time until **C** Remediation activities improve conditions and biotic communities to some **D** Current condition that is less functional than existed historically, with the difference termed as **E** Legacy. The goal of remediation is to maximize the return of ecosystem services termed **F** Eco (Ecological) lift. The easiest way to maximize lift should be to identify impaired sites that have the greatest likelihood of improving by considering the broader spatial context within which they exist (see text). Modeling expected lift (Watson et al. 2017) within the context of watershed-scale conditions could estimate **G** Expected in-stream conditions achievable, and thus provide a useful tool in the adaptive management cycle for managing watersheds heavily impacted by coal mining

Biotic response to remediation actions that improve water chemistry are much more variable. Generally, treated sites are compared to untreated impacted sites and un-impacted reference sites to evaluate recovery or ecological lift (Fig. 3) (DeNicola and Stapleton 2014). Occupancy, abundance, density, and community richness have been shown to improve with AMD treatment, but these measures poorly reflect recovery because they do not consider the ecology of specific species (DeNicola et al. 2012). Diatom assemblages improved significantly in AMD-treated streams, becoming more similar to un-impacted reference streams after 10 years (DeNicola and Stapleton 2014). Likewise, passive treatment significantly improved epilithic algae biomass and gross primary productivity but these were depressed compared to reference sites. Carlson (2013) found that all measures of benthic macroinvertebrate community structure did not improve one year after AMD treatment. Watson et al. (2017) found that macroinvertebrate richness noticeably increased and fishes recolonized AMD treatment sites located in the Abram Creek watershed in West Virginia, but overall biological conditions were much different and remained depressed compared to un-impacted reference sites, especially for fishes, Gunn et al. (2010) found rapid re-colonization of invertebrates in the first year after AMD treatment, but recovery was slow and the benthic community still lacked many expected taxa even after eight years. In a physical habitat restoration study, Louhi et al. (2011) found that macroinvertebrate communities did not improve after nearly 20 years.

Simon et al. (2006) found some recovery of benthic invertebrates downstream of passive wetland AMD treatment projects; additional un-treated AMD seeps apparently limited recovery. Similarly, fish communities have distinctly improved in the West Branch of the Susquehanna River after implementation of an extensive AMD treatment program, but are still below expectations compared to un-polluted sections in the watershed (Trout Unlimited 2011). Likewise, despite extensive reclamation of mined sites in Virginia over the last 20 years in the Powell River watershed in southwestern Virginia, the river's waters remain characterized by elevated concentrations of major ions and some trace elements (Mn, Ni, Se); and impacted mussel assemblages are not improving throughout the entire free-flowing section of river from Virginia to Tennessee (Zipper et al. 2016; Ahlstedt et al. 2016).

Typical hypotheses for lack of biotic recovery include lack of time elapsed for recovery to take place, and water quality can remain sub-optimal despite remediation efforts. In a review of stream mitigation results from >400 streams among 117 Appalachian mining permits, Palmer and Hondulu (2014) found that water chemistry stressors remained the key limiting factor to biotic recovery. Additionally, remediation designed without an ecosystem-based management perspective may limit recovery and ecological lift. For example, Miller (2015) found that efforts to restore stream channels degraded by erosive flows and sedimentation from MTR/VF that are not accompanied by improved water chemistry will not significantly improve biological conditions. In constructed stream channels on reclaimed mines in southern West Virginia, Petty et al. (2013) associated elevated TDS and conductance to a distinct shift to tolerant macroinvertebrates more typical of lentic environments. Collectors-gatherers replaced large shedders, thereby reducing organic matter processing rates compared to reference streams.

Likewise, recent sampling conducted in 2016 and 2017 in the Powell River watershed showed that mussel assemblages are still exposed to specific conductance on average at 768 μS/cm (Phipps 2019). Further, Evans et al. (2014) projected that it will require on average 20 years for specific conductance in Virginia's VFs to decline to <500 μS/cm, where many stream waters currently exceed 1000 μS/cm. This projection was based on study of older Virginia VFs, which are generally smaller than recent large MTR/VF operations in other states (Evans et al. 2014). Geochemical principles suggest that recovery times for waters emerging from larger mining disturbances and VFs will be longer. These headwater VFs are the likely main source of dissolved solids contributing to the high conductivity observed in downstream reaches of the Powell River where mussels remain (Cormier et al. 2013b).

Finally, impaired conditions in the broader, regional context may hamper ecological recovery at restoration sites. Re-colonization of aquatic organisms to improved local conditions may depend on conditions and movement ability in the surrounding network of streams (i.e., in the watershed). For example, Sundermann et al. (2011) found that restoration success depends on the quality of the species pool, and macroinvertebrate communities did not improve unless potential colonizers were within close range. Louhi et al. (2011) concluded that macroinvertebrate communities did not improve after 20 years because large-scale disturbance limited essential resources for macroinvertebrates. Watson et al. (2017) suggested that sub-optimal regional conditions in the North Branch of the Potomac River was a plausible explanation for the lack of fishery recovery with AMD treatment in a large tributary, Abram Creek. Thus, evidence indicates that regional conditions may dictate the success of river restoration projects (Petty and Thorne 2005; Merovich et al. 2013; Merriam and Petty 2016).

More research is needed to understand how conditions in surrounding streams mediate restoration potential of focal streams in mining landscapes, especially for imperiled taxa like mussels that have complex life histories and depend on fish and fish dispersal for reproduction. Fish assemblages that remain impaired in the watershed due to mining may constrain the ability for mussels to recover and disperse due to the lack of fish as hosts. It is possible that imperiled mussel populations will remain slow to improve at stream restoration sites until stream network connectivity and watershed-scale conditions improve. Research that links traits of local mussel populations and communities to conditions at the watershed-scale remain unexplored and represents an interesting avenue for future conservation studies on mussels.

Consequently, understanding how watershed-scale conditions and stream network topology mediate local recovery mechanisms could aid conservation strategies by identifying streams most likely to see maximum improvements in both water quality and biotic conditions (Merovich et al. 2013). For example, impaired sub-watersheds are expected to improve rapidly with remediation, if they are within good regions that supply colonizers. While full recovery of biological conditions is impossible, a priori tools that identify stream segments expected to have greatest ecological improvements (Fig. 3) could be developed (e.g., Watson et al. 2017).

10.2 Multiple Interacting Stressors

Mining represents only one of several dominant and pervasive land use activities within Appalachian watersheds. In field studies controlling for other land use activities and associated possible physiochemical impacts, increased ion concentrations from surface mining were strongly associated with altered macroinvertebrate assemblages in Appalachian streams (Timpano et al. 2018). However, there is considerable evidence suggesting stressors associated with coal mining combine with other land use activities and associated stressors to result in extensive physicochemical degradation and highly altered aquatic communities (Merriam et al. 2015). For example, nearly 50% of larger streams (i.e., basin areas > 5 km^2) within the MTR/VF region of West Virginia are characterized as having chemical contaminant mixtures indicative of multiple land use activities (i.e., surface mining, underground mining, and residential development) (Merriam et al. 2015). Within the same region, 67% of stream segments are predicted as impaired based on macroinvertebrate community index scores—40% of which are impacted by more than one land use (Merriam et al. 2015).

Multiple land use activities and associated stressors occurring throughout the Appalachian coalfields combine to degrade macroinvertebrate communities across multiple scales. At the local (i.e., stream segment) scale, mining-related stressors add to and interact with those of other land use activities to alter community composition. Merovich and Petty (2007) characterized a synergistic interaction between AMD and thermal effluent from a coal-fired power plant (i.e., their combined effect is greater than the sum of their individual effects). The combined effects of MTR/VF mining and residential development are generally additive (i.e., equivalent to the sum of their individual effects)—with macroinvertebrate community degradation being strongly associated with degraded water quality in mining-impacted streams and degraded physical habitat in development-impacted streams (Merriam et al. 2011). Studies have also documented a positive interaction between deep mining and residential development (i.e., their combined effect is less than the sum of their individual effects) on macroinvertebrate communities in the MTR/VF region, suggesting deep mine effluents may augment low flows and ameliorate the effects of untreated wastewater (Merriam et al. 2015). The lack of an equivalent interaction between residential development and MTR/VF mining—which has also been shown to augment summer low flows (Evans et al. 2015)—suggests differential effects of deep- and surface-mine effluents on invertebrate communities when occurring in combination with development-related effluents, further highlighting the need to better understand the toxicological effects of multi-stressor contaminant mixtures.

Ecotoxicological studies characterizing the effects of multiple chemical stressors on aquatic communities within Appalachian streams are rare. However, Kunz et al. (2013) observed differing toxicities associated with various ionic mixtures of reconstituted mine waters from Appalachian streams. Laboratory studies also have shown differing toxicities associated with major ions characteristic of both mining and residential development (Mount et al. 1997). Clements (2004) documented interactions between heavy metals (Zn, Cd, and Cu)—all of which are highly variable in mining-impacted streams throughout the Appalachian coalfields (Merriam et al. 2015). Results of these studies suggest it is likely that interactions among mining and non-mining stressors, including ions and metals, play an important role in modulating macroinvertebrate response to complex contaminant mixtures in mined Appalachian watersheds. Characterizing these interactions represents an important avenue for future research. For example, interactions between coal geology type and land use including instream distance from mining are important factors controlling biotic condition of streams. Petty et al. (2010) developed an index of mining that was correlated with water chemistry and the West Virginia Stream Condition Index (WV SCI), but their index indicated more degradation in watersheds with Freeport coal compared to Kittanning coal geology. Instream biotic conditions deteriorated rapidly with mining of only about 5% of maximum, but also impacts lessened with distance from mining activity. Thus, landscape-scale factors and underlying geology mediate the effect of mining intensity on stream biota (Petty et al. 2010).

Widespread physicochemical degradation of aquatic resources throughout the Appalachian coalfields resulting from multiple land use activities and associated stressors also has altered regional processes whereby biological communities interact via dispersal mechanisms. Even the most pristine streams within the MTR/VF region are characterized by degraded macroinvertebrate communities when isolated within heavily impacted regions and consequently cut off from source populations (Merriam and Petty 2016). Extensive mining also alters regional dispersal processes in fish communities in AMD-impacted stream networks (Martin 2010). Altered dispersal dynamics and reduced rescue processes result in extirpation and/or reduced abundance of sensitive taxa (e.g., brook trout) in even the most pristine streams, while tolerant or generalist taxa (e.g., creek chub) exhibit a concomitant increase as a result of decreased interspecific interactions and competition (Martin 2010). Given the extent to which multiple contemporary and historic land use activities have interacted and degraded these systems, the long-term persistence of many species, including unique mussels, and the potential recovery of associated communities is uncertain (Merriam and Petty 2016) and will require management at watershed scales.

10.3 Use of Bioindicators to Assess Aquatic Resource Status

Because aquatic biota respond to mining-related stressors in a consistent manner, they have been used as bioindicators of environmental quality for decades (Barbour et al. 1999). Ultimately, all aspects of the ecology of living organisms depend on their physico-chemical environment. Thus, monitoring biological communities can provide an integrative assessment of this abiotic template and biotic recovery.

In the US, use of aquatic life as bioindictors has focused on multimetric indexes of biotic integrity (IBI). These indexes are formulated with information on taxa richness, taxa composition, and trophic makeup that respond to stressor gradients, calibrated to conditions at the best available un-impacted sites. Bioassessment programs that use IBIs provide a standardized way to quantify ecological conditions. They are a basis for making management decisions, and are essential to carrying out the mandate of the Clean Water Act to protect and maintain water quality and designated aquatic life uses of the nation's waters.

Use of IBIs is commonplace; various US water-management agencies employ macroinvertebrate and fish IBIs to assess status of water resources (USEPA 2002). The use of fish communities to assess environmental condition of streams has rapidly accelerated and somewhat parallels the advancement of macroinvertebrate indices of ecosystem health (Rosenberg et al. 2008). As a consequence, a large number of fish IBIs have been developed regionally (e.g., McCormick et al. 2001) and locally (e.g., Kimmel and Argent 2006). Likewise, diatoms are widely used in bioassessment programs (Potapova et al. 2004). Recently, a periphyton IBI has been developed to detect AMD (Zalack et al. 2010).

Macroinvertebrates are considered excellent indicators of local stream conditions because of their sedentary nature. In contrast, fishes are more mobile than macroinvertebrates, can avoid local impairment via network dispersal in connected streams, and thus reflect regional conditions (Hitt and Angermeier 2008b). Freund and Petty (2007) found that the invertebrate WV SCI varied strongly with AMD in contrast to a significantly weaker response of the mid-Atlantic Highlands fish IBI (MAH-IBI) in the Cheat River watershed, likely due to influence of regional conditions on fish assemblages. They concluded that both macroinvertebrates and fish should be included in biomonitoring to quantify local and regional effects of mining on aquatic systems.

A disadvantage of current IBIs is their development to reflect conditions under a broad range of anthropogenic stressors. Consequently, IBI scores or assemblage metrics may be biased under certain pollution types. For example, caddisflies are important indicators of pollution but certain Hydropsychidae genera are tolerant of salt, acid, and aluminum exposure (Vuori 1996). Similarly, a challenge for fish IBI development has been the influence of network position on IBI response to environmental quality due fish movement dynamics (Hitt and Angermeier 2008a).

On the other hand, index development is often stratified by some important natural feature to control for known sources of variability in assemblage structure, such as coldwater streams for fishes (Lyons et al. 1996; Langdon 2001), limestone streams for benthic macroinvertebrates (Botts 2009), or specific regions including the mid-Atlantic Highlands (McCormick et al. 2001) and smaller watersheds (Kimmel and Argent 2006) for fishes. In a similar manner, mining-specific IBIs could be calibrated to detect and quantity mining impacts. A macroinvertebrate index of mining impact seems most feasible. As far as we know, only periphyton has been used specifically as an index for AMD. Zalack et al. (2010) recently developed and validated a diatom IBI to reflect AMD pollution that strongly correlated to water chemistry indicative of AMD.

Despite the assumption that macroinvertebrates are excellent indicators of local conditions, macroinvertebrate IBIs, like fishes, may also respond to watershed-scale conditions that are impaired by mining. For example, regional deflation or decline of macroinvertebrates species pools due to widespread mining reduces local macroinvertebrate quality despite good local conditions because of isolation from a source of colonizing individuals. Likewise, streams in poor physical or chemical condition may contain better quality macroinvertebrates than expected due to high colonization potential from nearby stream locations in good condition. Recent work in AMD (Merovich and Petty 2010) and MTR/VF (Pond et al. 2014; Merriam and Petty 2016) settings suggests that macroinvertebrate communities reflect regional conditions via drift and flight dispersal mechanisms. Traditional macroinvertebrate IBI applications may not reflect these processes. More research could investigate ways to represent regional-scale conditions that determine local ecological integrity.

Additional considerations to improve the use of IBIs in the management of mining influence include the following. Multivariate approaches such as similarity analyses (e.g., Bray–Curtis index) offer the possibility of incorporating taxon-specific information. Richness, although a highly responsive metric to AMD, merely counts the number of different taxa; in reality two assemblages with the same richness could have entirely different composition. Similarity analysis, unlike richness measures, can quantify the difference from reference communities. Thus, similarity analysis can provide a measure of impairment or, in contrast, a degree of recovery to reference condition. Similarity analysis can even be used to diagnose sources of impairment from multiple stressors (Rosenberg et al. 2008). For example, Merovich and Petty

(2007) used similarity analysis to partition the influence of AMD and thermal effluent on macroinvertebrate assemblages in the Cheat River.

Research also could revisit the use of RIVPACs-type (River Invertebrate Prediction and Classification System) (Clarke et al. 2003) modeling. This procedure predicts taxa expected to exist in the absence of stressors using discriminate functions developed from environmental variables. Observed taxa are then compared to those expected (Clarke et al. 2003). Observed to expected ratios are informative and intuitive measures of ecological condition and could be modified to reflect stream network dynamics that may influence local assemblage structure. In central Appalachian streams, Pond and North (2013) found that an observed to expected predictive model indicated that taxa loss strongly corresponded with surface mining and specific conductance.

Finally, IBIs developed at the genus- (or finer) level of taxonomic resolution are more sensitive to detecting compositional changes and may offer better evaluation of the ecological impacts of mines (Pond et al. 2008). Loss of sensitive genera will not be detected if organisms are tallied at the family level, for instance (Boehme et al. 2016). Fish IBIs that include data on functional attributes among fish species and information on riverine network topology (Hitt and Angermeier 2008a) are areas to improve on fish IBI response to mining influence as well.

11 Summary and Conclusions

Coal mining alters the physical and chemical conditions of aquatic ecosystems, and produces stressors that significantly affect aquatic life. Toxicity of mining effluent, and loss of stream and benthic habitat, directly affect aquatic organisms that persist after mining has ceased. Sensitive individuals are lost, and species richness is reduced to a few that are tolerant and that become more abundant. Simplification of community structure is consistent across all major aquatic groups discussed, compared to communities from streams unaffected by mining. Impacts persist due to continuing water-quality effects for years and decades following mine completions and closure.

Indirect impacts of mining on biota arise through altered food webs, altered biotic interactions, and disrupted dispersal dynamics. For instance, loss of herbivorous macroinvertebrates may explain the abundance of acidophilic algae in AMD waters. Regardless of the proximate cause, these changes make aquatic life effective indicators of mining impacts. However, the effects of mining-induced structural simplifications on aquatic ecosystem functions and services that are vital to human welfare remain unknown.

One important finding to date includes evidence that mining may have watershed-scale impacts, even to aquatic assemblages typically believed to be largely sessile (i.e., macroinvertebrates). Mining not only directly impacts local stream segments and aquatic life, but also indirectly impacts biota in unaffected streams not receiving mine runoff; these communities can show reduced richness, reduced IBI scores, and simpler food webs, because they can be cut off from the larger species pool by chemical barriers to dispersal (e.g., Freund and Petty 2007). Consequently, isolated stream reaches within a mosaic of mining impacts can lose biodiversity or struggle to recover from disturbance.

Much remains to be learned about how aquatic ecosystems recover from mining effects in Appalachia. While water chemistry remediation may improve water quality, biotic response seems to lag in streams within mining landscapes. This lag may relate to the persistence of mining effects on water quality and to downstream physical habitats, the lack of access to streams within mining-influenced landscapes by larger species pools. Consequently, ecological recovery may be a function of both local improvements in physico-chemical conditions and the condition of the watershed within which remediation and biological recovery take place. Therefore, impaired streams within watersheds of relatively good condition could have the best chance of experiencing biological recovery with a given unit of remediation effort. These streams can themselves become assets that further aid biological recovery of additional stream reaches to achieve larger watershed-scale improvements.

Acknowledgements Comments and suggestions from two anonymous reviewers and the editors greatly improved the content and organization of the manuscript. We especially thank Greg Pond for providing invaluable input and resources that immensely aided this production.

The views presented in the chapter are those of the authors and do not necessarily represent the views of the U.S. Department of Defense, the U.S. Fish and Wildlife Service, or their components.

References

Abell R, Thieme ML, Revenga C, Bryer M, Kottelat M et al (2008) Freshwater ecoregions of the world: a new map of biogeographic units for freshwater biodiversity conservation. Bioscience 58:403–414

Ahlstedt SA, Fagg MT, Butler R, Connell JF, Jones J (2016) Quantitative monitoring of freshwater mussel populations from 1979–2004 in the Clinch and Powell Rivers of Tennessee and Virginia, with miscellaneous notes on the fauna. Freshw Mollusk Biol Conserv 19:1–18

Allakhverdiev SI, Sakamoto A, Nishiyama Y, Inaba M, Murata N (2000) Ionic and osmotic effects of NaCl-induced inactivation of photosystems I and II in Synechococcus sp. Plant Physiol 123:1047–1056

Allan JD (1995) Stream ecology: structure and function of running waters. Chapman and Hall, London

Aluma E, Johnson KS, Hassett P (2017) Mercury bioaccumulation in crayfish in acid mine-impaired Appalachian streams. Water Air Soil Poll 228:200

Arnold MC, Bier RL, Lindberg TT, Bernhardt ES, Di Giulio RT (2017) Biofilm mediated uptake of selenium in streams with mountaintop coal mine drainage. Limnologica 65:10–13

Arnold MC, Friedrich LA, Lindberg TT, Ross M, Halden NM et al (2015) Microchemical analysis of selenium in otoliths of two West Virginia fishes captured near mountaintop removal coal mining operations. Environ Toxicol Chem 34:1039–1044

Arnold MC, Lindberg TT, Liu YT, Porter KA, Hsu-Kim H et al (2014) Bioaccumulation and speciation of selenium in fish and insects collected from a mountaintop removal coal mining-impacted stream in West Virginia. Ecotoxicology 23:929–938

Barbour MT, Gerritsen J, Snyder BD, Stribling JB (1999) Rapid bioassessment protocols for use in streams and wadeable rivers: periphyton, benthic macroinvertebrates, and fish, 2nd edn. U.S. Environmental Protection Agency, Washington DC

Battaglia M, Hose GC, Turak E, Warden B (2005) Depauperate macroinvertebrates in a mine affected stream: clean water may be the key to recovery. Environ Pollut 138:132–141

Bernhardt E, Palmer M (2011) The environmental costs of mountaintop mining valley fill operations for aquatic ecosystems of the central Appalachians. Science 1223:39–57

Berrill M, Hollett L, Margosian A, Hudson J (1985) Variation in tolerance to low environmental pH by the crayfish Orconectes rusticus, *O. propinquus*, and *Cambarus robustus*. Can J Zool 63:2586–2589

Bier RL, Voss KA, Bernhardt ES (2015) Bacterial community responses to a gradient of alkaline mountaintop mine drainage in Central Appalachian streams. ISME J 9:1378–1390

Black TR, Jones B, Mattingly H (2013) Development and validation of habitat models for the threatened Blackside Dace, *Chrosomus cumberlandensis*, at two spatial scales. Southeast Nat 12(sp4):27–48

Boehme EA, Zipper CE, Schoenholtz SH, Soucek DJ, Timpano AJ (2016) Temporal dynamics of benthic macroinvertebrate communities and their response to elevated specific conductance in Appalachian coalfield headwater streams. Ecol Indic 64:171–180

Botts W (2009) An index of biological integrity (IBI) for "true" limestone streams. PA Dep Environ Protection, Harrisburg, PA

Brandt JE, Bernhardt ES, Dwyer GS, Di Giulio RT (2017) Selenium ecotoxicology in freshwater lakes receiving coal combustion residual effluents: a North Carolina example. Environ Sci Technol 51:2418–2426

Bray JP, Broady PA, Niyogi DK, Harding JS (2008) Periphyton communities in New Zealand streams impacted by acid mine drainage. Mar Freshw Res 59:1084–1091

Brittain JE, Eikeland TJ (1988) Invertebrate drift—a review. Hydrobiologia 166:77–93

Buchwalter DB, Clements WH, Luoma SN (2017) Modernizing water quality criteria in the United States: a need to expand the definition of acceptable data. Environ Toxicol Chem 36:285–291

Campbell Grant EH (2011) Structural complexity, movement bias, and metapopulation extinction risk in dendritic ecological networks. J North Am Benthol Soc 30:252–258

Cannon WE, Kimmel WG (1992) A comparison of fish and macroinvertebrate communities between an unpolluted stream and the recovery zone of a stream receiving acid mine drainage. J Pa Acad Sci 66:58–62

Carline RF, DeWalle DR, Sharpe WE, Dempsey BA, Gagen CJ, Swistock B (1992) Water chemistry and fish community responses to episodic stream acidification in Pennsylvania, USA. Environ Pollut 78:45–48

Carlson B (2013) Water chemistry and benthic macroinvertebrate ecology in response to acid mine drainage. Master's Thesis, West Virginia University, Morgantown

Cashman MJ, Wehr JD, Truhn K (2013) Elevated light and nutrients alter the nutritional quality of stream periphyton. Freshw Biol 58:1447–1457

Cianciolo TR, McLaughlin DL, Zipper CE, Timpano AJ, Soucek DJ et al (2020) Selenium bioaccumulation across trophic levels and along a longitudinal gradient in headwater streams. Environ Toxicol Chem 39:692–704

Ciparis S, Phipps A, Soucek DJ, Zipper CE, Jones JW (2015) Effects of environmentally relevant mixtures of major ions on a freshwater mussel. Environ Pollut 207:280–287

Claiborne JB, Edwards SL, Morrison-Shetlar AI (2002) Acid-base regulation in fishes: cellular and molecular mechanisms. J Exp Zool 293:302–319

Clark EV, Daniels WL, Zipper CE, Eriksson K (2018) Mineralogical influences on water quality from weathering of surface coal mine spoils. Appl Geochem 91:97–106

Clark EV, Zipper CE, Soucek DJ, Daniels WL (2021) Contaminants in Appalachian water resources generated by non-acid-forming coal-mining materials. Chapter 9 In: Zipper CE, Skousen JG (eds) Appalachia's coal-mined landscapes. Springer

Clarke RT, Wright JF, Furse MT (2003) RIVPACS models for predicting the expected macroinvertebrate fauna and assessing the ecological quality of rivers. Ecol Model 160:219–233
Clements WH (2004) Small-scale experiments support causal relationships between metal contamination and macroinvertebrate community responses. Ecol Appl 14:954–967
Conley J, Funk DH, Buchwalter D (2009) Selenium bioaccumulation and maternal transfer in the mayfly Centroptilum triangulifer in a life-cycle, periphyton-biofilm trophic assay. Environ Sci Technol 43:7952–7957
Cook LJ, Francoeur SN (2013) Effects of simulated short-term road salt exposure on lotic periphyton function. J Freshw Ecol 28:211–223
Cormier S, Suter G, Zheng L, Pond G (2013a) Assessing causation of the extirpation of stream macroinvertebrates by a mixture of ions. Environ Toxicol Chem 32:277–287
Cormier SM, Wilkes SP, Zheng L (2013b) Relationship of land use and elevated ionic strength in Appalachian watersheds. Environ Toxicol Chem 32:296–303
Costello DM, Kulacki KJ, McCarthy ME, Tiegs SD, Cardinale BJ (2018) Ranking stressor impacts on periphyton structure and function with mesocosm experiments and environmental-change forecasts. PLoS ONE 13:e0204510. https://doi.org/10.1371/journal.pone.0204510
Covich AP, Palmer MA, Crowl TA (1999) The role of benthic invertebrate species in freshwater ecosystems: zoobenthic species influence energy flows and nutrient cycling. Bioscience 49:119–127
Crandall KA, Buhay JE (2008) Global diversity of crayfish (Astacidae, Cambaridae, and Parastacidae—Decapoda) in freshwater. Hydrobiologia 595:295–301
Creed RP, Reed JM (2004) Ecosystem engineering by crayfish in a headwater stream community. J N Am Benthol Soc 23:224–236
Das BK, Roy A, Koschorreck M, Mandal SM, Wendt-Potthoff K, Bhattacharya J (2009) Occurrence and role of algae and fungi in acid mine drainage environment with special reference to metals and sulfate immobilization. Water Res 43:883–894
Davic RD, Welsh HH (2004) On the ecological roles of salamanders. Annu Rev Ecol Evol S 35:405–434
Dean JC, Temple AJ, Smith JW (2019) Output performance of U.S.-built backpack electrofishers. N Am J Fish Manage https://doi.org/10.1002/nafm.10370
DeForest DK, Brix KV, Adams WJ (2007) Assessing metal bioaccumulation in aquatic environments: the inverse relationship between bioaccumulation factors, trophic transfer factors and exposure concentration. Aquat Toxicol 84:236–246
DeNicola DM, Layton L, Czapski TR (2012) Epilithic community metabolism as an indicator of impact and recovery in streams affected by acid mine drainage. Environ Manag 50:1035–1046
DeNicola DM, Lellock AJ (2015) Nutrient limitation of algal periphyton in streams along an acid mine drainage gradient. J Phycol 51:739–749
DeNicola DM, Stapleton MG (2002) Impact of acid mine drainage on benthic communities in streams: the relative roles of substratum vs. aqueous effects. Environ Pollut 119:303–315
DeNicola DM, Stapleton MG (2014) Benthic diatoms as indicators of long-term changes in a watershed receiving passive treatment for acid mine drainage. Hydrobiologia 732:29–48
Dillon BA, Lee KT (2002) Effects of abandoned mine drainage on crayfish distribution along a pH gradient. In: Escobar-Briones E, Alvarez F (eds) Modern approaches to the study of Crustacea. Kluwer Academic, New York, pp 275–279
Donaldson M, Cooke S, Patterson D, Macdonald J (2008) Cold shock and fish. J Fish Biol 73:1491–1530
Downey DM, French CR, Odom M (1994) Low cost limestone treatment of acid sensitive trout streams in the Appalachian Mountains of Virginia. Water Air Soil Poll 77:49–77
Drerup SA, Vis ML (2016) Responses of stream biofilm phospholipid fatty acid profiles to acid mine drainage impairment and remediation. Water Air Soil Poll 227:159

Drover DR, Zipper CE, Soucek DJ, Schoenholtz SH (2019) Using density, dissimilarity, and taxonomic replacement to characterize mining-influenced benthic macroinvertebrate community alterations in central Appalachia. Ecol Indic 106:105535. https://doi.org/10.1016/j.ecolind.2019.105535

Echols BS, Currie RJ, Cherry DS (2010) Preliminary results of laboratory toxicity tests with the mayfly, Isonychia bicolor (Ephemeroptera: Isonychiidae) for development as a standard test organism for evaluating streams in the Appalachian coalfields of Virginia and West Virginia. Environ Monit Assess 169:487–500

Eggleton MA, Morgan EL, Pennington WL (1996) Effects of liming on an acid-sensitive southern Appalachian stream. Restor Ecol 4:247–263

Ensign SH, Doyle MW (2006) Nutrient spiraling in streams and river networks. J Geophys Res-Biogeo 111. https://doi.org/10.1029/2005JG000114

Evans DM, Zipper CE, Donovan PF, Daniels WL (2014) Long-term trends of specific conductance in waters discharged by coal-mine valley fills in Central Appalachia, USA. J Am Water Resour 50:1449–1460

Evans DM, Zipper CE, Hester ET, Schoenholtz SH (2015) Hydrologic effects of surface coal mining in Appalachia (U.S.). J Am Water Resour 51:1436–1452

Frantz MW, Wood PB, Merovich GT Jr (2018) Demographic characteristics of an avian predator, Louisiana Waterthrush (Parkesia motacilla), in response to its aquatic prey in a Central Appalachian USA watershed impacted by shale gas development. PLoS ONE 13:e0206077. https://doi.org/10.1371/journal.pone.0206077

Freund JG, Petty JT (2007) Response of fish and macroinvertebrate bioassessment indices to water chemistry in a mined Appalachian watershed. Environ Manage 39:707–720

Fritz KM, Fulton S, Johnson BR, Barton CD, Jack JD et al (2010) Structural and functional characteristics of natural and constructed channels draining a reclaimed mountaintop removal and valley fill coal mine. J N Am Benthol Soc 29:673–689

Gillespie RB, Baumann PC (1986) Effects of high tissue concentrations of selenium on reproduction by bluegills. T Am Fish Soc 115:208–213

Gillis PL (2011) Assessing the toxicity of sodium chloride to the glochidia of freshwater mussels: implications for salinization of surface waters. Environ Pollut 159:1702–1708

Griffith M (2014) Natural variation and current reference for specific conductivity and major ions in wadeable streams of the conterminous USA. Freshw Sci 23:1–17

Griffith M (2017) Toxicological perspective on the osmoregulation and ionoregulation physiology of major ions by freshwater animals: Teleost fish, Crustacea, aquatic insects, and Mollusca: Osmoregulation and ionoregulation physiology of major ions. Environ Toxicol Chem 36:567–600

Griffith M, Barrows EM, Perry SA (1998) Lateral dispersal of adult aquatic insects (Plecoptera, Trichoptera) following emergence from headwater streams in forested Appalachian catchments. Ann Entomol Soc Am 91:195–201

Griffith MB, Norton SB, Alexander LC, Pollard AI, LeDuc SD (2012) The effects of mountaintop mines and valley fills on the physicochemical quality of stream ecosystems in the central Appalachians: a review. Sci Total Environ 2012:1–12

Grippo R, Dunson WA (1996) The body ion loss biomarker. 1. Interactions between trace metals and low pH in reconstituted coal mine-polluted water. Environ Toxicol Chem 15:1955–1963

Gunn J, Sarrazin-Delay C, Wesolek B, Stasko A, Szkokan-Emilson E (2010) Delayed recovery of benthic macroinvertebrate communities in Junction Creek, Sudbury, Ontario, after the diversion of acid mine drainage. Hum Ecol Risk Assess 16:901–912

Hamilton SJ, Muth RT, Waddell B, May TW (2000) Hazard assessment of selenium and other trace elements in wild larval Razorback Sucker from the Green River, Utah. Ecotox Environ Safe 45:132–147

Hartman KJ, Horn CD, Mazik PM (2010) Influence of elevated temperature and acid mine drainage on mortality of the crayfish Cambarus bartonii. J Freshw Ecol 25:19–30

Hartman KJ, Kaller JD, Howell JW, Sweka JA (2005) How much do valley fills influence headwater streams? Hydrobiologia 532:91–109

Herlihy A, Kaufmann PR, Church M Jr, Wigington PJ, Webb JR, Sale MJ (1993) The effects of acidic deposition on streams in the Appalachian Mountain and Piedmont region of the Mid-Atlantic United States. Water Resour Res 29:2687–2704

Hill TD, Willis DW (1994) Influence of water conductivity on pulsed AC and pulsed DC electrofishing catch rates for Largemouth Bass. N Am J Fish Manage 14:202–207

Hitt NP, Angermeier PL (2008a) River-stream connectivity affects fish bioassessment performance. Environ Manage 42:132–150

Hitt NP Angermeier PL (2008b) Evidence for fish dispersal from spatial analysis of stream network topology. J N Am Benthol Soc 27:304–320

Hitt NP, Chambers DB (2014) Temporal changes in taxonomic and functional diversity of fish assemblages downstream from mountaintop mining. Freshw Sci 33:915–926

Hitt NP, Floyd M, Compton M, McDonald K (2016) Threshold responses of blackside dace (Chrosomus cumberlandensis) and Kentucky Arrow Darter (Etheostoma spilotum) to stream conductivity. Southeast Nat 15:4160

Hitt NP, Roberts JH (2012) Hierarchical spatial structure of stream fish colonization and extinction. Oikos 121:127–137

Hitt NP, Smith DR (2015) Threshold-dependent sample sizes for selenium assessment with stream fish tissue. Integr Environ Asses 11:143–149

Hodson P, Reash R, Canton SP, Campbell P, Delos C et al (2010) Selenium risk characterization. In: Chapman PM et al (eds) Ecological assessment of selenium in the aquatic environment. CRC Press, pp 233–256

Hogsden KL, Harding JS (2012) Consequences of acid mine drainage for the structure and function of benthic stream communities: a review. Freshw Sci 31:108–120

Hogsden KL, Winterbourn MJ, Harding JS (2013) Do food quantity and quality affect food webs in streams polluted by acid mine drainage? Mar Freshw Res 64:1112–1122

Jacobus LM, Wiersema NA (2014) The genera Anafroptilum Kluge, 2011 and Neocloeon Traver, 1932, reinstated status, in North America, with remarks about the global composition of Centroptilum Eaton, 1869 (Ephemeroptera: Baetidae). Zootaxa 3814:385–391

Janz DM, DeForest DK, Brooks ML, Chapman PM, Gilron G et al (2010) Selenium toxicity to aquatic organisms. In: Chapman PM et al (eds) Ecological assessment of selenium in the aquatic environment. CRC Press, Boca Raton FL, pp 141–231

Jelks HL, Walsh SJ, Burkhead NM, Contreras-Balderas S, Diaz-Pardo E et al (2008) Conservation status of imperiled North American freshwater and diadromous fishes. Fisheries 33:372–407

Johnson BR, Fritz KM, Price R (2013) Estimating benthic secondary production from aquatic insect emergence in streams affected by mountaintop removal coal mining, West Virginia, USA. Fund Appl Limnol 182:191–204

Jones J (2015) Freshwater mussels of Virginia (Bivalvia: Unionidae): an introduction to their life history, status and conservation. Va J Sci 66:309–331

Jones J, Lane T, Otsby B, Beaty B, Ahlstedt SA et al (2018) Collapse of the Pendleton Island mussel fauna in the Clinch River, Virginia: setting baseline conditions to guide recovery and restoration. Freshw Mollusk Biol Conserv 21:36–56

Jones TG, Channel KB, Collins SE, Enz J, Stinson CM (2010) Possible extirpation of Cambarus veteranus (Big Sandy Crayfish) from West Virginia. Southeast Nat 9:165–174

Kanno Y, Pregler KC, Hitt NP, Letcher BH, Hocking DJ, Wofford JEB (2016) Seasonal temperature and precipitation regulate brook trout young-of-the-year abundance and population dynamics. Freshw Biol 61:88–99

Kefford BJ (2019) Why are mayflies (Ephemeroptera) lost following small increases in salinity? Three conceptual osmophysiological hypotheses. Philos Trans R Soc B 374:20180021. https://doi.org/10.1098/rstb.2018.0021

Keitzer SC, Goforth RR (2013) Spatial and seasonal variation in the ecological significance of nutrient recycling by larval salamanders in Appalachian headwater streams. Freshw Sci 32:1136–1147

Kennedy AJ, Cherry DS, Currie RJ (2004) Evaluation of ecologically relevant bioassays for a lotic system impacted by a coal-mine effluent, using Isonychia. Environ Monit Assess 95:37–55
Kennedy AJ, Cherry DS, Zipper CE (2005) Evaluation of ionic contribution to the toxicity of a coal-mine effluent using Ceriodaphnia dubia. Arch Environ Contam Toxicol 49:155–162
Kimmel WG, Argent D (2010) Stream fish community responses to a gradient of specific conductance. Water Air Soil Poll 206:49–56
Kimmel WG, Argent DG (2006) Development and application of an index of biotic integrity for fish communities of wadeable Monongahela River tributaries. J Freshw Ecol 21:183–190
King A, Gwinn D, Tonkin Z, Mahoney J, Raymond S, Beesley L (2015) Using abiotic drivers of fish spawning to inform environmental flow management. J Appl Ecol 53:34–43
Kozak KH, Wiens JJ (2010) Does niche conservatism promote speciation? a case study in North American salamanders. Evolution 60:2604–2621
Kruse Daniels N, LaBar J, McDonald LM (2021) Acid mine drainage in Appalachia: Sources, legacy, and treatment. Chapter 8 In: Zipper CE, Skousen JG (eds) Appalachia's Coal-mined landscapes. Springer
Kunz JL, Conley JM, Buchwalter DB, Norberg-King TJ, Kemble NE et al (2013) Use of reconstituted waters to evaluate effects of elevated major ions associated with mountaintop coal mining on freshwater invertebrates. Environ Toxicol Chem 32:2826–2835
Langdon RW (2001) A preliminary index of biological integrity for fish assemblages of small coldwater streams in Vermont. Northeast Nat 8:219–232
Leibold MA, Holyoak M, Mouquet N, Amarasekare P, Chase JM, Hoopes MF, Holt RD, Shurin JB, Law R, Tilman D, Loreau M, Gonzalez A (2004) The metacommunity concept: a framework for multi-scale community ecology. Ecol Lett 7:601–613
Lemly AD (1997) A teratogenic deformity index for evaluating impacts of selenium on fish populations. Ecotox Environ Safe 37:259–266
Lindberg TT, Bernhardt ES, Bier R, Helton AM, Merola RB et al (2011) Cumulative impacts of mountaintop mining on an Appalachian watershed. P Natl Acad Sci USA 108:20929–20934
Lituma CM, Cox JJ, Spear S, Edwards JW, De La Cruz JL, Muller LI, Ford WM (2021) Terrestrial wildlife in the Appalachian post-mined landscape: Status and opportunities. Chapter 6 In: Zipper CE, Skousen JG (eds) Appalachia's coal-mined landscapes. Springer
Loughman ZJ, Fagundo RA, Lau E, Welsh SA, Thoma RF (2013) Cambarus (C.) hatfieldi, a new species of crayfish (Decapoda:Cambaridae) from the Tug Fork River Basin of Kentucky, Virginia and West Virginia, USA. Zootaxa 3750:223–236
Loughman ZJ, Welsh SA (2010) Distribution and conservation standing of West Virginia crayfishes. Southeast Nat 9:63–78
Loughman ZJ, Welsh SA, Sadecky NM, Dillard ZW, Scott RK (2017a) Evaluation of physicochemical and physical habitat associations for Cambarus callainus (Big Sandy crayfish), an imperiled crayfish endemic to the Central Appalachians. Aquat Conserv 27:755–763
Loughman ZJ, Welsh SA, Thoma RF (2017b) Cambarus (C.) appalachiensis, a new species of crayfish (Decapoda: Cambaridae) from the New River Basin of Virginia and West Virginia, USA. Zootaxa 4243:432–454.
Louhi P, Mykra H, Paavola R, Huusko A, Vehanen T et al (2011) Twenty years of stream restoration in Finland: little response by benthic macroinvertebrate communities. Ecol Appl 21:1950–1961
Luís AT, Teixeira P, Almeida SFP, Ector L, Matos JX, Ferreira da Silva EA (2009) Impact of acid mine drainage (AMD) on water quality, stream sediments and periphytic diatom communities in the surrounding streams of aljustrel mining area (Portugal). Water Air Soil Poll 200:147–167
Lynch AJ, Myers BJE, Eby LA, Falke JA, Kovach RP et al (2016) Climate change effects on North American inland fish populations and assemblages. Fisheries 41:346–361
Lyons J, Wang L, Simonson TD (1996) Development and validation of an index of biotic integrity for coldwater streams in Wisconsin. N Am J Fish Manage 16:241–256
MacCausland A, McTammany ME (2007) The impact of episodic coal mine drainage pollution on benthic macroinvertebrates in streams in the Anthracite region of Pennsylvania. Environ Pollut 149:216–226

Martin RW (2010) Stream fish metacommunity and metapopulation ecology in a mined Appalachian watershed. Ph.D. Dissertation, West Virginia University, Morgantown

McClurg SE, Petty JT, Mazik PM, Clayton JL (2007) Stream ecosystem response to limestone treatment in acid impacted watersheds of the Allegheny Plateau. Ecol Appl 17:1087–1104

McCormick FH, Hughes RM, Kaufmann PR, Peck DV, Stoddard JL, Herlihy AT (2001) Development of an index of biotic integrity for the Mid-Atlantic Highlands region. T Am Fish Soc 130:857–877

McDonnell TC, Sloat MR, Sullivan TJ, Dolloff CA, Hessburg PF et al (2015) Downstream warming and headwater acidity may diminish coldwater habitat in southern Appalachian mountain streams. PLoS ONE 10:e0134757

McGarvey DJ, Johnston JM (2013) 'Fishing' for alternatives to mountaintop mining in southern West Virginia. Ambio 42:298–308

McGinley EJ, Raesly RL, Seddon WL (2013) The effects of embeddedness on the seasonal feeding of mottled sculpin. Am Midl Nat 170:213–228

McGregor S, Wynn EA, Garner JT (2013) Results of a survey of the mussel fauna at selected stations in the Black Warrior River system, Alabama, 2009–2014. Open-file Report 1301, Geol Surv Ala

Merovich GT Jr, Petty JT (2007) Interactive effects of multiple stressors and restoration priorities in a mined Appalachian watershed. Hydrobiologia 575:13–31

Merovich GT Jr, Petty JT (2010) Continuous response of benthic macroinvertebrate assemblages to a discrete disturbance gradient: consequences for diagnosing stressors. J N Am Benthol Soc 29:1241–1257

Merovich GT Jr, Petty JT, Strager MP, Fulton JB (2013) Hierarchical classification of stream condition: a house-neighborhood framework for establishing conservation priorities in complex riverscapes. Freshw Sci 32:874–891

Merovich GT Jr, Stiles JM, Petty JT, Fulton J, Ziemkiewicz PF (2007) Water chemistry based classification of streams and implications for restoring mined Appalachian watersheds. Environ Toxicol Chem 26:1361–1369

Merriam ER, Petty JT (2016) Under siege: Isolated tributaries are threatened by regionally impaired metacommunities. Sci Total Environ 560–561:170–178

Merriam ER, Petty JT, Merovich GT Jr, Fulton JB, Strager MP (2011) Additive effects of mining and residential development on stream conditions in a central Appalachian watershed. J N Am Benthol Soc 30:399–418

Merriam ER, Petty JT, Strager MP, Maxwell AE, Ziemkiewicz PF (2013) Scenario analysis predicts context-dependent stream response to landuse change in a heavily mined central Appalachian watershed. Freshw Sci 32:1246–1259

Merriam E, Petty T, Strager M, Maxwell A, Ziemkiewicz P (2015) Complex contaminant mixtures in multistressor Appalachian riverscapes. Environ Toxicol Chem 34:2603–2610

Miller EM (2015) Evaluating and prioritizing stream restoration in mined Appalachian watersheds. Ph.D. Dissertation, West Virginia University, Morgantown

Mitchell J, Pauley T, Withers D, Roble S, Miller B, Braswell A, Cupp P, Hobson C (1999) Conservation status of the southern Appalachian herpetofauna. Va J Sci 50:13–37

Mount DR, Erickson RJ, Highland TL, Hockett JR, Hoff DJ, Jenson CT, Norberg-King TJ, Peterson KN, Polaske ZM, Wisniewski S (2016) The acute toxicity of major ion salts to Ceriodaphnia dubia: I. influence of background water chemistry. Environ Toxicol Chem 35:3039–3057

Mount DR, Gulley DD, Hockett JR, Garrison TD, Evans JM (1997) Statistical models to predict the toxicity of major ions to Ceriodaphnia dubia, Daphnia magna and Pimephales promelas (fathead minnows). Environ Toxicol Chem 16:2009–2019

Mueller M, Pander J, Wild R, Lueders T, Geist J (2013) The effects of stream substratum texture on interstitial conditions and bacterial biofilms: Methodological strategies. Limnologica 43:106–113

Muncy BL, Price SJ, Bonner SJ, Barton CD (2014) Mountaintop removal mining reduces stream salamander occupancy and richness in southeastern Kentucky (USA). Biol Conserv 180:115–121

Mynsberge AR, Strager MP, Strager JM, Mazik PM (2009) Developing predictive models for freshwater mussels (Mollusca: Unionidae) in the Appalachians: limitations and directions for future research. Ecoscience 16:387–398

Naslund LC, Gerson JR, Brooks AC, Walters DM, Bernhardt ES (2020) Contaminant subsidies to riparian food webs in Appalachian streams impacted by mountaintop removal coal mining. Environ Sci Technol 54:3951–3959

Neves RJ, Bogan AE, Williams JD, Ahlstedt SA, Hartfield PW (1997) Status of aquatic mollusks in the southeastern United States: a downward spiral of diversity. In: Benz GW, Collins DE (eds) Aquatic fauna in peril: the southeastern perspective. Special Publication 1. Southeast Aquatic Research Institute, Lenz Design & Communications, Decatur, GA, pp 43–85

Newbold JD, Elwood JW, O'Neill RV, Winkle WV (1981) Measuring nutrient spiralling in streams. Can J Fish Aquat Sci 38:860–863

Nielsen DL, Brock MA, Rees GN, Baldwin DS (2003) Effects of increasing salinity on freshwater ecosystems in Australia. Aust J Bot 51:655–665

Nippgen F, Ross MRV, Bernhardt ES, McGlynn BL (2017) Creating a more perennial problem? Mountaintop removal coal mining enhances and sustains saline baseflows of Appalachian watersheds. Environ Sci Technol 51:8324–8334

Niyogi DK, Harding JS, Simon KS (2013) Organic matter breakdown as a measure of stream health in New Zealand streams affected by acid mine drainage. Ecol Indic 24:510–517

Ortmann AE (1909) The destruction of the fresh-water fauna in western Pennsylvania. P Am Philos Soc 48:90–110

Palmer MA, Hondula KL (2014) Restoration as mitigation: analysis of stream mitigation for coal mining impacts in southern Appalachia. Environ Sci Technol 48:10552–10560

Parkyn SM, Smith BJ (2011) Dispersal constraints for stream invertebrates: setting realistic timescales for biodiversity restoration. Environ Manage 48:602–614

Paybins KS, Messinger T, Eychaner JH, Chambers DB, Kozar MD (2000) Water quality in the Kanawha-new river basin West Virginia, Virginia, and North Carolina, 1996–98. US Geological Survey Circular 1–26

Peake DR, Pond GJ, McMurray SE (2004) Development of tolerance values for Kentucky crayfishes. Kentucky Environmental and Public Protection Cabinet, Ky Dep Environ Protection, Frankfort

Pericak AA, Thomas CJ, Kroodsma DA, Wasson MF, Ross MRV et al (2018) Mapping the yearly extent of surface coal mining in Central Appalachia using Landsat and Google Earth Engine. PLoS ONE 13:e0197758. https://doi.org/10.1371/journal.pone.0197758

Petty JT, Barker J (2004) Water quality variability in tributaries of the Cheat River, a mined Appalachian watershed. Proc Am Soc Min Reclam 21:1484–1504

Petty JT, Fulton JB, Strager MP, Merovich GT Jr, Stiles JM, Ziemkiewicz PF (2010) Landscape indicators and thresholds of stream ecological impairment in an intensively mined Appalachian watershed. J N Am Benthol Soc 29:1292–1309

Petty JT, Gingerich G, Anderson JT, Ziemkiewicz PF (2013) Ecological function of constructed perennial stream channels on reclaimed surface coal mines. Hydrobiologia 720:39–53

Petty JT, Thorne D (2005) An ecologically based approach to identifying restoration priorities in an acid-impacted watershed. Restor Ecol 13:348–357

Phipps A (2019) Evaluation of the effects of mining related contaminants on freshwater mussels (Bivalvia: Unionidae) in the Powell River of Virginia and Tennessee. MS Thesis, Virginia Tech

Pond GJ (2004) Effects of surface mining and residential land use on headwater stream biotic integrity in the eastern Kentucky coalfield region. Ky Dep Environ Protection, Frankfort

Pond GJ (2010) Patterns of Ephemeroptera taxa loss in Appalachian headwater streams. Hydrobiologia 641:185–201

Pond GJ (2012) Biodiversity loss in Appalachian headwater streams (Kentucky, USA): Plecoptera and Trichoptera communities. Hydrobiologia 679:97–117

Pond GJ, North SH (2013) Application of a benthic observed/expected-type model for assessing Central Appalachian streams influenced by regional stressors in West Virginia and Kentucky. Environ Monit Assess 185:9299–9320

Pond GJ, Passmore ME, Borsuk FA, Reynolds L, Rose CJ (2008) Downstream effects of mountaintop coal mining: comparing biological conditions using family- and genus-level macroinvertebrate bioassessment tools. J N Am Benthol Soc 27:717–737
Pond GJ, Passmore ME, Pointon ND, Felbinger JK, Walker CA et al (2014) Long-term impacts on macroinvertebrates downstream of reclaimed mountaintop mining valley fills in Central Appalachia. Environ Manage 54:919–933
Potapova MG, Charles DF, Ponader KC, Winter DM (2004) Quantifying species indicator values for trophic diatom indices: a comparison of approaches. Hydrobiologia 517:25–41
Presser TS (2013) Selenium in ecosystems within the mountaintop coal mining and valley-fill region of southern West Virginia: Assessment and ecosystem-scale modeling. U.S. Geological Survey Prof Paper 1803, Reston, VA
Railsback SF, Rose KA (1999) Bioenergetics modeling of stream trout growth: temperature and food consumption effects. T Am Fish Soc 128:241–256
Rainbow PS (2002) Trace metal concentrations in aquatic invertebrates: why and so what? Environ Pollut 120:497–507
Rogers JJ, Henley WF, Weberg AG, Jones JW, Cope WG (2018) Assessment of growth, survival, and organ tissues of caged mussels (Bivalvia: Unionidae) in a river-scape influenced by coal mining in the southeastern USA. Sci Total Environ 645:1273–1286
Rose S, Ghazi AM (1998) Experimental study of the stability of metals associated with iron oxyhydroxides precipitated in acid mine drainage. Environ Geol 36:364–370
Rosenberg DM, Resh VH, King RS (2008) Use of aquatic insects in biomonitoring. In: Merritt RW et al. (eds) An introduction to the aquatic insects of North America. Kendall/Hunt, pp 123–137
Ross MRV, McGlynn BL, Bernhardt ES (2016) Deep impact: effects of mountaintop mining on surface topography, bedrock structure, and downstream waters. Environ Sci Technol 50:2064–2074
Schlosser IJ (1991) Stream fish ecology: a landscape perspective. Bioscience 41:704–712
Schlosser IJ, Angermeier PL (1995) Spatial variation in demographic processes of lotic fishes: conceptual models, empirical evidence, and implications for conservation. Am Fish S S 17:392–401
Schofield CL, Driscoll CT (1987) Fish species distribution in relation to water quality gradients in the North Branch of the Moose River Basin. Biogeochemistry 3:63–85
Schorr MS, Dyson MC, Nelson CH, Van Horn GS, Collins DE, Richards SM (2013) Effects of stream acidification on lotic salamander assemblages in a coal-mined watershed in the Cumberland Plateau. J Freshw Ecol 28:339–353
Shirey P, Brueseke M, Kenny J B, Lamberti GA (2016) Long-term fish community response to a reach-scale stream restoration. Ecol Soc 21:11. https://doi.org/10.5751/ES-08584-210311
Silva EIL, Shimizu A, Matsunami H (2000) Salt pollution in a Japanese stream and its effects on water chemistry and epilithic algal chlorophyll-a. Hydrobiologia 437:139–148
Simmons JA (2012) Toxicity of major cations and anions (Na+, K+, Ca2+, Cl-, and SO4(2-)) to a macrophyte and an alga. Environ Toxicol Chem 31:1370–1374
Simmons JA, Currie WS, Eshleman KN, Kuers K, Monteleone S et al (2008) Forest to reclaimed mine land use change leads to altered ecosystem structure and function. Ecol Appl 18:104–118
Simon ML, Cherry DS, Currie RJ, Zipper CE (2006) The ecotoxicological recovery of Ely Creek and tributaries (Lee County, VA) after remediation of acid mine drainage. Environ Monit Assess 123:109–124
Skousen JG, Zipper CE (2021) Coal mining and reclamation in Appalachia. Chapter 3 in: Zipper CE, Skousen JG (eds) Appalachia's coal-mined landscapes. Springer
Smucker NJ, Drerup SA, Vis ML (2014) Roles of benthic algae in the structure, function, and assessment of stream ecosystems affected by acid mine drainage. J Phycol 50:425–436
Smucker NJ, Vis ML (2011) Acid mine drainage affects the development and function of epilithic biofilms in streams. Freshw Sci 30:728–738
Snyder CD, Hitt NP, Young JA (2015) Accounting for groundwater in stream fish thermal habitat responses to climate change. Ecol Appl 25:1397–1419

Soucek DJ (2007a) Bioenergetic effects of sodium sulfate on the freshwater crustacean, Ceriodaphnia dubia. Ecotoxicology 16:317–325
Soucek DJ (2007b) Comparison of hardness- and chloride-regulated acute effects of sodium sulfate on two freshwater crustaceans. Environ Toxicol Chem 26:773–779
Soucek DJ, Cherry DS, Currie RJ, Latimer HA, Trent GC (2000) Laboratory to field validation in an integrative assessment of an acid mine drainage–impacted watershed. Environ Toxicol Chem 19:1036–1043
Stewart R, Grosell M, Buchwalter D, Fisher N, Luoma S et al (2010) Bioaccumulation and trophic transfer of selenium. In: Chapman PM et al (eds) Ecological assessment of selenium in the aquatic environment. CRC Press, pp 93–140
Strayer DL, Downing JA, Haag WR, King TL, Layzer JB et al (2004) Changing perspectives on pearly mussels, North America's most imperiled animals. Bioscience 54:429–439
Sundermann A, Stoll S, Haase P (2011) River restoration success depends on the species pool of the immediate surroundings. Ecol Appl 21:1962–1971
Sutherland AB, Meyer JL, Gardiner E (2002) Effects of land cover on sediment regime and fish assemblage structure in four southern Appalachian streams. Freshw Biol 47:1791–1805
Timpano AJ, Schoenholtz SH, Soucek DJ, Zipper CE (2018) Benthic macroinvertebrate community response to salinization in headwater streams in Appalachia USA over multiple years. Ecol Indic 91:645–656
Trout Unlimited (TU) (2011) The West Branch Susquehanna recovery benchmark project. Trout Unlimited's Eastern Abandoned Mine Program, Lock Haven, PA
US Environmental Protection Agency (USEPA) (2002) Consolidated assessment and listing methodology. Office of Wetlands, Oceans, and Watersheds
US Environmental Protection Agency (USEPA) (2011) The effects of mountaintop mines and valley fills on aquatic ecosystems of the central Appalachian coalfields. Washington DC
US Environmental Protection Agency (USEPA) (2016) Aquatic life ambient water quality criterion for selenium—Freshwater 2016. EPA 822-R-16–006. Washington, DC
US Fish and Wildlife Service (USFWS) (2019a) Big Sandy crayfish (Cambarus callainus). In: Big Sandy crayfish (Cambarus callainus). https://ecos.fws.gov/ecp0/profile/speciesProfile?sId=8285
US Fish and Wildlife Service (USFWS) (2019b) Guyandotte River crayfish (Cambarus veteranus). In: Guyandotte River crayfish (Cambarus veteranus). https://ecos.fws.gov/ecp0/profile/speciesProfile?sId=10375
Vander Vorste RV, Timpano AJ, Cappellin C, Badgley BD, Zipper CE, Schoenholtz SH (2019) Microbial and macroinvertebrate communities, but not leaf decomposition, change along a mining-induced salinity gradient. Freshw Biol 64:671–684
Vannote RL, Minshall GW, Cummins KW, Sedell JR, Cushing CE (1980) The river continuum concept. Can J Fish Aquat Sci 37:130–137
Vaughn CC (2018) Ecosystem services provided by freshwater mussels. Hydrobiologia 810:15–27
Voss KA, Bernhardt ES (2017) Effects of mountaintop removal coal mining on the diversity and secondary productivity of Appalachian rivers. Limnol Oceanogr 62:1754–1770
Vuori K-M (1996) Acid-induced acute toxicity of aluminium to three species of filter feeding caddis larvae (Trichoptera, Arctopsychidae and Hydropsychidae). Freshw Biol 35:179–188
Wallace JB, Eggert SL (2009) Benthic invertebrate fauna, small streams. In: Encyclopedia of Inland Waters. Elsevier, pp 173–190
Wang N, Ivey CD, Dorman RA, Ingersoll CG, Steevens J et al (2018) Acute toxicity of sodium chloride and potassium chloride to a Unionid mussel (Lampsilis siliquoidea) in water exposures. Environ Toxicol Chem 37:3041–3049
Wang Z, Meador J, Leung K (2015) Metal toxicity to freshwater organisms as a function of pH: A meta-analysis. Chemosphere. https://doi.org/10.1016/j.chemosphere.2015.10.032
Ward JV, Stanford JA (1982) Thermal responses in the evolutionary ecology of aquatic insects. Annu Rev Entomol 27:97–117
Warren ML, Haag WR (2005) Spatio-temporal patterns of the decline of freshwater mussels in the Little South Fork Cumberland River, USA. Biodivers Conserv 14:1383–1400

Warren ML Jr, Burr BM, Walsh SJ, Bart HL Jr, Cashner RC et al (2000) Diversity, distribution, and conservation status of the native freshwater fishes of the southern United States. Fisheries 25:7–31

Watson AS, Merovich GT Jr, Petty JT, Gutta JB (2017) Evaluating expected outcomes of acid remediation in an intensively mined Appalachian watershed. Environ Monit Assess 189:339. https://doi.org/10.1007/s10661-017-6036-x

Welsh SA, Loughman ZJ (2015) Physical habitat and water quality correlates of crayfish distributions in a mined watershed. Hydrobiologia 745:85–96

Whitmore KM, Schoenholtz SH, Soucek DJ, Hopkins WA, Zipper CE (2018) Selenium dynamics in headwater streams of the central Appalachian coalfield. Environ Toxicol Chem 37:2714–2726

Wiley JB, Evaldi RD, Eychaner JH, Chambers DB (2001) Reconnaissance of stream geomorphology, low streamflow, and stream temperature in the mountaintop coal-mining region, southern West Virginia, 1999-2000. U.S. Geological Survey, Charleston, WV

Williams J, Bogan A, Butler R, Cummings K, Garner J, Harris J, Johnson N, Watters G (2017) A revised list of the freshwater mussels (Mollusca: Bivalvia: Unionida) of the United States and Canada. Freshw Mollusk Biol Conserv 20:33–58

Wood PB, Williams JM (2013) Impact of valley fills on streamside salamanders in southern West Virginia. J Herpetol 47:119–125

Worischka S, Schmidt S, Hellmann C, Winkelmann C (2015) Selective predation by benthivorous fish on stream macroinvertebrates—the role of prey traits and prey abundance. Limnologica 52:41–50

Zalack JT, Smucker NJ, Vis ML (2010) Development of a diatom index of biotic integrity for acid mine drainage impacted streams. Ecol Indic 10:287–295

Zipper CE, Beaty B, Johnson G, Jones J, Rapp J, Ostby B, Wolfe W, Donovan P (2014) Freshwater mussel population status and habitat quality in the Clinch River, Virginia and Tennessee, USA: A featured collection. J Am Water Resour 50:807–819

Zipper CE, Donovan PF, Jones J, Li J, Price J, Stewart RE (2016) Spatial and temporal relationships among watershed mining, water quality, and freshwater mussel status in an eastern USA river. Sci Total Environ 541:603–615

Economic Trends in Appalachian Coalfield Counties

George D. Santopietro and Carl E. Zipper

Abstract An economic profile of the Appalachian coalfield counties is presented including trends in population, income, employment, labor force participation and unemployment, educational attainment, and poverty. Appalachian coalfield counties with relatively low populations (less than 100,000) are categorized into county types based on historical coal production. Economic characteristics of these county types are compared with those of other Appalachian counties, with counties outside Appalachia, and with high-population counties in Kentucky, Maryland, Ohio, Pennsylvania, Tennessee, Virginia, and West Virginia. Results show declining populations and labor forces, and slower rates of income growth for rural high-coal-production Appalachian counties; these same county types have lower income levels, higher unemployment, and higher poverty rates compared to the non-Appalachian counties in the seven states. These differences have been evident for decades. As the coal-mining industry has experienced declining production and employment in recent years, economic conditions in rural high-coal-production counties have declined. Formal educational attainment rates in rural coal-production counties are lower than in the more prosperous non-Appalachian and high-population counties of the seven states.

Keywords Coal · Miners · Personal income · Population · Poverty · Unemployment

1 Introduction

Over more than a century, coal has provided economic support for the Appalachian counties where it is mined. Coal mining, however, is a cyclical industry and has not enabled many coal-production areas to achieve sustainable economic growth.

G. D. Santopietro (✉)
Radford University, Radford, VA, USA
e-mail: gsantopi@radford.edu

C. E. Zipper
Virginia Tech, Blacksburg, VA, USA

C. E. Zipper and J. Skousen (eds.), *Appalachia's Coal-Mined Landscapes*,
https://doi.org/10.1007/978-3-030-57780-3_11

Appalachian areas have historically fared less well on many measures of economic achievement than other counties in their respective states (ARC 1964). Many of the counties performing lowest on such measures are in the central Appalachian region of southern West Virginia, eastern Kentucky, southwestern Virginia, and northeastern Tennessee where much of Appalachia's coal is produced; though some counties in other coal-production areas also experience economic difficulties (Pollard and Jacobsen 2019). Differences in measures of economic well-being between rural Appalachian counties and other areas in the mid-Atlantic region have persisted for decades (ARC 1964; Isserman 1996; Pollard and Jacobsen 2019). This chapter illustrates the persistence of these economic and employment differences through comparisons of rural Appalachian coalfield counties to other areas of the states where those counties are located (Kentucky, Maryland, Ohio, Pennsylvania, Tennessee, Virginia, and West Virginia) and to national averages for some data.

Independent cities in Virginia and in Maryland (Baltimore) were considered in a manner identical to counties in the analysis and are termed as counties in the text that follows. Each county in the seven states was classified as one of eight county types using the following criteria (Table 1 and Fig. 1):

- Cumulative coal production, 1980–2017, based on data from EIA (2019b):
  - > 300 million tons (Coal > 300 Mt);
  - 100–300 million tons (Coal 100–300 Mt);
  - 10–100 million tons (Coal 10–100 Mt);
  - < 10 million tons (Coal < 10 Mt).
- Coalfield No Production (Coalfield 0 t): Counties for which some segment of Appalachian coal-bearing geological formations (including Pennsylvania

Table 1 Numbers of counties, population and cumulative coal production (1980–2017) in the county-type groupings used for analysis

County type[a]	Count	Population, 2018	Coal mined (million tons)
Population < 100,000			
Coal > 300 Mt[b]	12	331,061	6256
Coal 100–300 Mt	23	889,407	4351
Coal 10–100 Mt	45	1563,081	1840
Coal < 10 Mt	50	1536,595	174
Coalfield 0 t	35	1152,010	–
Other ARC	89	2565,865	<0.1[c]
Non-Coal/ARC	218	7122,236	–
Pop > 100,000	112	36,940,894	1379
Total	584	52,101,149	14,000

[a]For a listing of counties by county-type designation, see Zipper (2020)
[b]The City of Norton, Virginia was placed within the Coal > 300 Mt county type because it is located within a Coal > 300 Mt county (Wise, Virginia)
[c]Giles and Montgomery Counties, Virginia

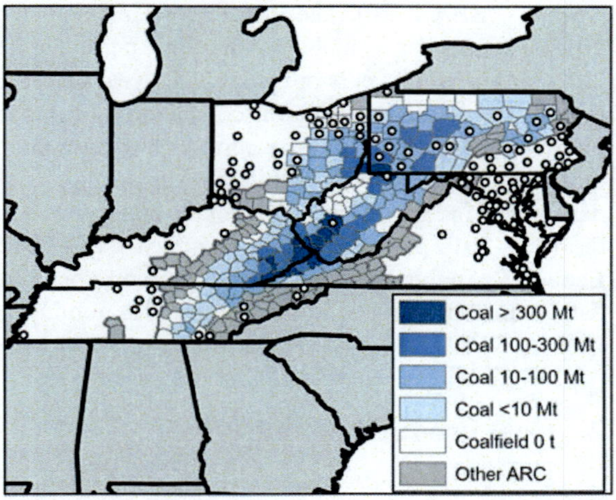

Fig. 1 Map designating county types used in analysis. Boundaries for counties outside of the Appalachian coalfield and the Appalachian Regional Commission service area (Non-Coal/ARC) are not shown. White circles designate counties with populations > 100,000 in 2018 (Pop > 100,000), a county type that supersedes other designations

anthracite) are located within their borders but with no recorded 1980–2017 coal production.

- Other ARC: Counties located within the Appalachian Regional Commission service area but not classified as coal-production or coalfield counties.
- Non-Coal/ARC: Other counties in the seven states.
- Pop > 100,000: Counties with populations greater than 100,000 in 2018 (this classification supersedes those above).

The population criterion was selected because it helps to separate counties with both high population and high coal production; the high population often results from socioeconomic drivers other than coal production which influence the economic metrics. For example, the following counties have high coal production but have high populations with diversified economies. Kanawha County, West Virginia, includes the state capital, Charleston, and its suburbs; Monongalia County, West Virginia, includes an urban center, Morgantown and West Virginia University; and Washington County, Pennsylvania, includes prosperous Pittsburgh suburbs. As such, all three of these counties with 2018 populations greater than 100,000 have economic-activity drivers that are not characteristic of other high-coal-production counties and, as a result, their socioeconomic measures also differ. In the text that follows, we use the term "high-coal-production counties" to describe rural counties with cumulative coal production greater than 100 million tons, the "coal-production counties" to designate the four-county types with post-1980 coal production; the term "rural" to designate counties with populations less than 100,000, and the term "high-population" to designate counties with populations greater than 100,000.

Economic and demographic measures were acquired for each county within the seven states and for the United States (US) as a whole where available within the same database. As appropriate, values were calculated for each county type on a population-weighted basis using annual populations from the Bureau of Economic

Analysis (BEA 2020). The measures were calculated annually from 1969 through 2018 or 2019 when available for the entire period, or from 2000/2001 through 2018 or 2019, or for recent years or recent multi-year periods when longer-period data were not available. The analysis was conducted by comparing aggregates of county-type metrics over the relevant time-period. Because coal mining is an important contributor to economic activity within the Appalachian coalfield, we also present regional data on that industry as a context for the county-type comparisons. Economic data for the 1969–2019 period are inflation-adjusted to a 2012 basis using the US Bureau of Economic Analysis Personal Consumption Expenditure price index while economic data for more recent periods are presented as nominal (non-inflation adjusted) dollars.

2 Coal Production and Prices

Coal has been an important driver of economic activity in Appalachia's coalfield but its economic contributions have been cyclical (Fig. 2). While annual production varied from year to year from 1969 to 2009, it was usually within a range of 350–450 million tons. National-average-coal prices peaked in 1979 during the energy crisis caused by sudden disruptions in the supply of imported oil. As a substitute for oil in heating and electrical energy production at the time, coal's prices increased. Despite coal's role in supporting the nearly continuous expansion of the US economy from 1983 to 2008 (NBER 2020), the inflation-adjusted price of coal steadily declined from this peak until the end of the twentieth century. The coal industry continued production as prices declined by increasing mining efficiency, including mechanization that increased labor productivity measured as tons per hour of labor (Fig. 2) (CRS 2017). Declining prices also contributed to changing industry structure with larger companies gaining greater market shares of total production (EIA 1993).

During the early 2000s, opportunities for efficiency gains dissipated as remaining reserves became more costly to mine, and surface miners faced new regulatory restrictions (Hook and Aleklett 2010; Copeland 2015). As mining costs increased, firms were able to maintain production because prices began to trend upward. With the Great Recession of 2007–09 production began to decline, even though prices were still rising, and then developed into a sustained downward trend. From 2012 on with both price and production falling and despite a return to increasing labor productivity, the coal industry experienced economic difficulty including multiple bankruptcies (CRS 2017; Krauss 2019; Macey and Salovaara 2019). A major reason was and continues to be the supply of lower-cost natural gas which caused electric power producers to retire multiple coal-fired electric generating plants (Bowen et al. 2018; EIA 2019c).

A slight reversal in price and production began in 2016 attributable to increased demand, mainly from overseas, and improved pricing for metallurgical coal. Central Appalachian mines produce high-quality metallurgical coals which generally command higher prices than the steam coal sold to the electric power markets. The increasing share of production devoted to overseas metallurgical markets

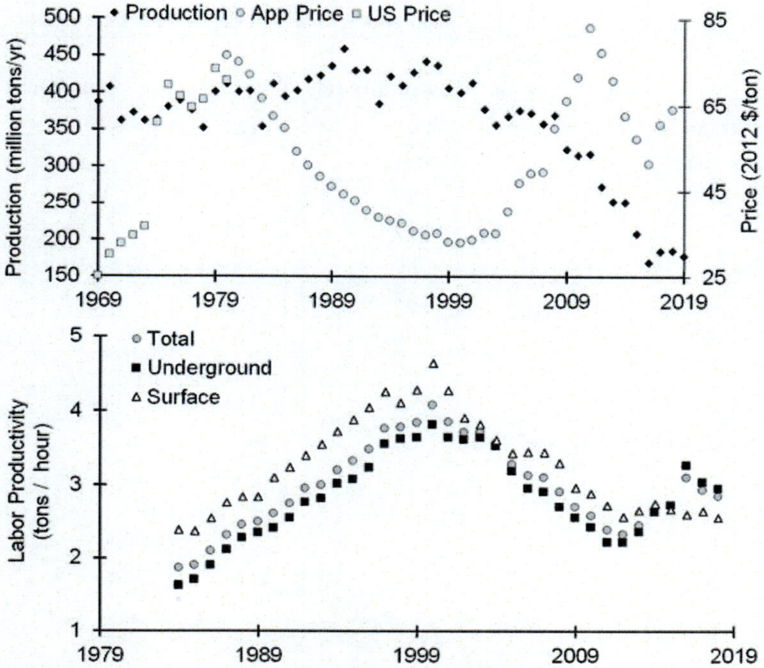

Fig. 2 (Above) Appalachian coal production and estimated average prices in seven states (from Ohio and Pennsylvania extending south through Tennessee). 1980–2018 production from EIA (2019b); 2019 production from EIA (2020c); pre-1980 production from Milici (1997). Appalachian regional average pricing (1980–2018) is estimated by calculating production-weighted annual averages state-level mine-mouth prices using EIA (2019b) data. US national average price for bituminous coal (1969–1980) from EIA (2012). (Below) Average labor productivity by mine type, 1983–2018, for the Appalachian region as defined by EIA (2019b) (includes Alabama). Data from EIA (2019b)

is a major cause for the increasing prices and production seen for 2016–2018, despite Appalachian shipments to domestic markets declining by an additional 10% (EIA 2019a).

3 Population

3.1 Total Population

Population levels in the Coal > 300 Mt and Coal 100–300 Mt counties are now lower than in 1969 although levels in other coalfield counties grew and Other ARC and Non-Coal/ARC counties experienced even more rapid growth (Table 2 and Fig. 3). This overall decline in rural high-coal-production counties was punctuated by population

Table 2 Population change in county groupings of 7 states and US for three time periods

County type	Population change			
	1969–2018 (%)	1990–2018 (%)	2001–2018 (%)	2011–2018 (%)
Coal > 300 Mt	−14.9	−18.4	−12.7	−9.0
Coal 100–300 Mt	−8.7	−9.5	−7.5	−5.4
Coal 10–100 Mt	4.7	−0.5	−2.5	−2.9
Coal < 10 Mt	26.4	9.2	2.3	−1.5
Coalfield 0 t	16.4	7.0	0.9	−1.1
Other ARC	47.6	21.8	6.8	0.5
Non-coal/ARC	45.0	22.6	8.7	1.7
Pop > 100,000	32.7	20.0	10.7	3.5
US	62.5	31.1	14.8	5.0

Data from BEA (2020, data series CAINC1)

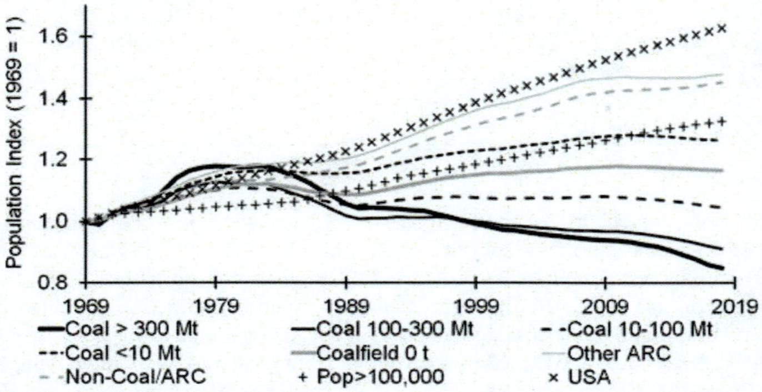

Fig. 3 Population trends, 1969–2019 by county type, indexed such that 1969 population levels = 1. Data from BEA (2020, data series CAINC1)

increases of about 8% during the "coal boom" from 1974 to 1980 when the population expanded in Appalachia and in other rural areas of the seven states. Appalachian population numbers remained relatively stable through 1984. Population declines in the high-coal-production counties thereafter were associated with the decline in coal prices from their 1979 peak and reductions in employment by the mining industry. By 1990, the high-coal-production counties' population was 9% below 1984. Other coalfield areas also lost population though more slowly as non-coalfield rural county populations continued to rise. Since 1990, a clear pattern of population change associated with historical levels of coal mining becomes evident. The higher production areas experienced the highest levels of population declines. These population losses occurred as the coal-mining industry dramatically improved labor productivity. By 2000, Appalachian coal mines had reduced employment to 48,021 from the > 130,000 employed in 1980 (EIA 2019b). This reduced demand for labor in Appalachian coal mines likely was a proximate cause for the population declines up until then.

Since 2001, population declines continued to be most significant in rural high-coal-production counties. On average, these areas suffered population declines of nearly 10%. Every county with a population of less than 100,000 with coal production ≥ 100 Mt and 2018 population < 100,000 experienced declining population over this period. Among counties within the seven-state region and the ≥ 100 Mt coal-production types, the greatest population loss was experienced by McDowell County, West Virginia (−31%). Other high-coal production counties experiencing > 15% losses were Buchanan County Virginia, Wyoming County West Virginia, and Breathitt, Harlan, and Leslie Counties in Kentucky. Two other coal-production counties and only two of more than 300 non-coal-production counties within the seven-state region also experienced > 15% population losses.

3.2 *Population Age Groups*

The population shares of two age groups are examined to provide a profile of changing demographics. The prime working-age group is from 25 to 54 years of age; the second age group is 65 years and older. One trend for both age groups, both men and women, is similar in all county types: the proportion of the population of prime working age (25–54) is declining while the age-65-and-over population proportions are increasing (Table 3). The prime-working-age population declines are happening more rapidly in rural high-coal-production counties than in other segments of the study area and more rapidly for women than for men in these counties. High-coal-production counties' prime-working-age population shares were noticeably higher

Table 3 Shares of population (expressed as %) represented by prime working-age (age 24–54) males and females and by age-65-and-over males and females in 2000 and 2018

	Age 25–54				Age 65 and over			
	Male		Female		Male		Female	
	2000	2018	2000	2018	2000	2018	2000	2018
Coal > 300 Mt	23.5	19.6	23.2	18.2	5.7	8.7	8.4	10.6
Coal 100–300 Mt	21.9	18.9	21.6	17.5	6.6	9.3	9.8	11.2
Coal 10–100 Mt	20.7	18.4	20.7	17.7	6.4	9.2	9.3	11.2
Coal < 10 Mt	19.9	18.4	20.0	17.7	6.0	9.2	8.2	10.8
Coalfield 0 t	20.2	18.0	20.4	17.6	6.0	8.9	8.6	10.8
Other ARC	19.9	18.4	19.7	17.9	5.5	9.1	7.7	10.8
Non-coal/ARC	19.9	18.7	19.8	18.4	5.2	8.1	7.4	10.0
Pop > 100,000	20.3	19.5	21.2	20.0	4.8	6.9	7.2	9.0

Data from BOC (2020b, c)

than those of comparison areas in 2000 and remained moderately higher in Coal > 300 Mt counties than in other coalfield areas in 2018. The availability of good-paying jobs in the coal industry is likely a factor keeping people in this age group in the high-coal-production counties and perhaps attracting workers from other areas. This distinction though appears to be thinning as coal jobs disappear. High-coal-production counties have maintained relatively high working-age population shares as overall populations declined by nearly 10% on average (Table 2).

Rural Appalachian counties (all coalfield county types and Other ARC) have moderately higher fractions of population 65 years of age and older than the Non-Coal/ARC and high-population (Pop > 100,000) counties. This effect is evident for both men and women. Within Appalachian areas, the 65-and-older population share is moderately lower in the Coal > 300 Mt counties than in other areas. With more of the population in the working-age group, fewer are in the older age group. This could also be the result of higher levels of mortality in the high-coal-production counties (Gohlke 2021), including mortality associated with coal mining employment if such mortality is affecting older residents preferentially.

4 Personal Income

Income is derived from two general sources: payments for labor, i.e., wages and salaries, and returns to investments in physical capital, natural resources, and savings. Payments for wages and salaries could also be thought of as a return on investment in human capital—skills and abilities developed by workers through education, training, and experience. The monetary payments may reflect not just the value produced by the worker but also the riskiness of the job to the worker's health. Higher-risk jobs often pay more than jobs with similar requirements but lower risk. In addition to wages and salaries, returns to the worker may also include non-monetary benefits such as job satisfaction and living in the desired location. The dilemma facing many coal miners is that the relatively high wage reflects both considerable human capital acquired through experience and training that cannot easily be transferred to other employment and a higher risk level than most other jobs (Murray and Schaeffer 2018). The next best alternative job therefore may not pay anywhere near as much as coal mining or may require relocation to a less desired area.

In this section, several measures of income and wages are examined. Some measures are aggregates for the county, i.e., a total value for all residents; others are on a per-capita basis reflecting average values for individuals. Aggregate values will be higher for the types with more counties and therefore more people; because raw data of this type cannot be represented well graphically, indexed values are presented leaving selected specific values for tables.

4.1 Total and Per-Capita Personal Income

Personal income as reported by BEA (2020) is an aggregate of all household income including wages, salaries, investment income and government transfer payments such as social security. Inflation-adjusted aggregate personal income has risen over the past five decades for all county types, although differences among rates of growth are apparent (Fig. 4). The coal-industry led prosperity and increasing populations (Fig. 3) during the energy crisis of the 1970s is reflected in the growth of income for the high-coal-production counties up until 1979. For the rest of the century, personal income in these counties declined and did not reach the 1979 level again in Coal > 300 Mt counties until the late 1990s. Personal income grew along with coal pricing until the next downturn after 2011. The lack of recent growth corresponds with the severe decline in coal production and pricing (Fig. 2).

Regardless of the time period analyzed, growth in personal income is consistently lower in the coal-production counties than in the Other ARC and Non-Coal/ARC counties and the nation as a whole (Table 4). Income growth up till 2018 is broken down into four time periods. For the coal-production counties, the rate of growth is

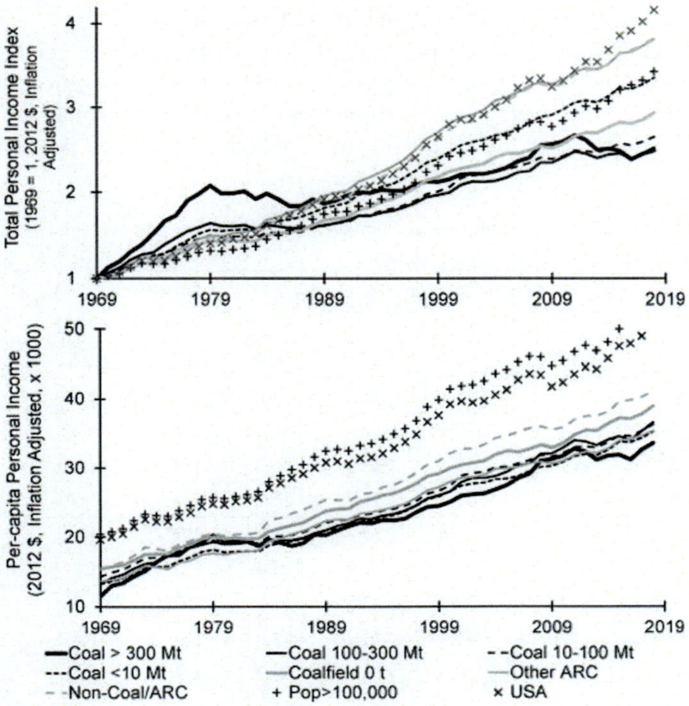

Fig. 4 Aggregate personal income index (1969 = 1, upper) and per-capita personal income (lower) by county type, 1969–2018. Data from BEA (2020), data series CAINC1

Table 4 Aggregate personal income annual growth rates for selected periods by county type

County type	Personal income total, inflation adjusted Average annual rate of growth (%)			
	1969–2018	1990–2018	2001–2018	2011–2018
Coal > 300 Mt	1.9	0.9	0.7	−1.0
Coal 100–300 Mt	1.9	1.5	1.1	0.3
Coal 10–100 Mt	2.0	1.7	1.2	0.9
Coal < 10 Mt	2.5	2.2	1.6	1.3
Coalfield 0 t	2.2	2.0	1.5	1.5
Other ARC	2.8	2.4	1.7	1.6
Non-coal/ARC	2.8	2.5	1.9	1.8
Pop > 100,000	2.5	2.4	2.0	2.2
USA	3.0	2.7	2.2	2.8

Data from BEA (2020), CAINC1 and CAINC5N, 2012-base, inflation-adjusted dollars

highest for the longest time period, 1969–2018, as it was buoyed by the boom of the 1970s. Since 2011, personal income has fallen by 1% in the Coal > 300 Mt counties, and growth was negligible in the Coal 100–300 Mt counties.

Per-capita personal income trends exhibit similar patterns while increasing for all county types (Fig. 4), as aggregate personal income trends also reflect changes in population. National average and high-population county per-capita incomes were greater than those of the more rural counties throughout the study period (Table 5).

Table 5 Average annual per-capita personal income, selected years, and average annual per-capita personal income growth rates, selected periods

County type	Per-capita income (×1000 2012-$)				Growth rate (%)		
					1969–2018	2001–2018	2011–2018
	1969	2001	2011	2018			
Coal > 300 Mt	11.5	25.9	32.8	33.6	2.2	1.5	0.3
Coal 100–300 Mt	13.2	27.9	34.0	36.6	2.1	1.6	1.0
Coal 10–100 Mt	14.3	28.8	33.0	36.3	1.9	1.4	1.3
Coal < 10 Mt	13.3	27.4	31.7	35.2	2.0	1.5	1.5
Coalfield 0 t	15.4	30.4	34.8	38.9	1.9	1.5	1.6
Other ARC	13.7	28.5	31.9	35.4	2.0	1.3	1.5
Non-coal/ARC	15.6	32.7	37.0	41.1	2.0	1.3	1.5
Pop > 100,000	20.4	41.9	46.7	52.6	2.0	1.4	1.7
USA	19.6	39.6	43.5	50.3	1.9	1.4	2.1

Data from BEA (2020), CAINC1 and CAINC5N, 2012-base, inflation-adjusted dollars

Although growth rates were modestly higher in the high-coal-production counties than for most other county types over the 1969–2018 and 2001–2018 periods, substantial per-capita income differences persist.

4.2 Wages and Salaries

Wages and salaries are payments for labor services and the largest source of income for most US households. All areas exhibit rising trends over the study period (Fig. 5). Wages and salaries experienced declines during the Great Recession of 2007–09 in most areas, excepting the high-coal-production counties. Total wages and salaries rose faster from 2008 through 2011 for the Coal 100–300 Mt and Coal > 300 Mt counties than in other county types, after which a dramatic reversal occurred in association with trends in coal-production and pricing (Fig. 2). As the coal industry experienced modest recovery in 2017 and 2018, wages and salaries in the high-coal-production

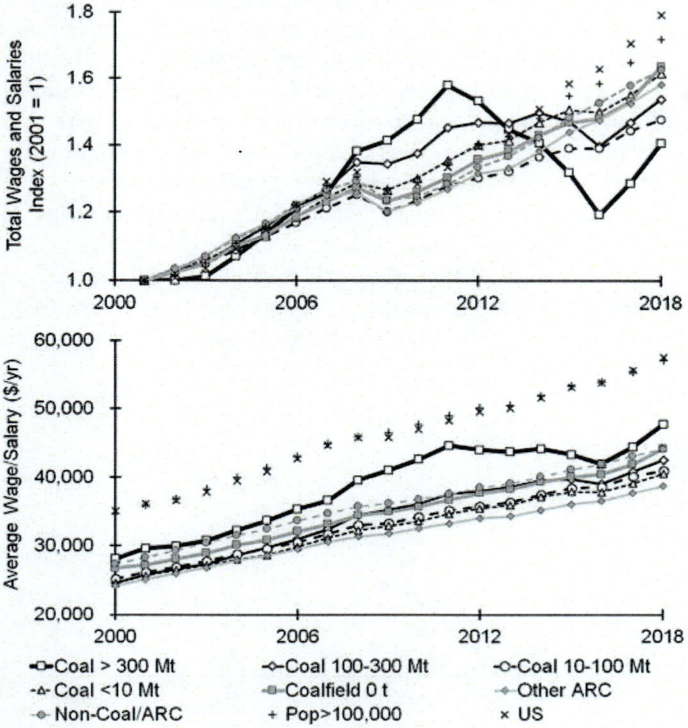

Fig. 5 Aggregate wages and salary index (2001 = 1) and average wages and salaries per employee (calculated as total wages and salaries/total wage and salary employees) by county type. Data from BEA (2020), CAINC4

counties again increased but did not return to 2011 levels in Coal > 300 Mt counties. For the full 2001–2018 period, all coal-production county types experienced slower wage-and-salary growth than Non-Coal/ARC counties, Pop > 100,000 counties, and the US as a whole.

Average wages and salaries are also presented, providing a view from the household perspective (Fig. 5). The Coal > 300 Mt counties have a higher average annual wage than all other rural county types throughout 2001–2018, a likely result of the relatively high rates of pay by the coal industry. While they escaped the stagnation of wage-growth experienced by other areas during the Great Recession, wages in these counties did decline from 2012 to 2016, likely reflecting the effects of reduced demand for mining labor given declines in coal production and pricing (Fig. 2).

4.3 *Wages and Salaries in Mining*

Average annual wages and salaries in coal mining in the eastern US rose steadily for most of the 2001–2018 period (Fig. 6). These values are aggregated at the state level because of data non-disclosures at the county level. They incorporate mining activity in western Kentucky in addition to Appalachian mines. The effects of this are minor, however, as annual western Kentucky mine employment ranged from 5 to 11% of total mine employment in the seven states over the study period. Even at the state level, however, data were suppressed for some states in some years so Fig. 6 data represent employment-weighted averages of reported values.

During most years, Appalachian wages and salaries are slightly below US averages for coal miners but still well above average levels for all workers in Appalachia's coalfields. Over the 2001–2018 period, the average annual regional coal-mining wage ranged from 71 to 86% above the overall average wage for the Coal > 300 Mt counties, and from 92 to 113% for the Coal 100–300 Mt counties. Since coal-miners

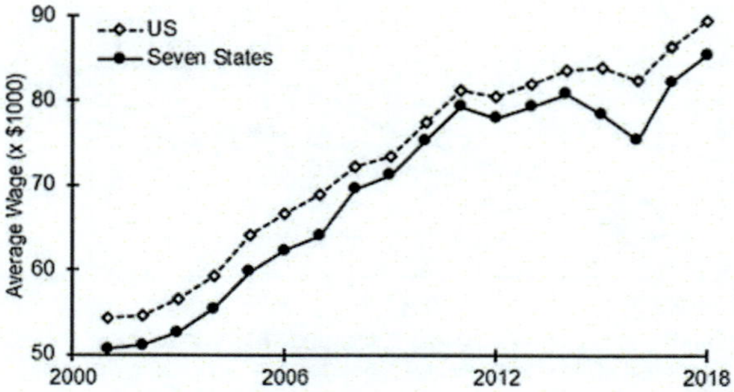

Fig. 6 Employment-weighted average annual wage for coal mining in seven Appalachian coalfield states including all of Kentucky, and US average. State-average wage data from BLS (2020a) were compiled to produce regional average using EIA (2019b) employment as state weightings

wages contribute to overall average wages within these counties, the actual wage increment for coal-miners' wages over other occupations' wages is higher.

5 Employment

5.1 Employment in Coal Mining

Appalachian employment in coal mining followed employment in other areas of the US in that industry since 2000 (Fig. 7). Employment started increasing in the mid-2000s and peaked in 2011 as mining firms sought to maintain or increase production in response to rising prices (Fig. 2) despite the declining quality of reserves and labor productivity. Employment fell precipitously from 2011 through 2016 with the fall in coal prices and the fall in production which reached its lowest level since the 1800s. The 25,168 coal-mining employees reported by EIA (2019b) for the seven Appalachian states (Tennessee northward but including only eastern Kentucky) in 2016 was less than half of the 55,513 employees reported for 2011. Employment rose slightly in the following two years, reaching 27,716 in 2018. In 2001 one-third of the jobs were in West Virginia; by 2018 one-half were there. Eastern Kentucky experienced the largest employment loss over this period as mine employment fell from 15,112 workers to 4121, a loss of 73%, and more than half of the region's total job loss.

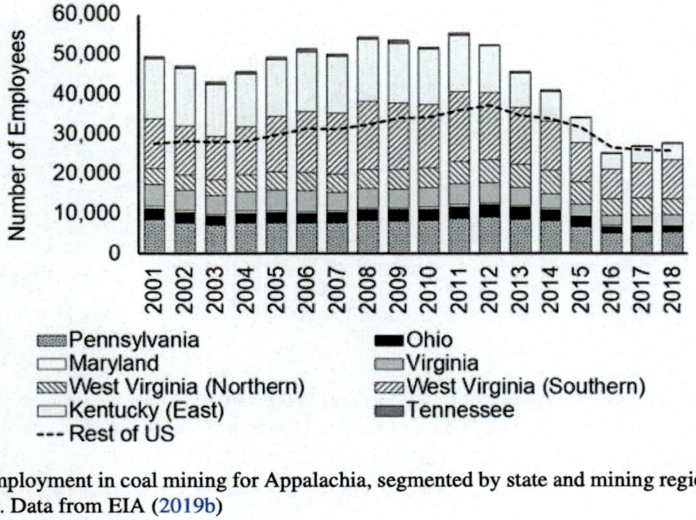

Fig. 7 Employment in coal mining for Appalachia, segmented by state and mining region, and for rest of US. Data from EIA (2019b)

5.2 Civilian Labor Force and Participation

The Civilian Labor Force (CLF) includes members of the population 16 years of age and older not institutionalized or in the military who are currently working or not working but actively seeking work. People not working and not looking for a job, such as many full-time students and retirees, do not count in this measure. All coal-production county types experienced a decline in the CLF over the 2001–2018 period (Fig. 8). Those declines occurred in an inverse relation to production levels and reflected both declining populations for the three highest coal-production county types and declining prime-working-age population shares (Tables 2 and 3). In contrast, the CLF in Non-Coal/ARC and Other ARC counties increased over this period despite some short-term declines at the time of the Great Recession, 2007–09, when some potential workers gave up looking for work. The CLF in high-population counties experienced continued growth over most of the period.

Civilian labor force incorporates both the working-age population and the labor force participation rate, i.e., the fraction of potential workers who are working or actively seeking work. Labor force participation has declined over the past decade in all county types (Fig. 9). These declines reflect ongoing national trends that have been evident since the 1990s arising from various factors including an aging population (Richter et al. 2018). Labor force participation rates have been lower in the coal-production counties than in the other county types and are lowest in the high-coal-production counties. Differences between Coal > 300 Mt and all non-coal-production county types exceeded 10% for both time periods.

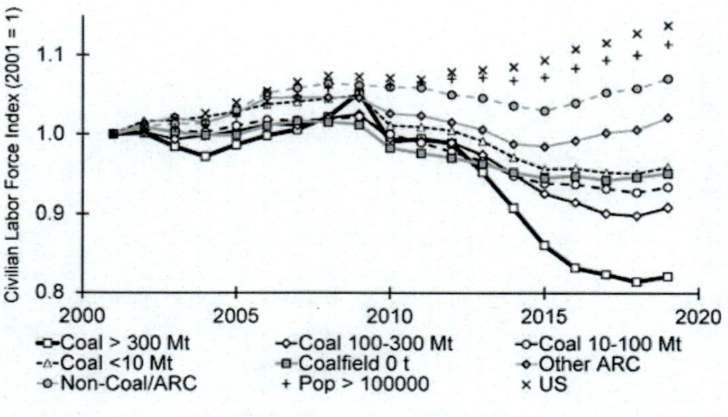

Fig. 8 Civilian labor force (CLF) by county type, 2001–2018, plotted as an index (2001 = 1) because of large differences among CLF magnitudes. Data from BLS (2020a), data code LAUCNTY

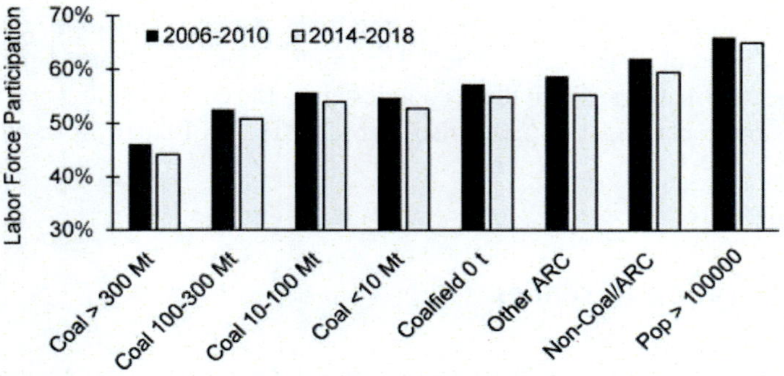

Fig. 9 Labor force participation rates, averages for five-year periods ending in 2010 and 2018. Vertical axis is truncated to emphasize differences. Data from BOC (2020a), data series ACSDP5

5.3 *Unemployment*

The official unemployment rate most commonly cited is U3, the percentage of the civilian labor force that is not working but actively seeking work. Unemployment rates varied widely throughout the study period for all county types, as all increased sharply during the Great Recession of 2007–09 (Fig. 10). Throughout most of the period, unemployment rates were lower in the rural non-coalfield and high-population counties than in coal-production areas. The primary exception to this pattern occurred from the Great Recession through 2011 for the Coal > 300 Mt counties as both coal pricing and coal-miners' wages were increasing (Figs. 2 and 6). From 2012–2016, coal pricing and production deteriorated even as the broader economy improved; and the coal-production counties resumed their status as having the highest unemployment rates within the study area.

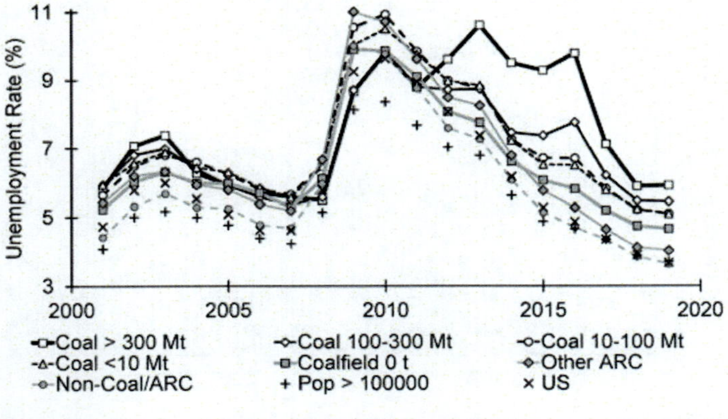

Fig. 10 Population-weighted average unemployment rates (U3) by county type, 2001–2018. Data from BLS (2020b)

Over most of the 2012–2018 period, the highest coal-production counties had the highest unemployment rates. The U3 unemployment rates likely understate the employment difficulties within these counties given the corresponding and relatively low labor force participation rates also recorded for 2014–2018 in these counties (Fig. 9).

6 Coal Mining Revenues

The economic value of an industry to an area can be assessed by considering how much revenue generated by firms in that industry remains in the area. Revenues from coal mining were estimated as annual county production quantities multiplied by average statewide price using data from EIA (2019b) and adjusted for inflation using the BEA Personal Consumption Expenditure price index (Fig. 11). An estimated total of $720 billion dollars in revenue, expressed on a 2012-dollar basis, was generated by coal mining in Appalachia from 1980 through 2018, equivalent to about $800 billion on a 2020-dollar basis. For the recent period of 2001–2018, that total is about $330 billion on a 2020-dollar basis.

The majority of coal revenues, about 75%, were generated in the two high-coal-production county types (> 100 Mt). Revenue does not necessarily follow the same trend as either price or production, nor does it necessarily track with profitability. If the price is falling because a decline in production costs leads to lower prices in a competitive market, production could be just as if not more profitable. The decline in prices from the peak of the energy crisis in 1979 to the early 2000s clearly had an impact on revenue even as production remained relatively stable. The increase in price from then until 2011 appears to have increased revenue as well. This all

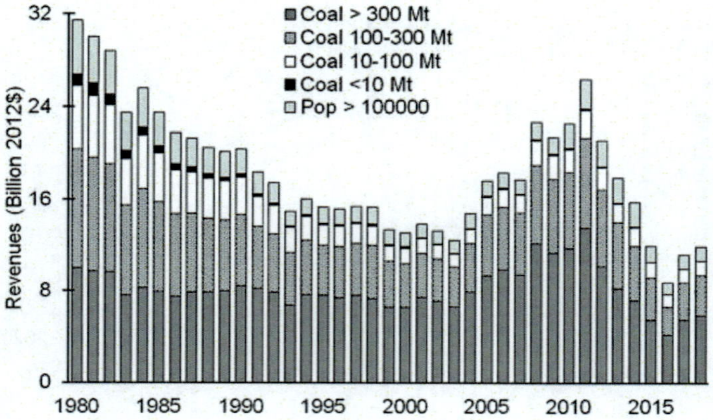

Fig. 11 Estimated annual revenue from coal mining, 1980–2018, by county type for mine locations; data from EIA (2019b) as compiled by Zipper (2020)

changed after 2011 when estimated revenues began to fall in association with price and production; all of these factors reversed to increase slightly in 2017–18, resulting in corresponding increases for regional coal-mining employment and wages (Figs. 5 and 7).

Given the data on revenue, employment (Fig. 7), and average wage in coal mining (Fig. 5), revenue per job can be estimated. Over the 2001–18 time period, wage expenses as a percent of revenue averaged 17%, only ranging between 15 and 18%. Non-wage portions of revenue may also remain in the regions via expenditures on other locally sourced inputs such as machinery and equipment, fuel, and via state and local taxes. Portions of revenue will also go to the owners of firms as resource rent and profit. Thus, local-area retained shares of revenues are greater where the coal being mined is owned by local entities, but large segments of land resources in some Appalachian coal-mining areas are owned by large firms based elsewhere (Spence et al. 2013).

7 Socio-economic Measures

7.1 Poverty Rates

The fraction of an area's population living on income below the federal poverty level as set by the US Department of Health and Human Services is known as the poverty rate. Within the seven-state area, the coal-production county types have the highest poverty rates (Fig. 12). Average poverty rates for the coal-production county types for 2018 ranged from 4.6 to 11.4% higher than the average for Non-Coal/ARC counties, which themselves have had poverty rates lower than the US average over the study period. Over the full period, annual average poverty rates in Coal > 300 Mt counties ranged from 4 to 10.1% above corresponding averages for the Coalfield 0

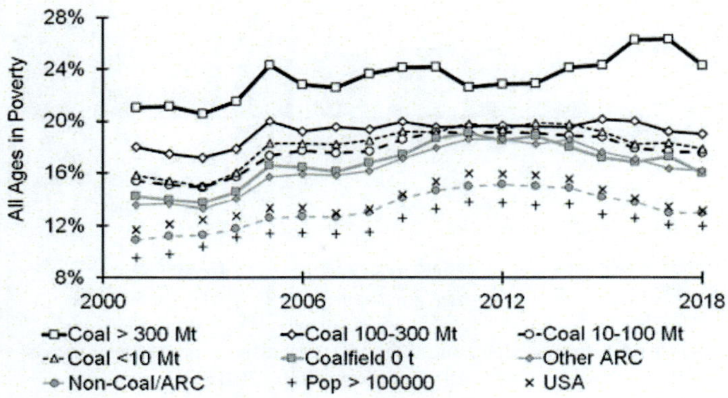

Fig. 12 Annual poverty rates by county type, 2001–2018. Data from BOC (2020d)

t and Other ARC county types. These figures indicate that a significant poverty gap does exist in the high-coal-production counties despite the relatively high wages for mining and that these gaps do not change much year to year.

The highest poverty rate is for Coal > 300 Mt counties even though this county type had an average wage higher than all other rural county types throughout the study period. The high wage rates in coal mining do not appear to translate into lower poverty rates. Coal > 300 Mt counties also had a low average CLF participation rate, as less than half the adult population was in the workforce. While residents of these areas working in the coal industry did well, many others, likely including those not in the workforce, did not share in that good economic fortune.

7.2 *Formal Education Attainment*

Rates of formal education attainment also varied by county type (Fig. 13); they were lower in coal-production counties than in other county types, lowest in the high-coal-production county types, and highest in the high-population counties. The 12.5% of rural Coal > 300 Mt county residents who have achieved a Bachelor's degree or higher level of formal education was exceeded by all other county types, including Coalfield 0 t (20.4%), Non-Coal/ARC (22.0%), and Pop > 100,000 (34.8%). In contrast, the share of the population that had not attained high school graduation or equivalency in Coal > 300 Mt counties (21.9%) exceeded that of all other county types by 6% or more.

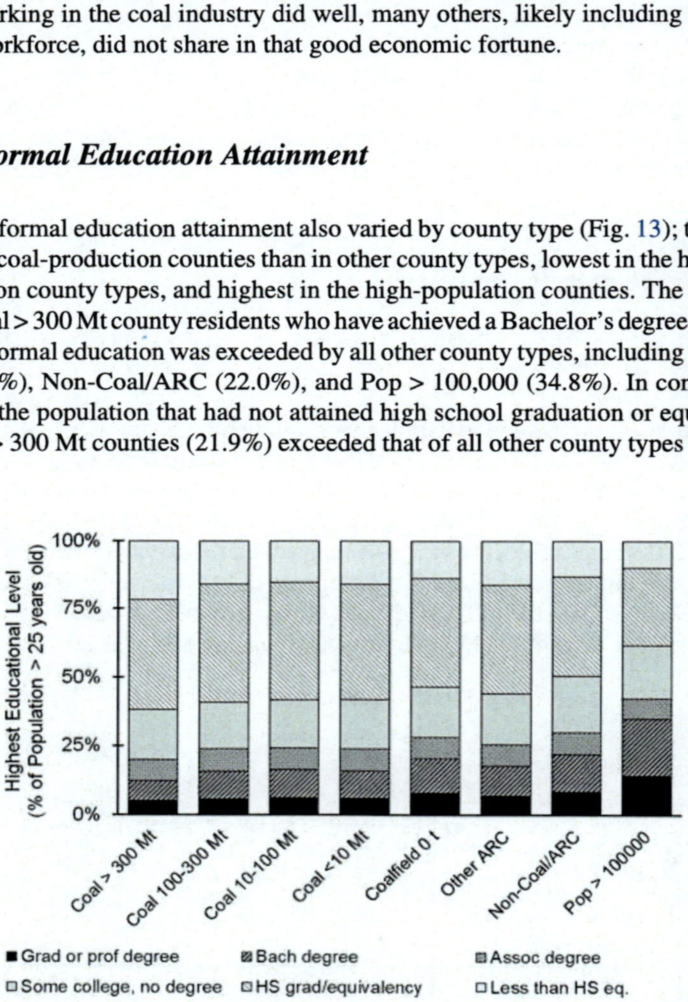

Fig. 13 Highest level of formal education attainment, expressed as percentages of age-25-and-older populations, five-year average (2014–18). Data from BOC (2020a)

8 Discussion

Measures of economic activity and human well-being in rural Appalachian counties lag behind those of many other US areas. This has been well-known since at least the 1960s and led to the formation of the Appalachian Regional Commission, a federally funded partnership of government agencies that aims to strengthen Appalachia's economy (ARC 2020). While ARC programs, state and local governments, community groups and others working toward these goals have achieved much success, recent analyses found that economic disparities remain (Pollard and Jacobsen 2019).

Analyses here have focused on Appalachia's coal-production areas. Economic indicators for county types within those areas are stratified by coal production levels, and contrasts are made with other areas of a seven-state region. These analyses found that measures of economic activity and well-being within rural coal-production counties are often lower than corresponding measures for non-coal-production areas, both within and beyond Appalachia, and for high-population areas; and that such disparities are greatest for the rural high-coal-production areas. Results of such comparisons vary through time, often driven by the economic fortunes of the coal-mining industry that is a significant source of income and employment for the local economies of coal-production areas.

In the United States, depopulation is occurring across many rural areas with natural-resource-dependent economies, including midwestern and southeastern areas that are economically dependent on agriculture, as well as mining-dependent areas in Appalachia and elsewhere (Johnson and Lichter 2019). Depopulation problems can be exacerbated by localized factors, such as the rising incidence of "deaths of despair" that has been noted in areas of chronic unemployment (Johnson and Lichter 2019) and which are evident in Appalachia's rural counties; while the elevated mortalities that have been noted in Appalachia's high-coal-production counties (Gohlke 2021) are also likely contributing.

Depopulation negatively impacts economic well-being and the general quality of life for remaining residents. Much has been written about the difficulties of maintaining public infrastructure in depopulating cities (e.g. Hoornbeek and Schwarz 2009) but we are aware of no similar studies for rural areas. As the population declines, the costs of maintaining infrastructures such as schools, roads, and water systems fall on fewer residents, in some cases resulting in failure to continue adequate maintenance and/or disrepair. Local businesses also struggle to survive with fewer customers.

The aging population, declining working-age population shares, and predominance of females within higher-age categories present in Appalachia are similar to national trends; however, in Appalachia, such trends are occurring more rapidly than in many other US areas (Mather et al. 2015). These national trends are driven by extended life expectancies enabled by improved healthcare, declining fertility and immigration, and in rural areas outmigration of younger residents. Aging populations typically require access to specialized business services such as healthcare but rural areas with declining and aging populations often experience a declining presence of local service businesses (Thiede et al. 2017).

County-mean per-capita income differences are apparent within the rural county-types of our study area. However, differences between the rural county types of Appalachia and the more urbanized counties with populations greater than 100,000 are far greater. This difference reflects national trends, as urban areas have generally fared far better economically than rural areas over recent decades. Convergence Theory predicts that lower income area economies will grow faster than higher income areas over the long-run and will eventually catch-up in terms of per-capita income (Barro and Sala-i-Martin 1995). The theory is based on the idea that lower-income areas are less industrialized and have less physical capital, and therefore will attract more capital investment and thereby grow faster until this difference is eliminated. In contrast to that theory, the data reviewed here indicate divergence over the time periods analyzed as differences among county types expanded.

The wavering but usually high unemployment rates in the coal-production counties reflect in part dependence on coal mining and the boom/bust cycles of coal markets. While the U3 unemployment rate in coalfield county types usually trends with the national rate, exceptions were evident and coalfield rates remain much above the national average. The persistence of high unemployment could be traced to several possible causes. As with many professions in natural resource extraction, the hope of another boom on the horizon keeps workers attached to their current profession even though not working, thereby extending the period or duration of unemployment. Even if an unemployed miner is willing to work at another job at a substantially lower wage, employers may be reluctant to hire the worker if the downturn is perceived as temporary and the worker is likely to go back soon to the higher-paying job. Workers 50 years of age and older also will face employers generally reluctant to hire in this age group; and those workers are less likely to seek educational and retraining programs, adding to the duration problem (Bersin and Chamorro-Premuzic 2019).

Miners may find that their job has few or no equivalents outside the coal industry (Murray and Schaeffer 2018). Their capabilities and tolerance for risk may not have a similar value in other industries. For displaced workers, miners in particular, one path to a similarly valued level of employment may be more formal educational or vocational training. Engineers, management, and administrative staff may have an easier time finding similarly valued work. Additionally, workers living in rural areas may want to remain there; however, employment opportunities of any kind may be few, and new business investment, whether local or attracted from outside the region, is scarce. One alternative is to commute but unless the county is in close proximity to an urban area, the level of competition for available jobs increases as more workers are willing to drive longer distances to workplaces (Murray and Schaeffer 2018).

Among the challenges faced by high-coal-production counties are the relatively low levels of formal education achievement by its populations and, hence, the potential workforce. The economics literature is replete with studies on the economic value of educational attainment, oftentimes referred to as investment in human capital (Belfield and Levin 2007; Checchi 2006; Tamborini et al. 2015). This extends as well to the inter-generational transmission of the benefits of education (Becker et al.

2018). Bureau of Labor Statistics data indicates that for the US as a whole, the median weekly earnings in 2017 for workers with a Bachelor's degree but no further formal education exceed those whose formal education terminates with a high-school diploma by more than 60% (Torpey 2018). While formal education attainment rates fail to capture the capabilities and skills of many workers, potential employers evaluate those statistics when considering locations for business expansion and investment. When coal miners could earn a relatively high wage based on industry-specific skills acquired by learning and by doing, other forms of investment in education and training were not perceived as offering as great a return. Thus, it is not surprising to see lower levels of educational attainment in the higher coal-production county types than in the Non-Coal/ARC and urban areas.

The argument has been made in the economics literature that economies with a high proportion of income and jobs related to natural resource extraction suffer from a "resource curse" that inhibits long-run economic growth (Van der Ploeg 2011). Numerous explanations as to how and why this would result appear in the literature. Some standard reasons offered include declining terms of trade, heavy reliance on taxes on the resource, crowding out of manufacturing industries, disincentives to invest in other forms of human capital, and slower rates of technological innovation. The price of coal is more dependent on the price of substitutes than on terms of trade. Technological innovation has been shown to be a driver of production even as prices have fallen (Fig. 2). Despite a declining civilian labor force, unemployment remains high indicating that workers are available. In recent studies of the "resource curse" effect on economic growth in Appalachia, Betz et al. (2015) only found links to population growth and entrepreneurship; while some effect of low formal educational attainment was found by Douglas and Walker (2017).

The rural coal-production counties, especially those with the greatest levels of historical mining, face a challenging future. Although the data show periods of improving and deteriorating economic conditions for coal producers ("booms and busts", as per Black et al. 2005), the coal-mining industry that has provided significant support for the good times faces declining prospects going forward. Future projections of Appalachian coal production with current policies in place estimate a 25% decline by 2030 relative to 2019 levels as electric power producers continue to take coal-powered electric generators out of service (EIA 2019c, 2020a, b). Projections of US coal use under a condition of policy change for climate mitigation show more drastic declines (IEA 2019; Mendelevitch et al. 2019). Many area residents are aware of these realities, and although remaining supportive of mining (Poudyal et al. 2019), are preparing for the transition to a future with far less reliance on coal (Carley et al. 2018).

High-coal-production areas are characterized by declining and aging populations, a shrinking civilian labor force, low labor force participation rates, high rates of poverty and unemployment, and low levels of formal education attainment relative to other areas. Certain area assets are not well captured by these statistics, however. These include a work-force segment that has proved capable of mining coal, a challenging endeavor that requires high levels of physical stamina, persistence, risk tolerance, and technological skill; communities and families with strong links to the region's traditions; and a natural environment with biodiverse ecosystems, clean water, and natural beauty where not impaired by past mining.

9 Conclusions

Economic growth results from any of four sources: increased investments in human capital, in physical capital, and in discovery and extraction of natural resources, and advances in the technologies using these three groups of resources to produce output. Economic growth in the coalfields has been primarily the result of the latter two—continued extraction and improved extraction methods. The potential for these to promote future growth now appears limited. Demand for coal is falling even as the geological resources are diminishing and becoming more costly to mine.

Opportunities to reinvest income generated from coal mining so as to put high-coal production counties on a sustainable path of economic growth and population stability have likely passed. Given the historical significance of income sources from coal mining for many of these counties, development of other sources of employment and income has been slow. The economic trend data reviewed here reveal no evidence that the high-coal-production counties have entered a sustainable growth path. Public policy could complement community-based and private sector efforts to achieve sustainable growth. This may include support for education and training, for improving health outcomes, for restoration and renewal of public infrastructure and environmental quality, and for promotion of entrepreneurial activity.

Acknowledgements Thanks to Patrician Donovan, Virginia Tech, for preparing Fig. 1.

References

Appalachian Regional Commission (ARC) (1964) Appalachia. A Report by the President's Appalachian Regional Commission, Washington DC
Appalachian Regional Commission (ARC) (2020) Appalachian Regional Commission. https://www.arc.gov/
Barro RJ, Sala-i-Martin X (1995) Economic growth. McGraw-Hill, New York
Becker G, Kominers S, Murphy K, Spenkuch J (2018) A theory of intergenerational mobility. J Political Econ 126(S1):S7–S25
Belfield C, Levin H (2007) The price we pay. Brookings Institution Press, Washington, DC
Bersin J, Chamorro-Premuzic T (2019) The case for hiring older workers. Harv Bus Rev (26 Sept)
Betz MR, Partridge MD, Farren M, Loba L (2015) Coal mining, economic development, and the natural resource curse. Energy Econ 50:105–116
Black D, MacKinnish T, Sanders S (2005) The economic impact of the coal boom and bust. Econ J 115:449–476
Bowen E, Christiadi Deskins J, Lego B (2018) An overview of the coal economy in Appalachia. Appalachian Regional Commission, Washington, DC

Carley S, Evans TP, Konisky DM (2018) Adaptation, culture, and the energy transition in American coal country. Energy Res Soc Sci 37:133–139

Checchi D (2006) The economics of education: human capital, family background and inequality. Cambridge Univ Press, Cambridge UK

Congressional Research Service (CRS) (2017) The U.S. coal industry: historical trends and recent developments. R44922

Copeland C (2015) Mountaintop removal mining: background on current controversies. Congressional Research Service. RS21421

Douglas S, Walker A (2017) Coal mining and the resource curse in the eastern United States. J Reg Sci 57(4):568–590

Gohlke J (2021) Human health in coalfield communities of Appalachia. Chapter 12 in: Zipper CE, Skousen JG. Appalachia's coal-mined landscapes. Springer

Hook M, Aleklett K (2010) Trends in U.S. recoverable coal supply estimates and future production outlooks. Nat Resourc Res 19:189–208

Hoornbeek J, Schwarz T (2009) Sustainable infrastructure in shrinking cities. Kent State University

International Energy Agency (IEA) (2019) World energy outlook 2019. France, Paris

Isserman AM (1996) Socio-economic review of Appalachia. Report to Appalachian Regional Commission. https://www.arc.gov/research/ResearchReports.asp

Johnson KM, Lichter DT (2019) Rural depopulation: growth and decline processes over the past century. Rural Sociol 84:3–27

Krauss C (2019) Murray energy Is 8th coal company in a year to seek Bankruptcy. New York Times (29 Oct)

Macey J, Salovaara J (2019) Bankruptcy as bailout: coal company insolvency and the erosion of federal law. Stanf Law Rev 71:879–962

Mather M, Jacobsen LA, Pollard KM (2015) Aging in the United States. Popul Bull 70(2). Population Reference Bureau, Washington, DC

Mendelevitch R, Christian Hauenstein C, Holz F (2019) The death spiral of coal in the U.S.: will changes in U.S. Policy turn the tide? Clim Policy 19:1310–1324

Milici RC (1997) The CoalProd database: historical production data for the major coal-producing regions of the conterminous United States. US Geol Surv, Open File Report 97–447

Murray M, Schaeffer P (2018) An economic analysis of the Appalachian coal industry ecosystem: human capital and CIE. Appalachian Regional Commission, Washington, DC

National Bureau of Economic Research (NBER) (2020) US business cycle expansions and contractions. https://www.nber.org/cycles.html

Pollard K, Jacobsen LA (2019) The Appalachian region: a data overview from the 2013–2017 American community survey. Appalachian Regional Commission, Washington DC

Poudyal NC, Gyawali BR, Simon M (2019) Local residents' views of surface mining: perceived impacts, subjective well-being, and support for regulations in southern Appalachia. J Clean Prod 217:530–540

Richter AW, Chapman D, Mihaylov E (2018) Declining U.S. labor force participation rates stand out. Econ Letter 13(6):1–4

Spence B, Kunkel K, Schewel E, Boettner T, Martin L (2013) Who owns West Virginia? West Virginia Center on Budget and Policy, Charleston. https://wvpolicy.org/who-owns-west-virginia-in-the-21st-century-2/

Tamborini C, Kim C, Sakamoto A (2015) Education and lifetime earnings in the United States. Demography 52(4):1383–1407. https://doi.org/10.1007/s13524-015-0407-0

Thiede BC, Brown DL, Sanders SR, Glasgow N, Kulcsar LJ (2017) A demographic deficit? Local population aging and access to services in rural America, 1990–2010. Rural Sociol 82:44–74

Torpey E (2018) Data on display: Measuring the value of education. U.S. Bureau of Labor Statistics. https://www.bls.gov/careeroutlook/2018/data-on-display/education-pays.htm

U.S. Bureau of Economic Analysis (BEA) (2020) Regional economic accounts: downloads https://apps.bea.gov/regional/downloadzip.cfm

U.S. Bureau of Labor Statistics (BLS) (2020a) Current employment statistics. https://www.bls.gov/ces/data/home.htm
U.S. Bureau of Labor Statistics (BLS) (2020b) Local area unemployment statistics. https://www.bls.gov/lau/
U.S. Bureau of the Census (BOC) (2020a) American community survey. https://www.census.gov/programs-surveys/acs
U.S. Bureau of the Census (BOC) (2020b) County intercensal datasets: 2000–2010. https://www.census.gov/data/datasets/time-series/demo/popest/intercensal-2000-2010-counties.html
U.S. Bureau of the Census (BOC) (2020c) National population by characteristics: 2010–2019. https://www.census.gov/data/tables/time-series/demo/popest/2010s-national-detail.html
U.S. Bureau of the Census (BOC) (2020d) Small area income and poverty estimates
U.S. Energy Information Administration (EIA) (1993) The changing structure of the U.S. Coal industry: an update. DOE/EIA-0513(93)
U.S. Energy Information Administration (EIA) (2012) Annual energy review 2011. DOE/EIA-0384 (2011)
U.S. Energy Information Administration (EIA) (2019a) Annual coal distribution report (and predecessor publications). https://www.eia.gov/coal/distribution/annual/
U.S. Energy Information Administration (EIA) (2019b) Annual coal report (and predecessor publications). https://www.eia.gov/coal/annual/
U.S. Energy Information Administration (EIA) (2019c). More U.S. coal-fired power plants are decommissioning as retirements continue. Today in Energy, 26 July 2019
U.S. Energy Information Administration (EIA) (2020a) Annual energy outlook 2020 https://www.eia.gov/outlooks/aeo/
U.S. Energy Information Administration (EIA) (2020b) New electric generating capacity in 2020 will come primarily from wind and solar. Today in Energy, 14 January 2020
U.S. Energy Information Administration (EIA) (2020c) Quarterly coal report. https://www.eia.gov/coal/production/quarterly/
Van der Ploeg F (2011) Natural resources: curse or blessing? J Econ Lit 49:366–420
Zipper CE (2020) Appalachian coal data. University Libraries, Virginia Tech. https://doi.org/10.7294/BDA6-7R15

Human Health in Coalfield Communities of Appalachia

Julia M. Gohlke

Abstract Differences in population health metrics for residents of Appalachia's coal-mining counties compared to other areas have been previously documented. Research examining the effects of mining on health has demonstrated that environmental exposures lead to respiratory disease in occupational settings; however, it has been more difficult to ascertain direct or indirect factors behind health disparities at the community-wide level. While associations between economic status, education, health care access and quality, or sociocultural factors and community health metrics have been documented, the causal linkages and uniqueness of these associations within coalfield areas are not well understood. Trends in population health between the 1970s and present day highlight remarkable improvements at state, national, and global levels, while population health improvements in Appalachia coalfield communities have lagged. Evaluation of time trends in population health metrics in central Appalachian coalfield counties offer insights into the interrelationships between factors potentially contributing to disparities when compared to state, national and global trends. Disparities in health outcomes including respiratory disease, cancer mortality, and adverse birth outcomes are summarized. Initiatives building from advances in telehealth and the strong social networks in Appalachian coalfield communities show promise for addressing health disparities.

Keywords Disparities · Infant mortality · Life expectancy · Mortality · Preterm birth

1 Introduction

Recent reports highlight differences in health metrics for Appalachia, particularly central Appalachian counties, compared to other areas in the United States (U.S.) (Behringer and Friedell 2006; McGarvey et al. 2011; Borak et al. 2012; Krometis et al. 2017; Meit et al. 2017, 2019). These studies indicate overall rates of premature

J. M. Gohlke (✉)
Virginia Tech, Blacksburg, VA, USA
e-mail: jgohlke@vt.edu

C. E. Zipper and J. Skousen (eds.), *Appalachia's Coal-Mined Landscapes*,
https://doi.org/10.1007/978-3-030-57780-3_12

death, liver disease, motor vehicle crashes, obesity, and cancer rates are heightened in Appalachia. Some studies have found that coal-producing counties within central Appalachia exhibit poorer health outcomes than non-coal producing counties (Hendryx and Ahern 2009; Bowen et al. 2018). For example, mortality rates over the 2000–2015 period were higher in Appalachia's coal-producing counties relative to Appalachia's non-coal producing counties and were higher in both than in the rest of U.S. (Bowen et al. 2018). Also over 2000–2015 within Appalachia, mortality rates in central Appalachia were increasing and were higher than in Appalachia's southern and northern regions where mortality has exhibited little change. In central Appalachia, where most of the region's surface-mined coal is produced, mortality rates rose from 11.6 deaths per 1000 people in 2000 to 13.5 deaths per 1000 in 2015 (Bowen et al. 2018).

Health indicators are strongly correlated with economic indicators, both in the U.S. and globally (Dow and Rehkopf 2010; Knapp et al. 2019). Appalachian coal-mining counties have struggled economically, especially when compared to non-Appalachian counties (Santopietro and Zipper 2021); therefore, economic factors are important to consider when examining health outcomes in coalfield communities. This chapter summarizes and interprets population health trends in central Appalachia coalfield areas starting in 1969 when reliable estimates of population health became available (CDC 2020) until the present day. By the 1970s, coal mining had already shaped economic conditions, sociocultural attitudes, and health conditions of coalfield communities since many parts of central Appalachia had already been mined extensively; therefore these analyses of population health data are confined to a period of intensive coal production and a more recent period of mining decline (Zipper et al. 2021). The analysis focus is the central Appalachian region of southern West Virginia, eastern Kentucky, southwestern Virginia, and northeastern Tennessee; although some data for the entire Appalachian coalfield are also reviewed.

Common population health metrics for Appalachia's coal-producing counties are reviewed and, where possible, compared to metrics for other regions. Metrics reviewed are life expectancy, mortality, cause-specific mortality, and several measures of infant and maternal health. While such metrics are useful for estimating a population's overall health, limitations are acknowledged. In particular, when the metrics are calculated for aggregated groupings of geographic areas they may obscure major differences within those groupings. Placement of coalfield patterns and trends into a national and global population health context enables understanding of dependencies on national and global advances as well as state-level and local factors. Also, recent community-level initiatives implemented to mitigate health disparities are reviewed.

2 Life Expectancy

Life expectancy is the number of years, on average, a person is expected to live based on year of birth. Life expectancy is calculated from mortality data collected by hospitals and reported to local health departments. These data are collated at the state or regional level, and then at national levels by health agencies around the world, allowing for a standardized method of evaluating trends and comparing across geographic regions.

Life expectancy changes over a lifetime since mortality risk is higher at specific ages; therefore age-specific life expectancy is often used to identify ages in a given population most at risk of dying. When the term life expectancy is used alone, it typically refers to the life expectancy of a newborn in a given year and considers the average probability of surviving across all ages based on ages of deaths in that year. For example, for a person born in 1970 in the U.S., life expectancy was 70.8 years at birth (with 2041 the estimated average year of death); however, because risk of mortality is high between ages 0 and 29 and because mortality rates have declined since 1970, that person's average life expectancy at age 45 (in year 2015) was 36 years, with 2051 the estimated year of death.

Life expectancy between 1959 and 2016 within the states of West Virginia, Kentucky, Tennessee and Virginia, the U.S. as a whole, and the state with the longest life expectancy (Hawaii) rose over most of the 1959–2016 period (Fig. 1). The life-expectancy gap between the U.S. and Hawaii remained stable across these 56 years. In contrast, West Virginia, Kentucky, and Tennessee life expectancy, although similar

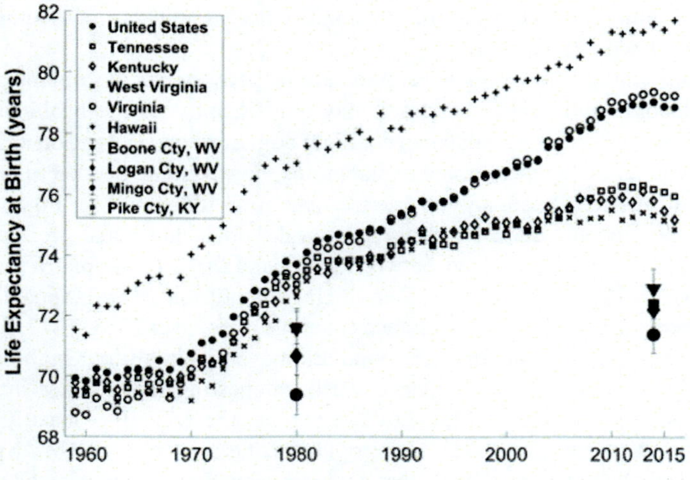

Fig. 1 Trends in life expectancy (years) in the U.S., states with partial areas in the central Appalachian coalfield (West Virginia, Kentucky, Tennessee, and Virginia), and the state with the highest life expectancy (Hawaii) over 1959–2016, and four high coal-production counties of central Appalachia for 1980 and 2014. Boone and Pike County have similar estimates in 1980. Data from Barbieri and Wilmouth (2019) and IHME (2019)

to the U.S. average through the early part of the analysis period, diverged beginning in the early 1980s. While average life expectancy at birth increased by 8.9 years, from 69.9 to 78.8 years, over the analysis period in the U.S. as a whole, life expectancy in West Virginia, with its large area of Appalachian coalfield, only increased by 5.5 years over that same period.

Another noticeable trend is the decreasing life expectancy between 2010 and 2016 in West Virginia with its large area of Appalachian coalfield, as well as Kentucky and Tennessee. Previous studies have attributed this effect primarily to increases in mid-life mortality caused by alcohol and drug overdoses, suicide, and alcoholic liver disease/cirrhosis of the liver (Meit et al. 2019), yet these causes still account for a relatively small proportion of overall deaths. Other studies have suggested the further decline in socioeconomic metrics in central Appalachian areas are driving up mortality due to metabolic disease, pulmonary disease, and cancer, which are no longer offset by the prominent decline in cardiovascular mortality seen over the 1980s and 1990s (Ezzati et al. 2008).

The lagging life expectancy improvements in central Appalachia's major coal-production counties relative to state and national averages is even more striking than the increasing health disparities at the state level. Over the 1980–2017 period, the four highest coal-production counties in central Appalachia (high-coal counties) were Pike County, KY, Boone County WV, Mingo County WV, and Logan County WV (Zipper 2020). Since 2001, coal mining was responsible for between 7 and 40% of total employment annually in individual counties (BEA 2019). Life expectancies in high-coal counties are considerably lower and have improved less in recent decades than state and national averages (Fig. 1). Mingo County life expectancy in 2014 (71.4 years) was similar to the average life expectancy in the U.S. in the early 1970s and Pike County has not experienced a statistically significant improvement in life expectancy from 1980 to 2014.

Previous studies have suggested associations between socioeconomic variables such as income and health outcomes may explain most of the health disparities characterized in central Appalachian coalfield communities (Halverson and Bischak 2008; Yao et al. 2012). Additional explanations, however, are needed to understand the pattern of life-expectancy divergence over time illustrated by Fig. 1. In fact, the per-capita income disparity between central Appalachia's coal-producing areas and the U.S. average narrowed between 1969 and 2018 (Santopietro and Zipper 2021; ARC 2015). In addition, coal-producing counties in central Appalachia have experienced increasing health-insurance coverage rates; over 2013–2017, those rates were similar or higher than the U.S. population average in all four high-coal counties of Fig. 1 (Pollard and Jacobsen 2019), suggesting that neither income nor health-insurance coverage differences can explain the widening disparities in life expectancy. A recent report analyzing population health metrics across Appalachia suggested that health-related behaviors, including inactivity, smoking, and teen pregnancies, have high explanatory power for poor health metrics, more so than healthcare access metrics such as the number of physicians within a county (Holmes et al. 2018).

Globally over the past two centuries, average life expectancy has increased from approximately 30 years to over 72 years, with some countries beginning this transition

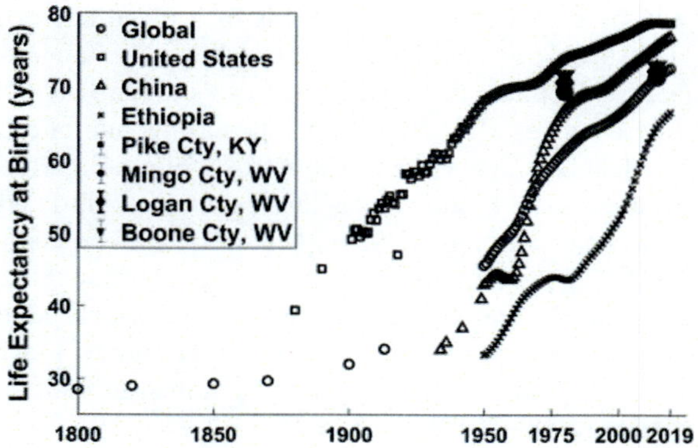

Fig. 2 Trends in life expectancy (years) globally and in the U.S., China, and Ethiopia; and in four high-coal Appalachian counties, 1980 and 2014. Data from (Riley 2005a, b) and United Nations Population Division, downloaded from Roser et al. (2019). County level data from IHME (2019) are also displayed in Fig. 1

in the 1800s and others not until the 1970s (Roser et al. 2019) (Fig. 2); this pattern reflects dramatic overall improvements in human health. This well-characterized transition, sometimes referred to as the demographic or epidemiologic transition, is marked by both reduced fertility (birth) rates and reduced mortality rates (McMichael et al. 2004; Riley 2005a, b; Roser et al. 2019).

Spatial disaggregation of life expectancy reveals different patterns of change over time across countries (Fig. 2). For example in the U.S., the transition started pre-1900, with life expectancy improving slowly into the 2000s, then leveling off in 2010; while in China, currently the world's largest coal producer, the epidemiologic transition occurred very rapidly in the 1960s and early 1970s, coincident with increased globalization and healthcare advances. For contrast to a country within sub-Saharan Africa where life expectancy has consistently been the lowest globally, the epidemiologic transition of Ethiopia occurred later and even more rapidly, over the late 1990s and 2000s, in association with rapid dissemination of health improvement strategies. Across the three countries shown in Fig. 2 and across 182 countries globally, disparities in life expectancy have been reduced between 1950 and the present day (Roser et al. 2019). This is in contrast to the increased disparity in life expectancy in three states with partial areas in the central Appalachian coalfield compared to U.S. averages over this same time period (Fig. 1). When comparing life expectancy change between 1980 and 2014 in the four high-coal counties to global data, it is striking to notice the lack of improvement compared to global trends, and that current life expectancies in the high-coal counties are lower than the average life expectancy in China and similar to the global average (Fig. 2).

3 Age-Adjusted Mortality

Death certificate data are often used to calculate age-specific mortality rates, which can then be used to estimate age-adjusted or age-standardized mortality rates for a given year. Common units for these metrics are deaths per 1000 persons. Mortality can be calculated for all persons of a given age or for populations by weighting age-specific mortality rates, which allows for standardized comparisons of mortality across populations with differing age structures.

Bowen et al (2018) found increasing mortality in central Appalachian coal-producing regions over 2000–2015 while mortality in Appalachia's northern and southern coal-producing regions experienced little change; however, differences in crude-mortality could be explained by differences in age structure. Bowen et al (2018) aggregates all coal-producing counties within central, northern, and southern Appalachia, but coal production varied widely among counties; hence, regional comparisons do not reveal patterns that may be specific to primary coal-producing counties. Finally, those authors examined mortality trends over a relatively short period (2000–2015).

To address these limitations, we have classified counties in Appalachia based on cumulative coal production between 1980–2017 (Table 1; Fig. 3) and analyzed mortality trends from the late 1960s. All mortality rates, as compared among these county types, are calculated as population-weighted means of the age-adjusted

Table 1 Classification of counties in seven Appalachian states by cumulative coal production, 1980–2017, population, and geographic location

County type[a]	Number of counties[b]		Total population	
	1968	2017	1968	2017
Coal > 300 Mt	11	12	369,657	334,998
Coal 100–300 Mt	23	23	945,721	896,754
Coal 10–100 Mt	45	45	1,455,510	1,569,282
Coal < 10 Mt	50	50	1,188,484	1,540,168
Coalfield 0 t[c]	35	35	974,998	1,155,154
Other ARC[d]	81	88	1,663,654	2,560,380
Non-coal/ARC[e]	195	217	4,533,756	7,088,176
Pop > 100,000[f]	111	112	28,214,794	36,827,461
Total	551	582	39,346,574	51,972,373

[a]Coal production data from US Department of Energy (Zipper 2020); Mt = million tons of coal production, 1980–2017

[b]Formation of independent cities in Virginia, and reversion of independent-city status by some localities in Virginia, caused county numbers to change over the analysis period. The City of Norton was tallied with Coal > 300 M tons counties following its formation since it is located physically within such a county (Wise, Virginia)

[c]Counties with geologic deposits of Appalachian bituminous or Pennsylvania anthracite coal located within their borders but no recorded coal production (1980–2017)

[d]Counties lacking recorded coal mining and located outside of the Appalachian coalfield but within the Appalachian Regional Commission (ARC) service territory

[e]Counties located outside of Appalachian coalfield areas and outside of Appalachian Regional Commission (ARC) service area

[f]Counties with 2017 populations greater than 100,000 were classified as this county type, regardless of coal production and geographic location

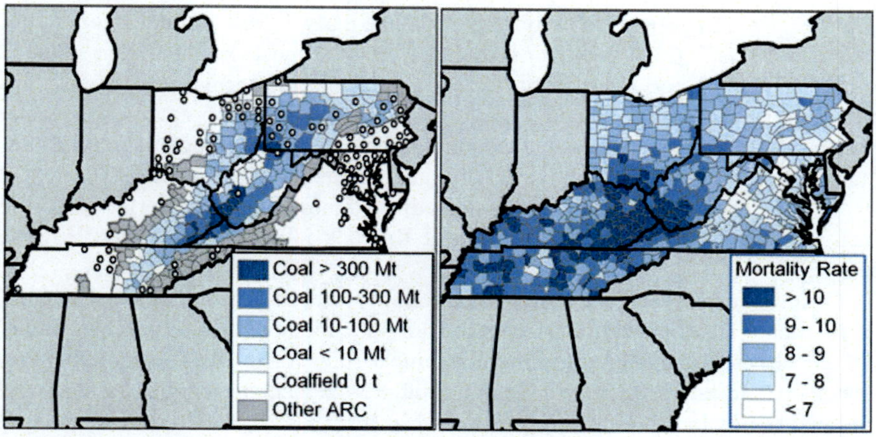

Fig. 3 (Left) Map of county classification based on coal production, population, and geography (Table 1). Counties of the seventh type, Non-Coal/ARC, are within the white areas of the analysis states with boundaries not shown. Open circles designate center points of counties with populations >100,000 in 2018 (Pop > 100,000). (Right) Map of 2009–2017 mean age-adjusted mortality rate per 1000 population. Data from CDC (2020)

rates published by CDC (2020). For these analyses, Virginia's independent cities are treated as if they are counties and are termed as such. Counties with population greater than 100,000 are in a separate category, noting that urbanization also impacts health metrics. This county-type designation supersedes coal-production and location-based designations in analyses presented. For a listing of counties by county-type designation, see Zipper (2020).

Mean age-adjusted mortality rates declined over 49 years, 1969 through 2017, for all county types (Fig. 4), following national and global trends. These comparisons, however, also reveal heightened mortality rates in the two highest coal-production county types through the analysis period, with widening disparities between 1995 and present day. The counties in the highest coal-production category have exhibited little

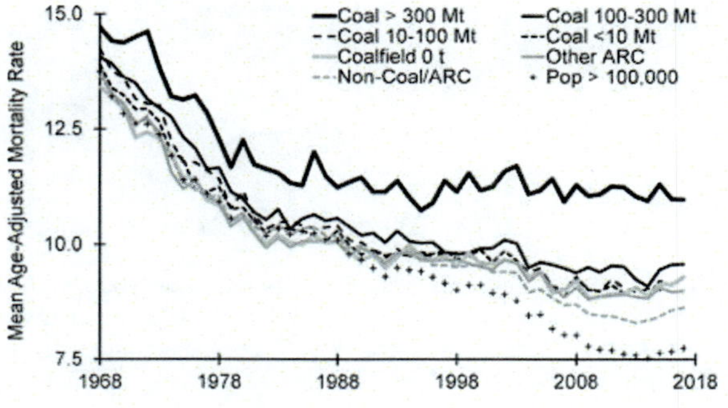

Fig. 4 Population-weighted age-adjusted mortality rates (per 1000 population) by county type. Data from CDC (2020)

improvement in age-adjusted mortality since the early 1990s. Counties with populations > 100,000 (high-population counties) saw the most decline in age-adjusted mortality over this period. At the beginning of the period, these high-population counties were similar in age-adjusted mortality to lower-population counties; however greater improvements in age-adjusted mortality are seen in high-population counties starting in the 1990s, with the widest disparity in the most recent years. The spatial distribution of high mean 2009–2017 age-adjusted mortality rates overlap with high coal-production areas near the junction of Kentucky, West Virginia, and Virginia (Fig. 3).

Confidence intervals for age-adjusted mortality rates have been estimated by CDC at the county, state, and national levels and provide a method for determining whether differences are statistically meaningful. Comparisons of age-adjusted mortality rate trends for the highest coal-production central Appalachian counties to state and national trends indicate differing experiences (Fig. 5). Logan Country started off with the highest age-adjusted mortality rate (15.8 per 1000 people), but then declined in the 2000s to match West Virginia average rates (9.6 per 1000 people in 2017). In contrast, age-adjusted mortality in Mingo County has been higher than state and national rates consistently for the full analysis period with the gap increasing slightly in recent years. Alternatively, neighbors in Boone County experienced mortality rates that did not differ from state or U.S. levels in 1969 and increased relative to state rates in the late 1970s and 1980s; and then experienced rapidly declining trends and mortality lower than state averages by 2000 and were statistically similar to the U.S. average in recent years. Finally, age-adjusted mortality in Pike County was similar to U.S. and state averages in the early 1970s, but was relatively constant from the

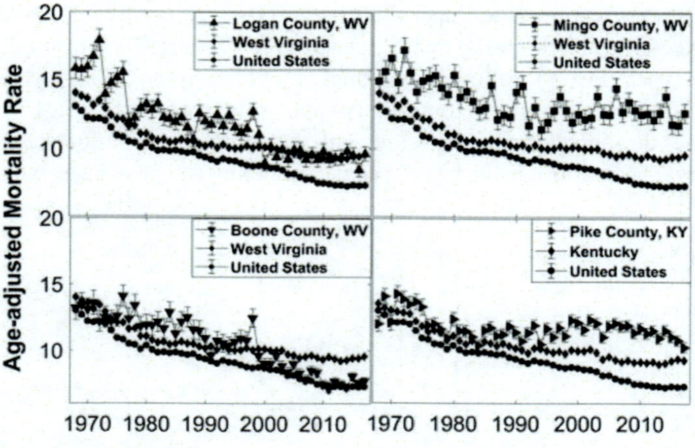

Fig. 5 Trends in age-adjusted mortality rates (per 1000 population) from 1968 to 2017 in the top four coal-production counties in central Appalachia compared to state and U.S. trends. Data from CDC (2020). Error bars represent 95% confidence intervals and are most evident for the county data. Mortality rates for West Virginia and Kentucky are higher than U.S. rates throughout the study period

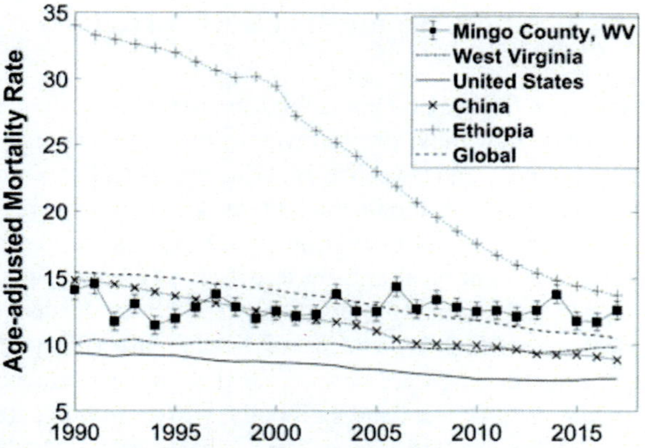

Fig. 6 Trends in age-adjusted mortality rate (per 1000 population) in a top coal-production county compared to state, U.S., China, Ethiopia, and global trends, 1990 through 2017. U.S., West Virginia, and Mingo County data from CDC (2020) and China, Ethiopia, and global trends from IHME (2019). Due to differences in CDC and IHME estimates, an adjustment factor (0.7) was applied to the IHME data based on matching West Virginia datasets across the two systems

1980s through 2015 while state and U.S. rates declined. Age-adjusted mortality rates in Pike County in the most recent years (2014–2017) have been declining, while state, U.S., and other high-coal county rates have experienced little change or increased slightly.

Across these four counties, Mingo County had the highest age-adjusted mortality rate and the least decline over this 48-year period. Figure 6 compares Mingo County trends in age-adjusted mortality to global and nationwide trends for selected countries over 28 years (1990–2017). While age-adjusted mortality rates in Ethiopia declined dramatically and while global, U.S., and China mortality also declined but more gradually, statewide West Virginia and Mingo County remained relatively stable over this period. Average age-adjusted mortality rates in China and globally have been lower than Mingo County rates since the mid-2000s. In 2017, overall age-adjusted mortality rates in Mingo County are similar to the rates seen in Ethiopia. The lack of improvement in Mingo County is in contrast to other world regions, highlighting the increased health disparities characterizing present-day central Appalachian coalfields and reflecting an overall stagnation in health improvement. Mingo County is not unique in this regard, as seven of the 38 counties with 1980–2017 coal production > 100 million tons also have mean age-adjusted mortalities greater than 12 per 1000 over the last five years of the analysis period.

4 Cause-Specific and Chronic-Disease Mortality

Health disparities characteristic of Appalachian coalfield areas are primarily chronic diseases. To understand how these disparities may have arisen, it is useful to review overall global and national trends in cause-specific mortality. Dissemination of effective methods for reducing communicable diseases has been a primary driver of improved life expectancy. The global health transition occurs as primary causes of death change from communicable diseases, such as measles, malaria, and diarrheal disease, to non-infectious chronic diseases such as cancer, cardiovascular, and pulmonary diseases which tend to occur later in life (Riley 2005a, b). Non-communicable, chronic diseases are now by far the primary cause of death worldwide, accounting for over 70% of deaths globally, with communicable causes dropping steadily from 33% in 1990 to 18% in 2017 (IHME 2018). The health transition in the U.S. was essentially complete by 1990 and remained stable through 2017, with close to 90% of deaths due to non-communicable causes over this period. Alternatively in China, deaths due to non-communicable disease rose from 71% in 1990 to 89% in 2017, with a coincident decrease in communicable disease deaths to below 10% of total. Deaths due to injuries remained relatively stable over this period and across countries, accounting for approximately 8% of deaths worldwide.

While the global health transition has resulted in longer lives, proportions of years lived in full health has been declining due to increasing chronic disease rates. The disability-adjusted life year (DALY) is a summative health metric based on mortality data and disease incidence or prevalence data; it is calculated by summing years lived with disease (YLD) and years of life lost (YLL). Each disease is given a disability weight to account for its severity based on health costs and other metrics (Murray 1994). Years lived with disability are higher as life expectancy increases due to the health transition and increased chronic diseases. Numerous studies across and within countries have characterized YLD increase during and following the health transition (Murray and Lopez 1997; Vos et al. 2016). In the U.S., increases in disabled life expectancy have outpaced disability free life expectancy over the past 40 years, particularly for women (Crimmins et al. 2016). These composite metrics are available for timeframes beginning in the 1990s.

These concepts can be applied to further understand underlying differences in age-standardized mortality rates. For example, neonatal disorders, lower respiratory infection and diarrheal disease in children due to malnutrition and unimproved water and sanitation infrastructure are still primary causes of death in Ethiopia in 2017; whereas primary causes of death in Mingo County are chronic diseases in older adults including ischemic heart disease, stroke, chronic obstructive pulmonary disease (COPD), lung cancer, and diabetes (IHME 2018). West Virginia had a much higher YLD rate (16,000 per 100,000 person-years) versus Ethiopia (9000 per 100,000 person-years) in 2017 (IHME 2018), even while total disability-adjusted life years (DALYs) are currently similar (approximately 42,000 per 100,000 person-years in both).

4.1 Previous Studies on Disease and Cause-Specific Mortality

Coal mining has directly contributed to mortality and morbidity in Appalachia communities, as the majority of over 100,000 miner fatalities occurring nationally on the job since 1900 having occurred in Appalachia (Zipper et al. 2021). On-the-job fatalities have declined dramatically over time, however, as 20 or fewer miners have died on the job nationally each year since 2011 (MSHA 2018). Many deaths are also recorded for former mine workers due to coal workers' pneumoconiosis (CWP) and progressive massive fibrosis (PMF), forms of the disease commonly known as "black lung." Recent increases in CWP prevalence among former miners, especially in Appalachia, are likely due to the increased excavation of silica-bearing rock required for mining of thin coal seams and consequent worker exposure to silica dust (Cohen et al. 2016; Blackley et al. 2018). Deaths from CWP in the U.S. between 1968 and 2018 totaled more than 30,000, the majority in Appalachia, with age-adjusted CWP death rates declining over this period (Mazurek et al. 2009; CDC 2020). While CWP death rates for those 65 years of age or older have continued to decline, CWP death rate in persons aged 25–65 years have remained unchanged but deaths have increased for younger adults (Mazurek et al. 2018). Although underground mine workers are most at-risk for CWP, a recent study found that respirable dust and quartz levels at surface mines sometimes exceeded permissible exposure limits and that drillers are the highest exposed group (Doney et al. 2020).

Community-level health effects due to coal production are more difficult to ascertain (Boyles et al. 2017; Krometis et al. 2017) although multiple studies have found elevated incidence of several negative health conditions in high-coal production Appalachian areas (Table 2). Rates of age-adjusted mortality due to cancer, particularly colorectal and lung cancer, heart disease, and chronic obstructive pulmonary disease (COPD) are heightened in central Appalachia and its coal-producing counties compared to U.S. averages (Krometis et al. 2017) and incidence of these diseases and other chronic diseases are also increased in central Appalachia overall (Barnett et al. 2000; Halverson et al. 2002, 2004; Paskett et al. 2011; Blackley et al. 2012).

Causes of these chronic diseases and related mortality are multi-factorial; as these diseases' incidence are associated with numerous socio-behavioral, economic, and demographic variables. A recent systematic review of observational epidemiological studies attributing coal mining as a potential cause of poor community-level health outcomes concluded most such studies were at risk of bias due to poor exposure characterization (typically at county level) and inadequate control of potential confounding variables, including smoking and sociodemographic factors (Boyles et al. 2017).

Table 2 Summary of published health studies conducted in coalfields of Appalachia

Health outcome measured, period evaluated	Exposure measurement	Results	References
All-cause and cause-specific mortality, 1950–2014	Coal production by county, per capita, by type, and method	Associations with all-cause and several chronic-disease-cause-specific mortality	**12 studies** Hendryx et al. (2008, 2010), Hendryx (2009), Hendryx and Ahern (2009), Hitt and Hendryx (2010), Esch and Hendryx (2011), Ahern and Hendryx (2012), Borak et al. (2012), Buchanich et al. (2014), Woolley et al. (2015b, a), Hendryx and Holland (2016), Shi et al. (2019)
Cancer, 1995–2012	Coal production by county and mining method	No significant associations with incidence, self-report, of hospitalizations	**5 studies** Hendryx et al. (2007, 2012), Hendryx and Ahern (2008), Christian et al. (2011), Hendryx (2013)
Cardiopulmonary disease, 2000–2014	Coal production by county and mining method and proximity to active surface mining	Associated with hospitalizations, symptoms, and diagnoses	**8 studies** Hendryx et al. (2007), Hendryx and Ahern (2008), Hendryx and Zullig (2009), Hendryx (2013), Brink et al. (2014), Hendryx and Entwhistle (2015), Hendryx and Luo (2015), Talbott et al. (2015)
Birth outcomes, 1996–2009	Coal production by county and mining method	Associated with low birth weight, but not birth defects	**3 studies** Ahern et al. (2011a, b), Lamm et al. (2015)
Mental health and substance use disorders, 2006–2008	Coal production by county, Coal mining permits by zip code	Associated with depression and substance use disorders	**2 studies** Hendryx and Innes-Wimsatt (2013), Canu et al. (2017)

(continued)

Table 2 (continued)

Health outcome measured, period evaluated	Exposure measurement	Results	References
Self-reported overall health, 2000–2012	Coal production by county and mining method; number of mining facilities near residence	Associated with poorer reported health status	**5 studies** Hendryx and Ahern (2008), Zullig and Hendryx (2010, 2011), Hendryx (2013), Woolley et al. (2015a, b)

Table based on Boyles et al. (2017) with additional studies added

4.2 Cause-Specific Mortality

Using the county-type groupings (Table 1; Fig. 3), Figs. 7 and 8 diagram time-trends of age-adjusted and population-weighted mortality rates for the primary causes of death in coalfield and non-coalfield counties in seven states. Consistent with state and national statistics, the highest cause-specific mortality rates in coalfield counties are for ischemic heart disease and cancer. Deaths due to ischemic heart disease have been decreasing steadily since 1969, while deaths due to cancer increased in the 1970s and 80s but have been decreasing since the 1990s. The highest coal-production areas show heightened mortality rates due to cancer. Mortality due to cancer was higher in counties with higher populations (> 100,000) in the earlier part of the timeframe analyzed but becomes lower in counties with high population later in the study period, similar to the pattern seen in overall age-adjusted mortality (Fig. 4). The age-standardized rate of mortality for tracheal, bronchus, and lung cancer for females in Mingo County was 94.5 per 100,000 (compared to 61.2 in West Virginia and 43.8 nationally) (IHME 2019).

Deaths due to respiratory causes show wide disparity across the county types analyzed, with the highest coal-production counties showing increasing mortality due to respiratory disease in contrast to stable or decreasing trends in non-coalfield areas from the 1990s to the present day. If surface coal production has a direct influence on chronic disease outcomes, a primary route of exposure would likely be via inhalational exposure of airborne particle emissions from mines or from mine-related transport, yet only a handful of studies have begun to delineate estimates of community-level exposure from these exposure pathways (Boyles et al. 2017; Krometis et al. 2017). Several studies have found areas near surface mines and coal haulage roads are subject to elevated dust exposures (Knuckles et al. 2013; Kurth et al. 2014, 2015; Aneja et al. 2017). Also, experimental studies have found inhalational exposure to a surface-coal mine particulate-matter mixture causes adverse cardiopulmonary outcomes in rodents (Nichols et al. 2015), a finding consistent with the broader literature on known adverse human health outcomes caused by

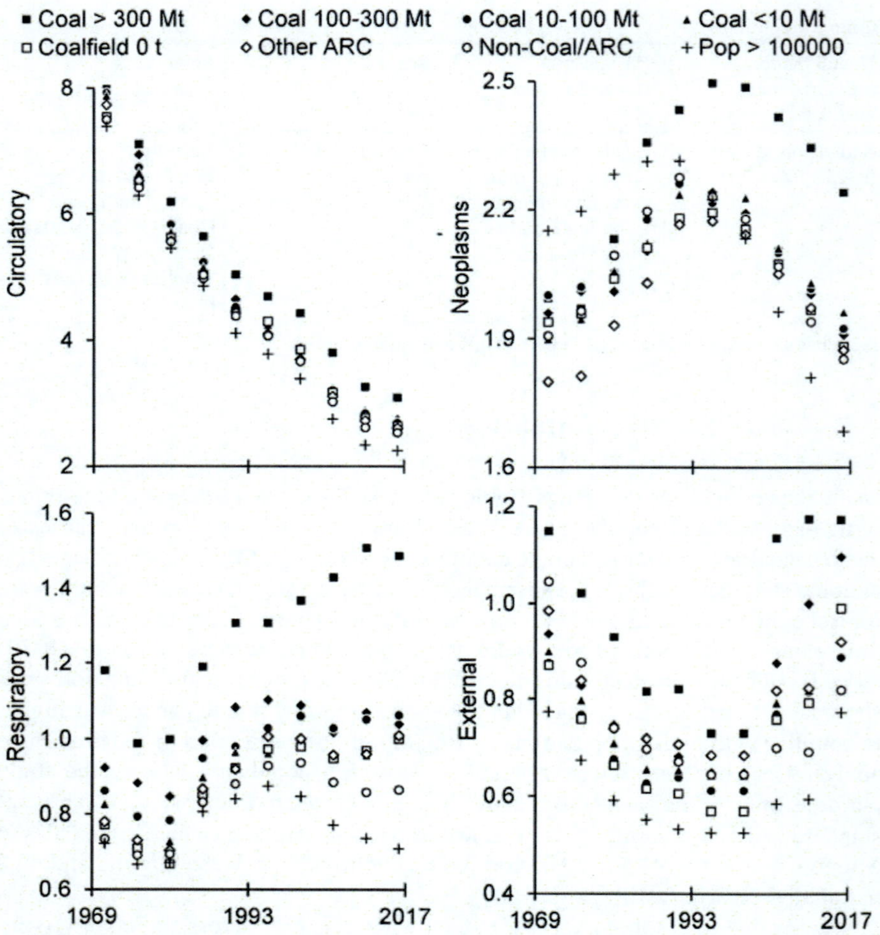

Fig. 7 Cause-specific mortality rates (per 1000 population) by county type. Population-weighted means were calculated for each county type using county-level cause-specific age-adjusted mortality rates from CDC (2020) for five-year periods. Primary causes are Circulatory (CDC codes I00-I99), Neoplasms (C00-D48), Respiratory (J00-J98), and External (V01-Y89)

exposure to particulate matter air pollution (Burnett et al. 2014). A recent observational study uses individual-level mortality records to adjust for confounding variables, and finds heightened mortality risk of non-malignant respiratory diseases in Virginia coal-mining counties (Shi et al. 2019).

While not the top killers in the coal-production counties of Appalachia, mortality disparities are particularly high for endocrine-related disorders, primarily diabetes, and digestive disorders, particularly liver disease, compared to other areas, and such disparities have been expanding in more recent decades. Deaths due to external causes, including intentional and unintentional injuries, have been consistently

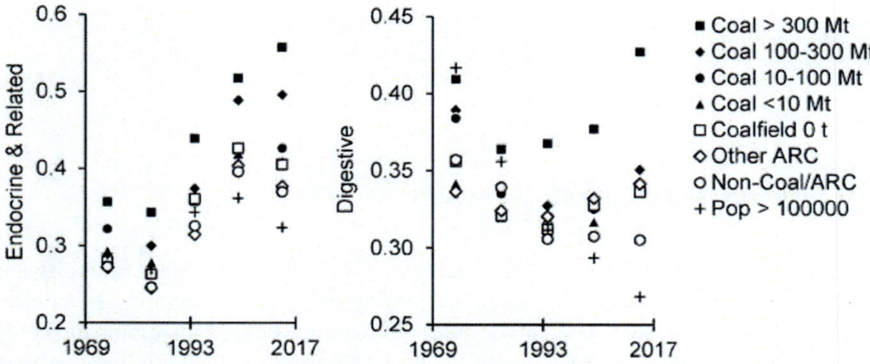

Fig. 8 Cause-specific mortality rates (per 1000 population) by county type. Population-weighted means were calculated for each county type using county-level cause-specific age-adjusted mortality rates from CDC (2020) for ten-year periods. Primary causes are Endocrine, Nutritional and Metabolic (CDC codes E00-E88) and Digestive (K00-K92)

greater in high coal-production counties than in comparison areas. Mortality from liver diseases, poisonings, and suicides (a category that includes drug overdose deaths) was also higher, on average, in high coal-production counties during the period of data availability, 1999–2017 (data from CDC 2020 but not shown). The age-standardized mortality rates for mental and substance use disorders for females in Mingo County, 38.1 per 100,000, was higher than statewide West Virginia (15.8) and national (8.2) averages. From 1980 to 2014, however, female deaths due to mental and substance use disorder increased more dramatically (3569%) than did female tracheal, bronchus, and lung cancer deaths (64%) in Mingo County (IHME 2019).

5 Maternal and Newborn Child Health

Birth outcomes, including preterm birth and low birth weight, are considered critical population health metrics because they are predictors of stunting, adult-onset chronic disease rates, productivity, and life expectancy. Data on health at birth are commonly collected and recorded in centralized databases by governments, allowing for standardized approaches for assessing trends over time and space. A shortened length of gestation (<37 weeks) and low weight at birth (<2500 g) are key indicators of survival and risk of disease later in life (Kramer et al. 2000; Johnson and Schoeni 2011; Howson et al. 2013). Infant mortality (age-specific mortality rate in the initial year of life) and maternal mortality (death of a woman while pregnant or within 42 days of having been pregnant) are substantial contributors to overall life expectancy, particularly in countries still in the health transition (Sartorius and Sartorius 2014). In the U.S., infant mortality rates have been stable or decreasing (Mathews and Driscoll 2017), but maternal mortality has been stable or increasing compared to decreases globally (WHO 2014); however, improved surveillance may contribute to the estimates of increased rates of maternal mortality in the U.S. (Joseph et al. 2017).

5.1 Infant and Maternal Mortality

Overall infant mortality has decreased dramatically in the U.S.; infant mortality has also been declining within Appalachia, 1990 to 2013, but not as rapidly as elsewhere (Singh et al. 2017). Another study showed no change in infant mortality rates in West Virginia and Kentucky between 2005 and 2014 while national and global rates decreased (Mathews and Driscoll 2017). Figure 9 shows infant mortality rates in states that incorporate central Appalachia between 1968 and 2017. Infant mortality was heightened in West Virginia, Tennessee, and Virginia compared to the U.S. average between 1969 and 1980. Between 1980 and 1999, rates in states with central Appalachian areas were more similar to national trends, with the exception of Tennessee. More recently, infant mortality has remained unchanged in West Virginia, Kentucky, and Tennessee while infant mortality in Virginia and nationally have continued to decrease.

Age-specific mortality rates, such as infant mortality, can be unstable over short periods in low-population areas; therefore, data were combined into multiple-year periods for the four high coal-production counties (Fig. 10). Infant-mortality rates were heightened during the 1970s in three of the four high coal-production counties (Logan, Mingo, and Pike). Infant mortality rates then declined in the 1980s and 1990s to become more similar to state and national rates. Starting in the 2000s, rates stabilized with some indication of increases, particularly in Pike County. Infant mortality in Boone County was below state and national averages through most of the period considered but has increased recently.

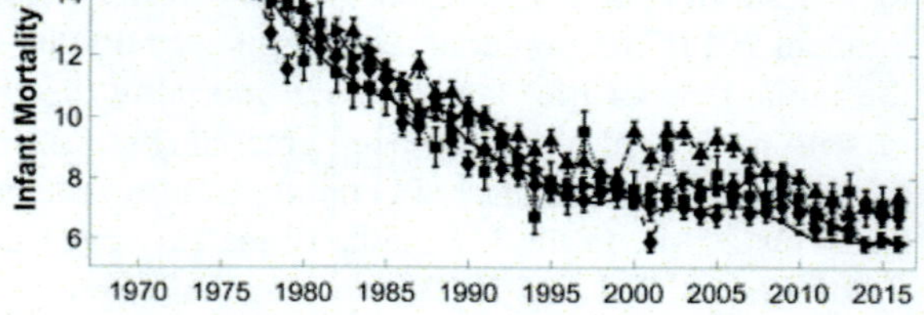

Fig. 9 Trends in infant mortality rate from 1968 to 2016 in states with central Appalachian counties and U.S. Inset shows same data for 2010–2016 with expanded vertical scale. Error bars represent 95% confidence intervals in estimates. Data from CDC (2020)

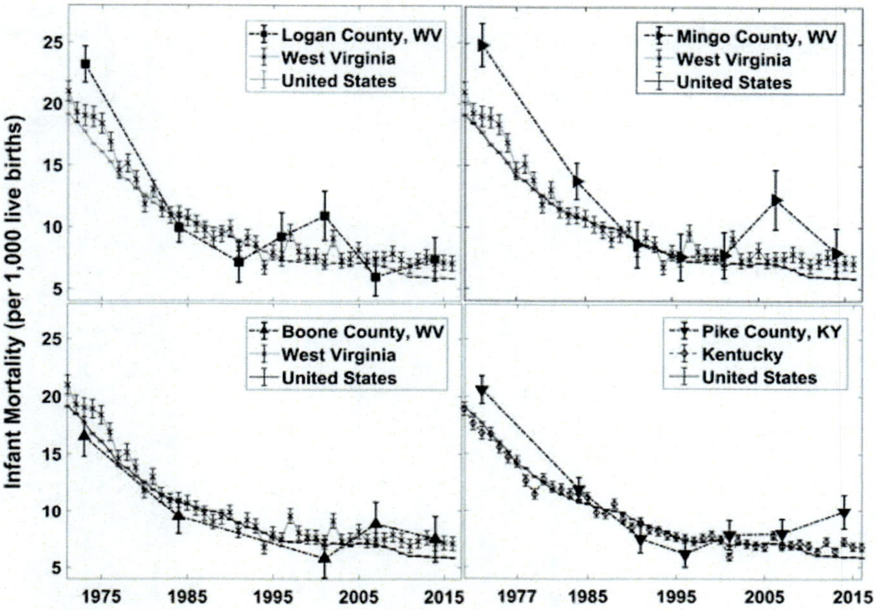

Fig. 10 Trends in infant mortality rates (per 1000 live births), 1973 through 2014, in the four high-coal counties compared to state and national averages, 1971 through 2016. County estimates are averaged over multiple-year periods but state and national estimates are annual. Error bars represent 95% confidence intervals in estimates. Data from CDC (2020)

In contrast to other population health metrics presented, maternal mortality in states with central Appalachian areas and the four high coal-production counties, 1969 through 2016, were consistent with national trends, decreasing steadily through the 1980s and stabilizing through the 1990s and 2000s. Starting in the late 1980s, rates in West Virginia and Kentucky were lower than the national average until about 2010, when rates became more similar to the national average (data not shown). Estimates of age-adjusted maternal mortality in the four high coal-production counties were similar to state trends, with averages of 12.1–14.4 per 100,000 females between 1968–1978 and dropping to 4.5–5.1 in 1999–2016 (CDC 2020).

5.2 Preterm Birth and Low Birth Weight

Preterm birth (PTB) and low birth weight (LBW) rates are heightened in West Virginia, Kentucky, and Tennessee compared to Virginia and the U.S. average between 1995 and 2018 (Fig. 11). PTB and LBW have been on the rise in the U.S. since 2014. Virginia preterm birth and low birth weight rates were more similar to other states with central Appalachian areas in the mid to late 1990s, but then declined

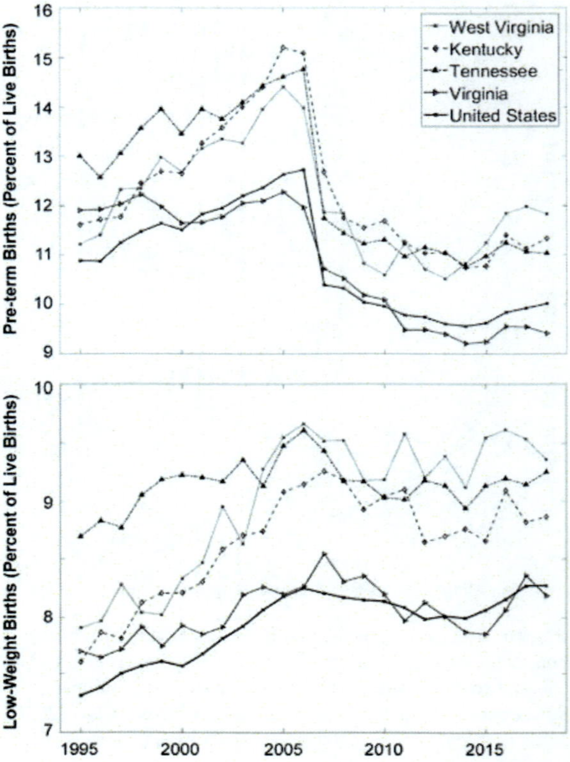

Fig. 11 Trends in preterm birth and low birth weight rates from 1995 to 2018 in states with central Appalachian areas compared to U.S. trends. Data from CDC (2020), Natality Files 1995–2002, 2003–2006, and 2007–2018. Note that the gestational age estimation from use of the last menstrual period estimate through 2006 to the obstetric estimate starting in 2007 partially explains the preterm birth decrease from 2006 to 2007 (Martin et al. 2015)

to become more similar to national averages over the 2000s. Conversely, preterm birth rates in West Virginia have risen since 2014 such that the rate in 2018 (11.8%) was higher than in 1995 (11.2%), even after accounting for the improvement in gestational age estimation that reduced preterm birth rate estimates in the U.S. between 2006 and 2007 (Martin et al. 2015). Low birth weight rates increased between 1995 and 2018 in the U.S. and states with central-Appalachian areas, particularly between 1995 and 2005, but leveled off between 2005 and 2018 (Fig. 11).

While county-level data enabling examination of trends over time are sparse, current preterm birth and low birth weight rates are considerably higher in the four major coal-production counties compared to state, national, and global averages. Preterm birth rate estimates for 2014–2017 range from 12% in Pike County to 14% in Boone County, with Logan and Mingo Counties at 13% (MOD 2018). For comparison, the global preterm birth rate estimate in 2014 was 11% of live births, with China at 7%, and Ethiopia at 12% (Chawanpaiboon et al. 2019), while the U.S. national average was 10% (Fig. 11). Low birth-weight rates in the high coal-production counties over 2011–2017 ranged from 10% of live births in Pike County to 12% in Mingo County, with Logan and Boone Counties at 11% (RWJF 2019), all higher than the U.S. national average of 8% (Fig. 11). The estimated worldwide low-birth-weight rate in 2015 was 15%, with China at 5%, and countries in Sub-Sahara Africa at 14% (Blencowe et al. 2019), indicating that these high coal-production counties have high preterm birth and low birth weight rates relative to both national and worldwide comparisons.

6 Health Initiatives in Coalfield Communities

Health disparities in Appalachia have been long-standing and have been documented previously. A recent analysis identified behaviors associated with chronic disease disparities seen across Appalachia by highlighting "bright spots," Appalachian counties experiencing reduced chronic disease rates relative to other counties in the region (Marshall et al. 2017; Holmes et al. 2018). The analysis found that such counties also report reduced teen pregnancy, higher rates of exercise, and lower rates of smoking. No county with substantial coal production was identified as a bright spot. A subset analysis such as presented here indicates differences among the four high coal-production counties, as Boone and Logan counties have lower age-adjusted mortalities than Mingo and Pike Counties and lower infant mortalities than Pike County (Figs. 5 and 9).

A variety of interventions intended to reduce health disparities in Appalachia have been implemented and evaluated; they focus on cancer screening, diabetes management, obesity prevention, and smoking cessation (Behringer et al. 2007; Jessee and Rutledge 2012; Schoenberg et al. 2012; Studts et al. 2012; Paskett et al. 2018). Interventions primarily tackle reduced access to medical care as well as sociocultural and environmental factors that may reduce the risk of disease or disease progression. Broadly, interventions can be classified into two groups: (1) improving rural health medical care models (e.g. telehealth initiatives such as phone or internet-based care, or coordination of care) to improve access and quality of care; and (2) utilizing social structure (e.g. church-based programs) to support behavioral change at the individual, family, and community levels, particularly for chronic disease prevention and management and for preventative healthcare (e.g. cancer screening).

Advances in telehealth and healthcare coordination have produced promising models of improved access and quality of care in Appalachia. Travel time to a healthcare provider is a primary barrier to access; therefore, healthcare coordination and telehealth maximize benefits of each in-person visit by addressing multiple health issues versus requiring multiple visits, while addressing issues that do not require in-person visits using phone and internet communication services. For example, Barker et al (2016) found educational telephone interventions by nurse practitioners to be as effective as in-person outpatient clinic education sessions for diabetes management in rural West Virginia. A nurse practitioner-coordinated group-visit approach improved diabetes management in southwest Virginia compared to usual care (Jessee and Rutledge 2012). Integration of behavioral health services at primary care clinics was an effective method for improving mental health care access in Appalachia (Correll et al. 2011). A stroke prevention education program was effective when delivered by phone versus in-person in Appalachian Virginia (Schweickert et al. 2011). Telehealth is also a promising delivery method for substance use disorder and mental health treatment services in central Appalachia (Moody et al. 2017). The primary barrier for further implementation of telehealth is limited reimbursement for service providers, which varies widely by state-level policies and health insurance plans (Dorsey and Topol 2016). For telehealth methods that require video conferencing, limited broadband access in Appalachia is an important barrier (Bauerly et al. 2019).

Incorporating health management into important sociocultural entities within communities, particularly within church communities, has led to improvements in chronic disease management and cancer screening in Appalachia. For example, church-based interventions to reduce obesity and cancer via improved physical activity and nutrition have proven effective (Martinez et al. 2016; Paskett et al. 2018). Training and supporting lay health advisors within churches is another model proven effective for improving cervical cancer screening (Studts et al. 2012) and for disease prevention efforts such as a smoking cessation programs (Schoenberg et al. 2015). Several analyses suggest long-term partnerships of health researchers and caregivers, local health departments and locally respected leaders can address place- and cultural-based barriers to improve health in Appalachia (Behringer and Friedell 2006; Schoenberg et al. 2012).

7 Conclusions

Population health trends demonstrate disparities between high coal-production counties and other areas for some health metrics in the 1970s, whereas for other health outcomes, disparities were not evident until the 1980s or 1990s. Age-adjusted mortality disparities have been evident over five decades, with residents of high coal-production counties typically experiencing greater mortality than residents of other Appalachian areas and of areas outside of Appalachia in those same states. Disparities are greatest for high coal-production counties and have increased with time, particularly between 1995 and present day, but they also vary among counties. Trends in cause-specific mortality for the primary causes of death (including cardiovascular, cancer, pulmonary, and diabetes) follow a divergent pattern, with increasing disparities in high coal-production counties relative to other county types, particularly for respiratory-related mortality. Trends in infant mortality, preterm birth, and low birth weight in states with central Appalachian areas and for some of Appalachia's four high coal-production counties became noticeably divergent from national and state trends in the 2000s. Determining whether these elevated mortalities in high coal-production areas are caused by common factors is an important consideration for future research and to inform more effective health interventions. For example, some studies have suggested maternal chronic disease status are determinants of adverse birth outcomes, and in turn, adverse birth outcomes increase the risk of chronic disease later in life (Barker 2012). While CWP is a directly attributable link between coal mining occupations and health that has been well-documented since the 1960s, it remains an ongoing problem for Appalachia's coal-mine workers even with government-mandated occupational protections.

While population health has improved dramatically across the globe and the US since the 1970s, health metrics in rural high coal-production counties have seen much less and in some cases no corresponding improvement. Dissemination of improvements in healthcare methods has been a major driver of population health trends nationally and globally. The fact that health metric improvements in high coal-production counties are lagging behind those of comparison areas suggests utilization of such methods may also be lagging and/or that factors unique to those counties may be serving to impair human health. Further, state- and national-level policies play an important role in determining population health; therefore, the differences in trends over time of high coal-production counties from state and national trends may indicate time and geographically specific events as potential additional explanatory factors for the health disparities characterized. Refinement of metrics for potential exposure factors that may be influential in coalfield areas, including factors directly related to mining such as mining-origin air particulate exposures or factors indirectly related via economic effects, are critical for determining causal pathways that may link living and working in coalfield communities with poor health outcomes. One potential linkage is the ultra-fine atmospheric particulates produced by operating coal surface mines; now that coal-surface mining is declining, researchers may wish to determine if community health conditions potentially linked to such exposures (e.g., respiratory, pulmonary) also follow that downward trend.

Community-based and health-infrastructure intervention research suggests promising avenues for addressing current health disparities in Appalachia; however, the interventions evaluated are generally focused on health-care access and behavior change solutions that are not specific to rural areas within the coalfields of Appalachia. Health intervention research that directly considers potential factors that may link the coal industry and coal mining to coalfield health disparities has been limited. By considering the coalfield-specific health disparities documented herein, future health intervention research targeting unique aspects of coalfield communities may lead to more effective interventions.

References

Ahern MM, Hendryx M (2012) Cancer mortality rates in Appalachian mountaintop coal mining areas. J Environ Occup Health 1:63–70

Ahern MM, Hendryx M, Conley J, Fedorko E, Ducatman A, Zullig KJ (2011) The association between mountaintop mining and birth defects among live births in central Appalachia, 1996–2003. Environ Res 111:838–846

Ahern MM, Mullett M, MacKay K, Hamilton C (2011) Residence in coal-mining areas and low-birth-weight outcomes. Matern Child Health J 15:974–979

Appalachian Regional Commission (2015) Appalachia then and now: examining changes to the Appalachian region since 1965. https://www.arc.gov/assets/research_reports/AppalachiaThenAndNowCompiledReports.pdf

Aneja VP, Pillai PR, Isherwood A, Morgan P, Aneja SP (2017) Particulate matter pollution in the coal-producing regions of the Appalachian Mountains: integrated ground-based measurements and satellite analysis. J Air Waste Manag Assoc 67:421–430

Barbier M, Wilmouth J (2019) United States Mortality Database. University of California, Berkeley. 16 Dec 2019. https://usa.mortality.org

Barker DJ (2012) Developmental origins of chronic disease. Public Health 126:185–189

Barker K, Mallow J, Theeke L, Schwertfeger R (2016) A telehealth rural practice change for diabetes education and management. J Nurse Pract 12:e225–e229

Barnett E, Halverson JA, Elmes GA, Braham VE (2000) Metropolitan and non-metropolitan trends in coronary heart disease mortality within Appalachia, 1980–1997. Ann Epidemiol 10:370–379

Bauerly BC, McCord RF, Hulkower R, Pepin D (2019) Broadband access as a public health issue: the role of law in expanding broadband access and connecting underserved communities for better health outcomes. J Law Med Ethics 47(2_suppl):39–42

Behringer B, Friedell GH (2006) Appalachia: where place matters in health. Prev Chronic Dis 3:A113

Behringer B, Friedell GH, Dorgan KA, Hutson SP, Naney C et al (2007) Understanding the challenges of reducing cancer in Appalachia. Calif J Health Promot 5(SI):040–049

Blackley D, Behringer B, Zheng S (2012) Cancer mortality rates in Appalachia: descriptive epidemiology and an approach to explaining differences in outcomes. J Commun Health 37:804–813

Blackley DJ, Halldin CN, Laney AS (2018) Continued increase in prevalence of coal workers' pneumoconiosis in the United States, 1970–2017. Am J Public Health 108:1220–1222

Blencowe H, Krasevec J, de Onis M, Black RE, An X et al (2019) National, regional, and worldwide estimates of low birthweight in 2015, with trends from 2000: a systematic analysis. Lancet Glob Health 7:e849–e860

Borak J, Salipante-Zaidel C, Slade MD, Fields CA (2012) Mortality disparities in Appalachia: reassessment of major risk factors. J Occup Environ Med 54:146–156

Bowen E, Christiadi DJ, Lego B (2018) An overview of the coal economy in Appalachia. Appalachian Regional Commission, Washington DC

Boyles AL, Blain RB, Rochester JR, Avanasi R, Goldhaber SB et al (2017) Systematic review of community health impacts of mountaintop removal mining. Environ Int 107:163–172

Brink LL, Talbott EO, Stacy S, Marshall LP, Sharma RK, Buchanich J (2014) The association of respiratory hospitalization rates in WV counties, total, underground, and surface coal production and sociodemographic covariates. J Occup Environ Med 56:1179–1188

Buchanich JM, Balmert LC, Youk AO, Woolley SM, Talbott EO (2014) General mortality patterns in Appalachian coal-mining and non-coal-mining counties. J Occup Environ Med 56:1169–1178

Burnett RT, Pope CA III, Ezzati M, Olives C, Lim SS et al (2014) An integrated risk function for estimating the global burden of disease attributable to ambient fine particulate matter exposure. Environ Health Perspect 122:397–403

Canu WH, Jameson JP, Steele EH, Denslow M (2017) Mountaintop removal coal mining and emergent cases of psychological disorder in Kentucky. Commun Ment Health J 53:802–810

Chawanpaiboon S, Vogel JP, Moller A-B, Lumbiganon P, Petzold, et al. (2019) Global, regional, and national estimates of levels of preterm birth in 2014: a systematic review and modelling analysis. Lancet Glob Health 7:e37–e46

Christian WJ, Huang B, Rinehart J, Hopenhayn C (2011) Exploring geographic variation in lung cancer incidence in Kentucky using a spatial scan statistic: elevated risk in the Appalachian coal-mining region. Public Health Rep 126:789–796

Cohen RA, Petsonk EL, Rose C, Young B, Regier M et al (2016) Lung pathology in U.S. coal workers with rapidly progressive pneumoconiosis implicates silica and silicates. Am J Respir Crit Care Med 193:673–680

Correll JA, Cantrell P, Dalton WT III (2011) Integration of behavioral health services in a primary care clinic serving rural Appalachia: reflections on a clinical experience. Fam Syst Health 29:291
Crimmins EM, Zhang Y, Saito Y (2016) Trends over 4 decades in disability-free life expectancy in the United States. Am J Public Health 106:1287–1293
Doney BC, Blackley D, Hale JM, Halldin C, Kurth L et al (2020) Respirable coal mine dust at surface mines, United States, 1982–2017. Am J Ind Med 63:232–239
Dorsey ER, Topol EJ (2016) State of telehealth. N Engl J Med 375:154–161
Dow WH, Rehkopf DH (2010) Socioeconomic gradients in health in international and historical context. Ann NY Acad Sci 1186:24–36
Esch L, Hendryx M (2011) Chronic cardiovascular disease mortality in mountaintop mining areas of central Appalachian states. J Rural Health 27:350–357
Ezzati M, Friedman AB, Kulkarni SC, Murray CJ (2008) The reversal of fortunes: trends in county mortality and cross-county mortality disparities in the United States. PLoS Med 5:e66
Halverson JA, Barnett E, Casper M (2002) Geographic disparities in heart disease and stroke mortality among black and white populations in the Appalachian region. Ethnic Dis 12(S3):82–91
Halverson JA, Bischak G (2008) Underlying socioeconomic factors influencing health disparities in the Appalachian region. Appalachian Regional Commission, Washington DC
Halverson JA, Ma L, Harner EJ (2004) An analysis of disparities in health status and access to health care in the Appalachian region. Appalachian Regional Commission, Washington, DC
Hendryx M (2009) Mortality from heart, respiratory, and kidney disease in coal mining areas of Appalachia. Int Arch Occup Environ Health 82:243–249
Hendryx M (2013) Personal and family health in rural areas of Kentucky with and without mountaintop coal mining. J Rural Health 29(s1):s79–s88
Hendryx M, Ahern MM (2008) Relations between health indicators and residential proximity to coal mining in West Virginia. Am J Public Health 98:669–671
Hendryx M, Ahern MM (2009) Mortality in Appalachian coal mining regions: the value of statistical life lost. Public Health Rep 124:541–550
Hendryx M, Ahern MM, Nurkiewicz TR (2007) Hospitalization patterns associated with Appalachian coal mining. J Toxicol Environ Health A 70:2064–2070
Hendryx M, Entwhistle J (2015) Association between residence near surface coal mining and blood inflammation. Extr Ind Soc 2:246–251
Hendryx M, Fedorko E, Anesetti-Rothermel A (2010) A geographical information system-based analysis of cancer mortality and population exposure to coal mining activities in West Virginia, United States of America. Geospat Health 4:243–256
Hendryx M, Holland B (2016) Unintended consequences of the Clean Air Act: mortality rates in Appalachian coal mining communities. Environ Sci Policy 63:1–6
Hendryx M, Innes-Wimsatt KA (2013) Increased risk of depression for people living in coal mining areas of central Appalachia. Ecopsychology 5:179–187
Hendryx M, Luo J (2015) An examination of the effects of mountaintop removal coal mining on respiratory symptoms and COPD using propensity scores. Int J Environ Health Res 25:265–276
Hendryx M, O'Donnell K, Horn K (2008) Lung cancer mortality is elevated in coal-mining areas of Appalachia. Lung Cancer 62:1–7
Hendryx M, Wolfe L, Luo J, Webb B (2012) Self-reported cancer rates in two rural areas of West Virginia with and without mountaintop coal mining. J Commun Health 37(2):320–327
Hendryx M, Zullig KJ (2009) Higher coronary heart disease and heart attack morbidity in Appalachian coal mining regions. Prev Med 49:355–359
Hitt NP, Hendryx M (2010) Ecological integrity of streams related to human cancer mortality rates. EcoHealth 7(1):91–104
Holmes GM, Lane NM, Arcury TA, Schwalbe ML, Ranolph R et al (2018) Identifying bright spots in Appalachian health: statistical analysis. Appalachian Regional Commission, Washington, DC
Howson CP, Kinney MV, McDougall L, Lawn JE (2013) Born too soon: preterm birth matters. Reprod Health 10(S1):S1

Institute for Health Metrics and Evaluation (IHME) (2018) Global burden of disease study. University of Washington, Seattle. https://ghdx.healthdata.org/gbd-results-tool
Institute for Health Metrics and Evaluation (IHME) (2019) US county profiles. University of Washington, Seattle. https://www.healthdata.org/us-county-profiles
Jessee BT, Rutledge CM (2012) Effectiveness of nurse practitioner coordinated team group visits for type 2 diabetes in medically underserved Appalachia. J Am Acad Nurse Prac 24:735–743
Johnson RC, Schoeni RF (2011) Early-life origins of adult disease: national longitudinal population-based study of the United States. Am J Public Health 101(12):2317–2324
Joseph K, Lisonkova S, Muraca GM, Razaz N, Sabr Y et al (2017) Factors underlying the temporal increase in maternal mortality in the United States. Obstet Gynecol 129:91–100
Knapp EA, Bilal U, Dean LT, Lazo M, Celentano DD (2019) Economic insecurity and deaths of despair in US counties. Am J Epidem 188:2131–2139
Knuckles TL, Stapleton PA, Minarchick VC, Esch L, McCawley M et al (2013) Air pollution particulate matter collected from an Appalachian mountaintop mining site induces microvascular dysfunction. Microcirculation 20:158–169
Kramer MS, Demissie K, Yang H, Platt RW, Sauvé R, Liston R (2000) The contribution of mild and moderate preterm birth to infant mortality. Fetal and Infant Health Study Group of the Canadian Perinatal Surveillance System. JAMA 284:843–849
Krometis LA, Gohlke J, Kolivras K, Satterwhite E, Marmagas SW, Marr LC (2017) Environmental health disparities in the Central Appalachian region of the United States. Rev Environ Health 32:253–266
Kurth LM, Kolker A, Engle M, Geboy N, Hendryx M et al (2015) Atmospheric particulate matter in proximity to mountaintop coal mines: sources and potential environmental and human health impacts. Environ Geochem Health 37:529–544
Kurth LM, McCawley M, Hendryx M, Lusk S (2014) Atmospheric particulate matter size distribution and concentration in West Virginia coal mining and non-mining areas. J Expo Sci Environ Epidemiol 24:405–411
Lamm SH, Li J, Robbins SA, Dissen E, Chen R, Feinleib M (2015) Are residents of mountain-top mining counties more likely to have infants with birth defects? The West Virginia experience. Birth Defects Res A Clin Mol Teratol 103:76–84
March of Dimes (MOD) (2018) March of Dimes PeriStats. March of Dimes Foundation. https://www.marchofdimes.org/Peristats/Peristats.aspx
Marshall J, Thomas L, Lane N, Holmes G, Arcury T et al (2017) Creating a culture of health in Appalachia: disparities and bright spots. Appalachian Regional Commission, Washington DC
Martin JA, Osterman M, Kirmeyer S, Gregory E (2015) Measuring gestational age in vital statistics data: transitioning to the obstetric estimate. US Centers for Disease Control and Prevention. Natl Vital Stat Rep 64(5)
Martinez DJ, Turner MM, Pratt-Chapman M, Kashima K, Hargreaves MK et al (2016) The effect of changes in health beliefs among African-American and rural White church congregants enrolled in an obesity intervention: a qualitative evaluation. J Commun Health 41:518–525
Mathews T, Driscoll AK (2017) Trends in infant mortality in the United States, 2005–2014. U.S. Centers for Disease Control and Prevention, NCHS Data Brief 279, March 2017
Mazurek J, Laney A, Wood J (2009) Coal workers' pneumoconiosis-related years of potential life lost before age 65 years-United States, 1968–2006. US Centers for Disease Control and Prevention. Morb Mortal Wkly Rep 58:1412–1416
Mazurek JM, Wood J, Blackley DJ, Weissman DN (2018) Coal workers' pneumoconiosis–attributable years of potential life lost to life expectancy and potential life lost before age 65 years—United States, 1999–2016. US Centers for Disease Control and Prevention. Morb Mortal Wkly Rep 67:819
McGarvey EL, Leon-Verdin M, Killos LF, Guterbock T, Cohn WF (2011) Health disparities between Appalachian and non-Appalachian counties in Virginia USA. J Commun Health 36:348–356
McMichael AJ, McKee M, Shkolnikov V, Valkonen T (2004) Mortality trends and setbacks: global convergence or divergence? Lancet Public Health 363:1155–1159

Meit M, Heffernan M, Tanenbaum E (2019) Investigating the impact of the diseases of despair in Appalachia. J Appalachian Health 1:7
Meit M, Heffernan M, Tanenbaum E, Hoffmann T (2017) Appalachian diseases of despair. Appalachian Regional Commission, Washington DC
Moody LN, Satterwhite E, Bickel WK (2017) Substance use in rural central Appalachia: current status and treatment considerations. J Rural Ment Health 41:123–135
Murray CJ (1994) Quantifying the burden of disease: the technical basis for disability-adjusted life years. Bull World Health Organ 72(3):429
Murray CJ, Lopez AD (1997) Regional patterns of disability-free life expectancy and disability-adjusted life expectancy: Global Burden of Disease Study. Lancet 349:1347–1352
Nichols CE, Shepherd DL, Knuckles TL, Thapa D, Stricker JC et al (2015) Cardiac and mitochondrial dysfunction following acute pulmonary exposure to mountaintop removal mining particulate matter. Am J Physiol Heart Circ Physiol 309:H2017–H2030
Paskett ED, Baltic RD, Young GS, Katz ML, Lesko SM et al (2018) A group randomized trial to reduce obesity among Appalachian church members: the walk by faith study. Cancer Epidemiol Biomarkers Prev 27:1289–1297
Paskett ED, Fisher JL, Lengerich EJ, Schoenberg NE, Kennedy SK et al (2011) Disparities in underserved white populations: the case of cancer-related disparities in Appalachia. Ooncologist 16:1072–1081
Pollard K, Jacobsen LA (2019) The Appalachian Region: a data overview from the 2013–2017 American Community Survey. Appalachian Regional Commission, Washington DC
Riley JC (2005a) Estimates of regional and global life expectancy, 1800–2001. Popul Dev Rev 31:537–543
Riley JC (2005b) The timing and pace of health transitions around the world. Popul Dev Rev 31:741–764
Robert Wood Johnson Foundation (RWJF) (2019) County Health Rankings Reports. Robert Wood Johnson Foundation and University of Wisconsin Population Health Institute. https://www.countyhealthrankings.org/explore-health-rankings
Roser M, Ortiz-Ospina E, Ritchie H (2019) Life expectancy. https://ourworldindata.org/life-expectancy. 15 Dec 2019
Santopietro G, Zipper CE (2021) Economic trends in Appalachian coalfield counties. Chapter 11 in: Zipper CE, Skousen JG (eds) Appalachia's coal-mined landscapes. Springer
Sartorius BK, Sartorius K (2014) Global infant mortality trends and attributable determinants—an ecological study using data from 192 countries for the period 1990–2011. Popul Health Metr 12:29
Schoenberg NE, Howell BM, Fields N (2012) Community strategies to address cancer disparities in Appalachian Kentucky. Fam Commun Health 35:31
Schoenberg NE, Bundy HE, Bispo JAB, Studts CR, Shelton BJ, Fields N (2015) A rural Appalachian faith-placed smoking cessation intervention. J Relig Health 54:598–611
Schweickert PA, Rutledge CM, Cattell-Gordon DC, Solenski NJ, Jensen ME et al (2011) Telehealth stroke education for rural elderly Virginians. Telemed E-Health 17:784–788
Shi R, Meacham S, Davis GC, You W, Sun Y, Goessl C (2019) Factors influencing high respiratory mortality in coal-mining counties: a repeated cross-sectional study. BMC Public Health 19:1484
Singh GK, Kogan MD, Slifkin RT (2017) Widening disparities in infant mortality and life expectancy between Appalachia and the rest of the United States, 1990–2013. Health Aff 36:1423–1432
Studts CR, Tarasenko YN, Schoenberg NE, Shelton BJ, Hatcher-Keller J, Dignan MB (2012) A community-based randomized trial of a faith-placed intervention to reduce cervical cancer burden in Appalachia. Prev Med 54:408–414
Talbott EO, Sharma RK, Buchanich J, Stacy SL (2015) Is there an association of circulatory hospitalizations independent of mining employment in coal-mining and non-coal-mining counties in West Virginia? J Occup Environ Med 57:e30–e36
U.S. Bureau of Economic Analysis (BEA) (2019) Total Full-Time and Part-Time Employment by NAICS Industry. https://apps.bea.gov/regional/downloadzip.cfm CAEMP25N

US Centers for Disease Control and Prevention (CDC) (2020) Wide-ranging Online Data for Epidemiologic Research (CDC WONDER) Underlying Cause-of-Death Data. December 2019 and January 2020. https://wonder.cdc.gov/mortSQL.html

U.S. Mine Safety and Health Administration (MSHA) (2018) Coal Fatalities for 1900 through 2018. United States Department of Labor. Washington, D.C. https://arlweb.msha.gov/stats/centurystats/coalstats.asp

Vos T, Allen C, Arora M, Barber RM, Bhutta ZA et al (2016) Global, regional, and national incidence, prevalence, and years lived with disability for 310 diseases and injuries, 1990–2015: a systematic analysis for the Global Burden of Disease Study 2015. Lancet 388:1545–1602

Woolley SM, Meacham SL, Balmert LC, Talbott EO, Buchanich JM (2015a) Comparison of mortality disparities in central Appalachian coal-and non-coal-mining counties. J Occup Environ Med 57:687–694

Woolley SM, Youk AO, Bear TM, Balmert LC, Talbott EO, Buchanich JM (2015b) Impact of coal mining on self-rated health among Appalachian residents. J Environ Public Health 2015:501837

World Health Organization (WHO) (2014) Trends in maternal mortality: 1990 to 2013. World Health Organization, Geneva

Yao N, Matthews SA, Hillemeier MM (2012) White infant mortality in Appalachian states, 1976–1980 and 1996–2000: changing patterns and persistent disparities. J Rural Health 28:174–182

Zipper CE (2020) Appalachian coal data. University Libraries, Virginia Tech. https://doi.org/10.7294/BDA6-7R15

Zipper CE, Adams MB, Skousen JG (2021) The Appalachian coalfield in historical context. Chapter 1 in: Zipper CE, Skousen JG (eds) Appalachia's coal-mined landscapes. Springer

Zullig KJ, Hendryx M (2010) A comparative analysis of health-related quality of life for residents of US counties with and without coal mining. Public Health Rep 125:548–555

Zullig KJ, Hendryx M (2011) Health-related quality of life among central Appalachian residents in mountaintop mining counties. Am J Public Health 101:848–853

The Appalachian Coalfield's Energy Transition and Prospects

Carl E. Zipper, Jeff Skousen, and Christopher D. Barton

Abstract The USA and world economies are experiencing an energy transition. The transition is occurring rapidly in eastern USA with consequent effects on Appalachian coal production and markets. Areas of Appalachia where coal mining was most intensive face the future with mining-impaired lands and waters and with economic and human health problems. The region's underutilized mined lands, most of which are not in productive use, present opportunities for productive use that will expand if current energy production and use trends continue as projected. Quality of life enhancements, including environmental restoration of mined areas and improved educational and health-care access, have potential to aid Appalachian coalfield communities' transition to a more prosperous future.

Keywords Coal legacy · Coal mining · Economic development · Environmental restoration · Quality of life · Renewable energy

1 Introduction

The world economy is experiencing an energy transition (IEA 2019; S&P Global 2020). Symptoms include widespread replacement of coal-fired electrical generation by natural gas and by renewable biofuels, solar and wind, and economic electrification. This transition is occurring rapidly in the USA and other advanced economies but is also occurring in other parts of the world. The transition has had a significant and rapid impact on the Appalachian coalfield (Santopietro and Zipper 2021), creating

C. E. Zipper (✉)
Virginia Tech, Blacksburg, VA, USA
e-mail: czip@vt.edu

J. Skousen
West Virginia University, Morgantown, WV, USA

C. D. Barton
University of Kentucky, Lexington, KY, USA

C. E. Zipper and J. Skousen (eds.), *Appalachia's Coal-Mined Landscapes*,
https://doi.org/10.1007/978-3-030-57780-3_13

risk that some parts of the region may be left behind as the world economy advances. Here we ask the question: What actions might ease the Appalachian coalfield's fate as the global energy transition continues? In response, we offer observations on possible paths forward.

2 Energy Transition

2.1 Historical Context

Energy use is fundamental to human life in the modern age. The global economy utilizes energy in abundant quantities. Humans use energy from physical, chemical, and biological sources to do what cannot be done using human bodies alone. Humans' abilities to direct such energies are essential to current society and to the planet's support of more than 7 billion people.

Energy transitions occur when an individual energy source increases in importance relative to economic activities and development over time. Energy transitions are often associated with increasing economic activity generally, although such changes do not affect all people equally. The rising energy source makes activities possible that were not possible before or makes known activities possible at reduced economic cost; but the shifting economic fortunes created by the transition may disadvantage people and industries formerly dependent upon the prior energy regime.

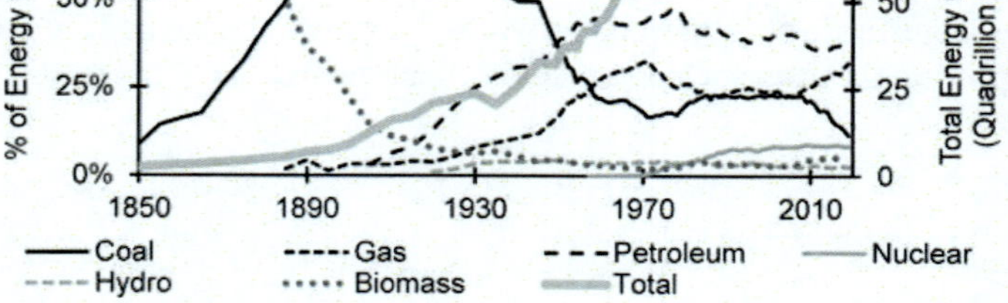

Fig. 1 USA energy consumption by source as % of total (left axis), and total consumption (right axis), 1850–2019 for major historical sources and nuclear; and 1970–2019 for geothermal, solar, and wind (inset above). Data from US EIA (2011b and 2020b)

Multiple energy transitions have occurred throughout world history (Smil 2016a, b) and in the USA (Fig. 1). Energy from biomass combustion was utilized by prehistoric people and then enabled development of early industry. In the USA, wood and charcoal fuels were used by the early iron industry and powered the nation's first railroads (Zipper et al. 2021a). Replacement of biomass by coal as an industrial energy source began while spawning the industrial revolution in England during the eighteenth century and then expanding to other parts of the world. Coal began widescale replacement of wood as an industrial fuel in the USA during the mid-nineteenth century, while its use in China and other Asian nations were expanding rapidly until recently. Industrial-scale production and use of liquid hydrocarbon fuels began in the latter part of the nineteenth century; liquid hydrocarbons replaced coal as the dominant energy source for the global economy and in the USA during early-to-mid-twentieth century. Today, essential societal functions are becoming more dependent upon electrical energy, continuing a process that has extended over more than a century (Rosenberg 1998); while the energy sources used to generate electricity continue to diversify (IEA 2019).

2.2 *Transition in Eastern USA*

Historically, energy transitions causing major economic transformations have occurred slowly, often over centuries, but they can occur more quickly within national and sub-national areas (Smil 2016a, b). Such is the case in eastern USA, the primary market region for Appalachian coal. Although liquid and gas fuels and coal from other USA regions eroded Appalachian coal markets over much of the twentieth century, Appalachian coal production remained at robust levels until early in the twenty-first century when production began a steady decline (Fig. 2) As rapid technology development expanded production and reduced prices for natural gas, an energy source that competes with Appalachian coal for electricity generation, that decline accelerated. Hydrofracturing technologies have expanded natural gas production from the Marcellus and Utica shales of northern Appalachia by more than 20 times from <1 trillion cubic feet per year in 2010 to more than 10 trillion cubic feet, 32% of US natural gas production, in 2019. Similarly, eastern USA electric generators, the predominant purchasers of Appalachian coal, halved their use of coal fuels while tripling their use of natural gas since the mid-2000s (Fig. 3).

The energy-technology developments causing such rapid effects in Appalachia are also affecting the global economy. Hydrofracturing has enabled the USA to become an exporter of both oil and gas, and no longer dependent on international markets for essential energy supplies as it had been since the 1950s (US EIA 2015a). The USA's increasing production and exports have been major contributors to declining prices for oil and gas in global markets since the early 2010s (BP 2020). Similarly, energy production from renewable energy sources (solar and wind) has been increasing rapidly in the USA and globally (IEA 2019).

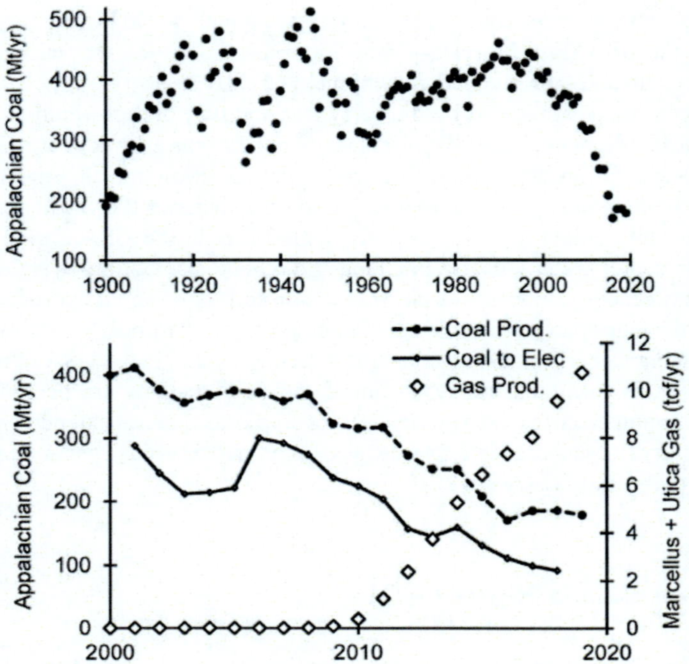

Fig. 2 Annual Appalachian coal production (million tons per year) from Tennessee northward (1900–2019 above; and 2000–2019 below); Appalachian coal shipments to electric power generators (million tons per year, below), and dry natural gas production (trillion cubic feet per year) from two northern-Appalachia geologic formations, the Marcellus and Utica shales (below) (Data from EIA 2019a, b, 2020c, e; Milici 1997)

Economics has been a major driver for this rapid change. Hydrofracturing technologies reduced eastern-USA market prices for natural gas by more than 2 times since the mid-2000s while coal prices rose slightly over that same period (Fig. 3). Similarly, costs for generating electricity with renewable energy sources, solar and wind, have declined to the point where these sources are now competitive with natural gas and capable of generating electricity for less cost than at new coal plants (US EIA 2020a). As those economics prevent construction of new coal plants, older coal plants are being retired at a rapid rate; more than one-third of the coal-fired generating capacity within the Appalachian coal market area of eastern USA has been retired since 2010 while additional retirements are planned through 2023 (US EIA 2020d).

While natural gas prices have been declining, coal users have been facing increasing costs and restrictions. These include limits on emissions of multiple atmospheric pollutants that are released by combustion of coal (e.g., Zhang 2016). New regulation of coal ash disposal (US EPA 2019) has also increased electric generators' costs of coal utilization. Although some regulatory restrictions established in past years have been proposed for rollback (Mendelevitch et al. 2019), electric power

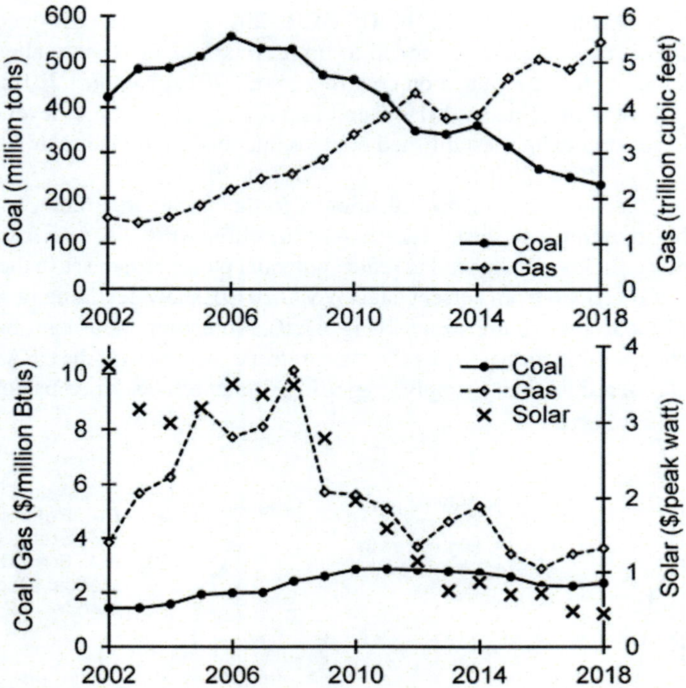

Fig. 3 Annual coal and gas shipments (above) and shipment-weighted average prices (below) for coal and gas received by electric generators within the Appalachian coal market area of eastern USA; and national average prices for photovoltaic modules shipped to USA purchasers (below) (Data sources: US EIA 2009, 2011a, 2019a, c, d)

generators have made long-term investment decisions that reduce usage of coal (US EIA 2019e).

Factors causing increased mining costs are also influencing coal markets. These include coal reserve depletion in Appalachia where coal has been mined for more than two centuries, which increases extraction costs as miners are forced to access less-desirable seams (Milici 2000; Hook and Aleklett 2010). Environmental restrictions include limits on valley fill construction and increased regulation of water effects (US EPA 2016; Skousen and Zipper 2021). Although mean coal prices paid by eastern USA electric generators declined slightly in recent years (Fig. 3), this has occurred as multiple coal producers were driven out of the market and some, presumably those with high costs, entered bankruptcy (NYT 2019, S&P Global 2019).

Also, other factors are driving the eastern USA energy transition. These include environmental concerns as coal-combustion air pollutants affect both human health and climate to a greater degree than natural gas, while wind and solar produce no such pollutants. Even with modern air-pollution controls in place, coal-fired electric generators release air pollutants that threaten human health (Sunderland et al. 2016; Clay and Muller 2019). Also, the climate-altering effect of coal-generated greenhouse gases is well established (IPCC 2014; US GCRP 2017). Some electric generators are making public commitments intended to limit emissions of climate-altering gases which include reduced reliance on coal (e.g., AEP 2019; NextEra 2018; Southern Company 2018; Duke Energy 2019; Dominion Energy 2020). State-level regulators of electric generators have established policies intended to limit climate-altering gas emissions (e.g., RGGI 2020).

Future projections provide no indication of forthcoming turnaround. Even the US Energy Information Administration (EIA) projections (Fig. 4), which incorporate only current policies and thus fail to reflect potential policy responses to the changing climate that is evident from current data (WMO 2019), show declining production of Appalachian coal over future decades (Fig. 4 left). Projections incorporating potential climate-change mitigation policies show even steeper declines for the USA and North American coal production, strongly suggesting similar effects for Appalachia (Fig. 4 right).

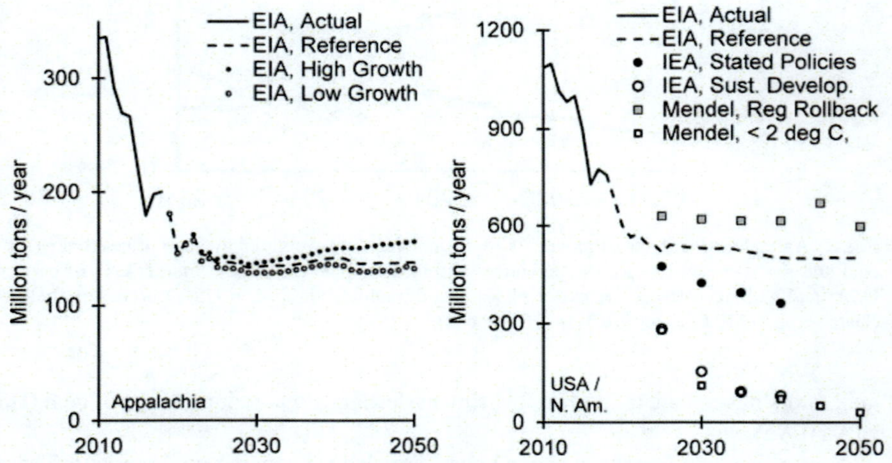

Fig. 4 Appalachian (left), USA, and North American coal production (right), actual (2010–2018), and projected through 2050. Actual data from US EIA (2019b). Projections from US EIA (2020a) for Reference Case and for high and low economic growth projections; IEA (2019) for Stated Policies Scenario (essentially, business as usual) and a Sustainable Development scenario with policies to limit greenhouse gas emissions; and by Mendelevitch et al. (2019) for USA steam coal if currently proposed policies rolling back environmental regulations are implemented (Reg Rollback); and with greenhouse gas-emission limits intended to limit climate warming to <2 °C. IEA (2019) projections are for North America; other data for the USA only. Appalachia data are for the region extending from Alabama northward

3 Prospects

Here, we ask the question: What will become of the Appalachian coalfield and its communities? The recent and rapid coal-production decline appears as both real and permanent but not yet finished. What can or should be done to ease consequent effects on people within a region that has made essential contributions to the nation's economic growth over more than two centuries (Zipper et al. 2021a)? That region now finds itself disadvantaged in the competitive global economy due to multiple factors: geographic isolation from the mainstream of today's commerce, impaired lands and waters that are a legacy of coal mining (Clark et al. 2021; Kruse Daniels et al. 2021; Sena et al. 2021), communities with diminished fiscal capacities due to declining populations and losses of coal-mining jobs (Santopietro and Zipper 2021), and populations with multiple health problems (Gohlke 2021).

We have no comprehensive answers to that question. But we do have some observations. Our perspective is limited by our status as natural resource scientists. Although with extensive experience working in the Appalachian coalfield, we claim no expertise in economic development, community planning, or similar fields. Nonetheless, we see the following as clear:

1. The Appalachian coalfield has assets that are becoming more valued in the global economy, some of which are unique to the region.

Those assets include a talented workforce, sufficiently so to have enabled a highly technological endeavor—coal mining—to thrive until the recent and globally significant turn in the region's fortunes; a natural environment that remains as highly biodiverse relative to other non-tropical world regions and with natural beauty where not impaired by mining; a humid and temperate climate regime that is perceived as pleasurable by many people and is conducive to biological productivity; extensive mined-land areas that are not in productive use but often accessible from roadways and available for development; and relatively unpopulated landscapes within an increasingly crowded world.

2. The energy transition that has affected the region's fortunes negatively is a reality.

Coal is not coming back. Coal-market changes to date have been massive and rapid. Given electric generators' decisions to retire coal-fired generating capacity and replace it with other non-coal generating units, those changes are likely irreversible. But they may not be finished, given rapidly advancing photovoltaic technologies and the human-induced climate change that is affecting the entire world.

Scientific evidence supporting the human-induced climate-change theory is clear and becoming more evident with time (IPCC 2014; US GCRP 2017; WMO 2019). Should national policy makers act to mitigate the changing climate, Appalachian coal's decline would likely continue (Fig. 4 right). But within such a policy environment, emerging renewable energy technologies could provide new opportunities for the Appalachian coalfield.

3. Some coalfield areas are currently benefiting from the intensive economic activity generated by hydrofracturing, but the future of hydrofracturing is not assured.

The primary locus of hydrofracturing of the Marcellus and Utica shales is northern Appalachia, not within the central Appalachian region where coal mining's legacy is most severe. Should the northern Appalachia hydrofracturing industry's recent expansion stall or reverse, the region may experience a shift of economic fortune.

Scientific research has found clear links between hydrofracturing and multiple environmental and human-health threats. Multiple studies have found leaks of methane, a potent greenhouse gas, to be common near hydrofracturing facilities (Lyon et al. 2016; Omara et al. 2016). Hydrocarbon gases with known capability to affect human health are associated with those methane leaks (Macey et al. 2014; Paulik et al. 2018) and appear to be causing negative human health effects to parties residing close to hydrofracturing sites (Jemielita et al. 2015; Stacy et al. 2015; Casey et al. 2016; Currie et al. 2017). Water-quality degradation has also been found in the vicinity of hydrofracturing (Vidic et al. 2013); while the methane leaks have significant climate implications (Alverez et al. 2018; Howarth 2019). Hence, the northern Appalachian hydrofracturing industry is subject to regulation risks.

Much of the hydrofracturing industry's recent activity is based on expectation of future profitability, as evidenced by the high levels of debt (WSJ 2019; IEEFA 2020). Some of the hydrofracturing industry's future projections are based on an expectation of rapidly expanding exports of liquified natural gas (LNG). However, natural gas production in other world regions is also expanding, as are other nations' LNG export expectations (IEA 2019). Also, significant shale-gas geologic resources, totaling >10 times those of the USA in total, occur in South America, China. and other parts of the world (US EIA 2015b).

4. Mined lands are an underutilized resource with potential for conversion into productive economic assets.

The global energy transition is providing opportunities for mined-land utilization that was not evident previously.

<u>Solar</u>: A constraint to photovoltaic development in eastern USA is availability of suitable land areas that are not in other uses (Zipper et al. 2021b). Appalachian mined lands are plentiful; in close proximity to major eastern USA population centers and energy grids; and, for the most part, not in other productive or managed uses and therefore potentially available for photovoltaic energy development.

<u>Wind</u>: Although some formerly prime wind-energy sites have been impaired by coal mining, many remain. Some mining operations have replaced former mountains with facsimiles comprised of mine spoils that provide less geotechnical stability than the former intact geology, while other former mountains have been reduced in elevation and, hence, wind energy accessibility. Despite those impairments, mined lands of the Appalachian coalfield may offer opportunities for wind energy development.

<u>Biomass</u>: The capability of mined-land areas to support biomass crops also offers opportunities, especially if policies to restrict greenhouse gas emissions are enacted.

Use of biofuels for climate mitigation is problematic when biofuel crops displace native ecosystems directly or if they displace agricultural crops and thus cause natural ecosystem disturbance elsewhere via agricultural expansion (Fargione et al. 2008). Growing deep-rooted biofuels on marginal land, however, can provide climate benefits both by substituting a biofuel product for fossil fuel and by rebuilding soil carbon stocks (Evans et al. 2015). Clearly, the extensive mined lands of the Appalachian coalfield offer such opportunities.

Enaction of climate-mitigation policies would likely expand demand for sustainably-produced liquid biofuels as substitutes for petroleum-based vehicle fuels. Given transportation costs for agricultural biomass, biorefineries for such conversion would likely be located close to biomass sources.

Also, published strategies for limiting climate warming to ≤ 1.5 °C employ bioenergy with carbon capture and sequestration (BECCS) for drawing down atmospheric carbon (IPCC 2018). Such facilities would burn renewable biomass to produce energy for electricity generation or other purposes, capture the carbon that would otherwise be emitted as combustion gases, and store or sequester that carbon stably in deep geologic formations (IEA 2019). In addition to mined lands with biofuel production capability (Brown et al. 2015; Scagline-Mellor et al. 2018; Zipper et al. 2021b), the Appalachian coalfield is underlain by depleted oil and gas reservoirs, unmineable coal seams, and deep saline aquifers capable of storing large quantities of inorganic carbon (US DOE 2015). Hence, the region has resources that could support BECCS deployment.

Rapidly emerging and increasingly cost-competitive battery storage technologies (US EIA 2019f) have potential for pairing with photovoltaic and wind generation to provide continuous renewable power energy for local consumers desiring such and to regional electric-power transmission grids. Establishment of renewable energy facilities on many mined lands would aid environmental restoration by displacing the invasive-plant dominated novel ecosystems that currently occupy many mined areas (Sena et al. 2021).

5. Restoration of natural ecosystems on mined lands would create multiple benefits.

Appalachia's natural ecosystems are highly valued globally (Ricketts et al. 1999; Pickering et al. 2003). Methods for restoring Appalachian forests and headwater streams on degraded mined lands are known and available (Zipper et al. 2021b).

Forest restoration has potential to create economic benefits by producing marketable timber with potential for eventual sale revenues exceeding restoration costs on a dollar-for-dollar basis. However, those revenues would occur far in the future, and conventional economic frameworks incorporating time-value-of-money considerations do not justify those restoration expenses. Nonetheless, forest restoration on mined lands would also sequester large quantities of carbon, removing it from the atmosphere via conversion to growing trees and soil organic matter; should national policies be enacted to address climate change, such sequestration may become more economically valued. The deep and often organic-matter deficient soils on mined lands offer capacity to sequester and store atmospheric carbon (Skousen et al. 2021).

Ecosystem restoration on mined lands would also produce non-market benefits. These include increased levels of ecosystem services. Strategic ecosystem restoration might include locational considerations, given the ecological value of large tracts of relatively undisturbed natural ecosystems; some such areas remain in the Appalachian coalfield, while additional areas with similar attributes could be created with ecological restoration on strategically located mined areas.

Another non-market benefit created by strategic mined-land ecological restoration would be quality-of-life enhancement in former coal-intensive communities. Some mined areas are both highly visible to local populations and clearly different from unmined Appalachian forest. These unique environments often support adaptive wildlife, some with conservation value, that do not thrive in the region's forests (Lituma et al. 2021). The presence of such wildlife enhances local quality of life, provides recreational opportunities and, in some cases, economic benefits. Environmental restoration approaches for displacing or controlling the highly-invasive exotic plant species that dominate such areas while maintaining suitable habit for valued wildlife are needed; development and widespread application of such approaches would create regional benefits.

Similarly, mined-land restoration projects carried out with community engagement could produce additional benefits. Many Appalachian areas retain a sense of community that is sorely lacking in some of the USA's more urbanized areas.

Aquatic resources have also been impacted by mining. Mining-origin water pollutants emerge from the huge volumes of geologic materials disturbed by past production of coal. Passage of time can mitigate those effects, although the time required for full mitigation is not clear (Clark et al. 2021; Kruse Daniels et al. 2021). Environmental restoration of stream corridors and headwater-stream watersheds in such areas has potential to aid recovery of aquatic life that has been impaired by mining (Merovich et al. 2021).

6. Enhancement of education opportunities can be an economic development strategy.

Education is essential to employment success in today's knowledge-based economy. Educational levels in coalfield counties are lower than in other eastern USA areas (Santopietro and Zipper 2021). This finding is similar to what other studies have found for European areas that have suffered fates similar to that of Appalachia, declining coal production, and consequent economic problems and job losses. Government investments in educational opportunities for the younger generations of current populations and for displaced miners have been seen as a response to coal-job losses in these other regions, although with recognition that positive effects are more likely over long than short terms (Caldecott et al. 2017). A regional development agency also recommends efforts to improve educational outcomes in Appalachia as an economic development strategy (ARC 2015).

Enhancement of educational opportunities may provide entrees to alternative employment for some displaced miners and other underemployed regional residents, while also providing the next generation with expanded employment opportunities. Also, higher-quality educational opportunities can enhance the quality of life available to local residents; and, hence, may also aid in drawing new industry and investment to the region.

7. Actions to enable population health improvements would also be a viable strategy for improving quality of life in coalfield communities.

Appalachian communities in high coal-production counties experience negative health outcomes (Gohlke 2021). The academic literature indicates limited health-care access as a contributing factor, at least in some coal-mining areas (Shi et al. 2019); while a regional agency also sees improvements in health-care infrastructure and outcomes as an important priority (ARC 2015). Because human behaviors also contribute to the region's health problems, initiatives building from social networks and utilizing communications technology show promise for addressing health disparities in Appalachian areas (Gohlke 2021).

8. The combination of quality-of-life enhancements, environmental amenities, and renewable energy resources could serve as a draw for new businesses, industry, and investment.

The rural lifestyle available in many areas of the Appalachian coalfield would likely prove more attractive if greater economic opportunity were available. Enhancement of educational opportunities and environmental quality via restoration on mined areas would likely increase that appeal.

Many corporations seek locations with high quality of life for new investments. Increasingly, many multinational corporations are also seeking to ensure that their operations are carbon neutral (Google 2019) and, in some cases, even carbon negative (Microsoft 2020). One major firm states that all of its own facilities operate on 100% renewable energy, and that it is seeking to ensure similar operations by its suppliers (Apple 2019). The corporate trend toward carbon neutrality is not restricted to the technology industry; other entities pledging to achieve carbon-neutral operations in their own facilities, encouraging carbon neutrality by their suppliers, or both include global manufacturers and retailers (Morgan 2019).

Given Appalachia's unique environment and storied history, coalfield communities offering essential infrastructure such as sites for facility development and renewable energy, along with educational, environmental, and other quality-of-life amenities, could find themselves in a unique position to attract employers.

4 Constraints

Some of the prospects observed above can be accessed by local entities but others would require state-level and national intervention. Clearly, some would require significant financial investment. Many environmental restoration projects, for example, do not pay for themselves when viewed through the conventional economic lens, but such projects often yield net benefits when considering indirect as well as direct economic effects. In states where public K-12 schools receive primary support from local taxes, improved educational opportunities would likely require state and/or federal investments to supplement the funds generated locally from tax bases depleted by losses of population and coal jobs; while mobilization of post-secondary educational entities typically requires resources well beyond those of local entities. Health-care access is a challenge nationally as well as in Appalachia.

5 Conclusions

We do not pretend that the potential remedies suggested by our observations, even if fully embraced and acted upon, would be adequate to address current conditions within the Appalachian coalfield. We expect that others, including residents, can develop other and more creative solutions to the region's problems. The region, its communities, and residents have experienced a rapid change of fortune in response to energy-related events of global significance. What can or should be done to ease the region's fate? We offer observations while recognizing that there are no easy answers. While bearing the legacy of past mining impacts, Appalachia's coal-mined landscapes offer opportunities in the new energy era.

References

Alvarez RA, Zavala-Araiza D, Lyon DR et al (2018) Assessment of methane emissions from the U.S. oil and gas supply chain. Science 361:186–188

American Electric Power (AEP) (2019) AEP's clean energy strategy will achieve significant future carbon dioxide reductions. https://www.aep.com/news/releases/read/1503/AEPs-Clean-Energy-Strategy-Will-Achieve-Significant-Future-Carbon-Dioxide-Reductions-

Appalachian Regional Commission (2015) Investing in Appalachia's Future, The Appalachian Regional Commission's Five-Year Strategic Plan for Capitalizing on Appalachia's Opportunities

Apple (2019) Making without taking sounds impossible. But it's our goal. https://www.apple.com/environment/our-approach/

British Petroleum (BP) (2020) Statistical review of world energy

Brown C, Griggs T, Keene T, Marra M, Skousen J (2015) Switchgrass biofuel production on reclaimed surface mines: I. Soil quality and dry matter yield. Bioenergy Res 9:31–39

Caldecott B, Sartor O, Spencer T (2017) Lessons from previous 'coal transitions' high-level summary for decision-makers. IDDRI and Climate Strategies. https://www.iddri.org/en/publications-and-events/report/lessons-previous-coal-transitions

Casey JA, Savitz DA, Rasmussen SG, Ogburn EL, Pollak J, Mercer DG et al (2016) Unconventional natural gas development and birth outcomes in Pennsylvania, USA. Epidemiology 27:163–172

Clark EV, Zipper CE, Soucek DJ, Daniels WL (2021) Contaminants in Appalachian water resources generated by non-acid-forming coal-mining materials. In: Zipper CE, Skousen JG (eds) Chapter 9, Appalachia's coal-mined landscapes. Springer

Clay K, Muller NZ (2019) Recent increases in air pollution: evidence and implications for mortality. Natl Bureau Econ Res, Working Paper 26381

Currie J, Greenstone M, Meckel K (2017) Hydraulic fracturing and infant health: new evidence from Pennsylvania. Sci Adv 3:e1603021

Dominion Energy (2020) Delivering Clean Energy. https://www.dominionenergy.com/ourpromise/clean-energy

Duke Energy (2019) Duke Energy aims to achieve net-zero carbon emissions by 2050. https://news.duke-energy.com/releases/duke-energy-aims-to-achieve-net-zero-carbon-emissions-by-2050

Evans SG, Ramage BS, DiRocco TL, Potts MD (2015) Greenhouse gas mitigation on marginal land: a quantitative review of the relative benefits of forest recovery versus biofuel production. Environ Sci Technol 49:2503–2511

Fargione J, Hill J, Tilman D, Polasky S, Hawthorne P (2008) Land clearing and the biofuel carbon debt. Science 319:1235–1238

Gohlke JM (2021) Human health in coalfield communities of Appalachia. In: Zipper CE, Skousen JG (eds) Chapter 12, Appalachia's coal-mined landscapes. Springer

Google Inc. (2019) 2019 Environmental Report. https://sustainability.google/reports/environmental-report-2019/

Hook M, Aleklett K (2010) Trends in U.S. recoverable coal supply estimates and future production outlooks. Nat Resour Res 19:189–208

Howarth RW (2019) Ideas and perspectives: is shale gas a major driver of recent increase in global atmospheric methane? Biogeosciences 16:3033–3046

Institute for Energy Economics and Financial Analysis (IEEFA) (2020) Bankruptcies in fracking sector mount in 2019. https://ieefa.org/wp-content/uploads/2020/01/Bankruptcies-in-Fracking-Sector-Mount-in-2019_January-2020.pdf

Intergovernmental Panel on Climate Change (IPCC) (2014) Climate change 2014. https://www.ipcc.ch/

Intergovernmental Panel on Clim Change (IPCC) (2018) Global warming of 1.5 °C

International Energy Agency (IEA) (2011) Combining bioenergy with CCS: reporting and accounting for negative emissions under UNFCCC and the Kyoto Protocol

International Energy Agency (IEA) (2019) World Energy Outlook 2019

Jemielita T, Gerton GL, Neidell M et al (2015) Unconventional gas and oil drilling is associated with increased hospital utilization rates. PLoS ONE 10:e0131093

Kruse Daniels N, LaBar J, McDonald LM (2021) Acid mine drainage in Appalachia: sources, legacy, and treatment. In: Zipper CE, Skousen JG (eds) Chapter 8, Appalachia's coal-mined landscapes. Springer

Lituma CM, Cox JJ, Spear S, Edwards JW, De La Cruz JL, Muller LI, Ford WM (2021) Terrestrial wildlife in the Appalachian post-mined landscape: status and opportunities. In: Zipper CE, Skousen JG (eds) Chapter 6, Appalachia's coal-mined landscapes. Springer

Lyon DR, Alvarez RA, Zavala-Araiza D, Brandt AR, Jackson RB, Hamburg SP (2016) Aerial surveys of elevated hydrocarbon emissions from oil and gas production sites. Environ Sci Technol 50: 4877−4886

Macey GP, Breech R, Chernaik M, Cox C, Larson D, Thomas D, Carpenter DO (2014) Air concentrations of volatile compounds near oil and gas production: a community-based exploratory study. Environ Health 13:82

Mendelevitch R, Christian Hauenstein C, Holz F (2019) The death spiral of coal in the U.S.: will changes in U.S. policy turn the tide? Clim Policy 19:1310–1324

Merovich GT, Hitt NP, Merriam ER, Jones JW (2021) Response of aquatic life to coal mining in Appalachia. In: Zipper CE, Skousen JG (eds) Chapter 10, Appalachia's coal-mined landscapes. Springer

Microsoft (2020) Microsoft announces it will be carbon negative by 2030. https://news.microsoft.com/2020/01/16/microsoft-announces-it-will-be-carbon-negative-by-2030/

Milici RC (1997) The CoalProd database: historical production data for the major coal-producing regions of the conterminous United States. USGS Open File Report 97–447

Milici RC (2000) Depletion of Appalachian coal reserves—how soon? Int J Coal Geol 44:251–266

Morgan B (2019) 101 companies committed to reducing their carbon footprint. Forbes, 26 August 2019. https://www.forbes.com/sites/blakemorgan/2019/08/26/101-companies-committed-to-reducing-their-carbon-footprint
NextEra (2018) NextEra Energy establishes carbon dioxide emissions rate reduction target as part of its commitment to creating a sustainable energy future. https://www.investor.nexteraenergy.com/news-and-events/news-releases/2018/09-06-2018-113047243
New York Times (NYT) (2019) Murray energy is 8th coal company in a year to seek Bankruptcy. https://www.nytimes.com/2019/10/29/business/energy-environment/murray-energy-bankruptcy.html
Omara M, Sullivan MR, Li X, Subramanian R, Robinson AL, Presto AA (2016) Methane emissions from conventional and unconventional natural gas production sites in the Marcellus shale basin. Environ Sci Technol 50:2099–2107
Paulik LB, Hobbie KA, Rohlman D et al (2018) Environmental and individual PAH exposures near rural natural gas extraction. Environ Poll 241:397–405
Pickering J, Kays R, Meier A, Andrew S and Yatskievych R (2003) The Appalachians. Wilderness Earth's Last Wild Places. Conservation International, Washington, DC
Regional Greenhouse Gas Initiative (RGGI) (2020) The regional greenhouse gas initiative an initiative of New England and Mid-Atlantics states of the US. https://www.rggi.org/
Ricketts TH, Dinerstein E, Olson DM, Loucks CJ, Eichbaum W, DellaSalla D, Kavanagh K, Hedao P, Hurley P, Carney K, Abell R, Walters S (1999) Terrestrial ecoregions of North America: a conservation assessment. Island Press, Washington
Rosenberg N (1998) The role of electricity in industrial development. Energy J 19:7–24
S&P Global (2019) Latest round of coal bankruptcies hits buyers of others' distressed assets
S&P Global (2020) What is energy transition. https://www.spglobal.com/en/research-insights/articles/what-is-energy-transition
Santopietro GD, Zipper CE (2021) Economic trends in Appalachian coalfield counties. In: Zipper CE, Skousen JG (eds) Chapter 11, Appalachia's coal-mined landscapes. Springer
Scagline-Mellor S, Griggs T, Skousen J, Wolfrum E, Holaskova I (2018) Switchgrass and giant miscanthus biomass and theoretical ethanol production from reclaimed mine lands. BioEnergy Res 11:562–573. https://doi.org/10.1007/s12155-018-9915-2
Sena K, Franklin JA, Swab RM, Hall SL (2021) Plant communities on Appalachian mined lands. In: Zipper CE, Skousen JG (eds) Chapter 5, Appalachia's coal-mined landscapes. Springer
Shi R, Meacham S, Davis GC, You W, Sun Y, Goessl C (2019) Factors influencing high respiratory mortality in coal-mining counties: a repeated cross-sectional study. BMC Public Health 19:1484
Skousen JG, Daniels WL, Zipper CE (2021) Soils on Appalachian coal-mined lands. In: Zipper CE, Skousen JG (eds) Chapter 4, Appalachia's coal-mined landscapes. Springer
Skousen JG, Zipper CE (2021) Coal mining and reclamation in Appalachia. In: Zipper CE, Skousen JG (eds) Chapter 3, Appalachia's coal-mined landscapes. Springer
Smil V (2016a) Energy transitions: global and national perspectives, 2nd edn. Praeger
Smil V (2016b) Examining energy transitions: a dozen insights based on performance. Energy Res Soc Sci 22:194–197
Southern Company (2018) Planning for a low-carbon future. https://www.southerncompany.com/content/dam/southern-company/pdf/corpresponsibility/Planning-for-a-low-carbon-future.pdf
Stacy SL, Brink LL, Larkin JC et al (2015) Perinatal outcomes and unconventional natural gas operations in southwest Pennsylvania. PLoS ONE 10:e0126425
Sunderland EM, Driscoll CT, Hammitt JK et al (2016) Benefits of regulating hazardous air pollutants from coal and oil-fired utilities in the United States. Environ Sci Technol 50:2117–2120
US Department of Energy (US DOE) (2015) Carbon storage atlas. Fifth Edition, Office of Fossil Energy
US Energy Information Administration (US EIA) (2009) Cost and quality of fuels for electric plants (and prior editions)
US Energy Information Administration (US EIA) (2011a) Annual Energy Review (Table 10.8)

US Energy Information Administration (US EIA) (2011b) History of energy consumption in the United States, 1775–2009. Today in Energy, 9 February 2011
US Energy Information Administration (US EIA) (2015a) U.S. energy imports and exports to come into balance for first time since 1950s. Today in Energy, 15 April 2015
US Energy Information Administration (US EIA) (2015b) World Shale Resource Assessments. https://www.eia.gov/analysis/studies/worldshalegas/
US Energy Information Administration (US EIA) (2019a) Annual Coal Distribution Report (and prior editions)
US Energy Information Administration (US EIA) (2019b) Annual Coal Report (and prior editions)
US Energy Information Administration (US EIA) (2019c) Annual Solar Photovoltaic Module Shipments Report
US Energy Information Administration (US EIA) (2019d) Electric Power Annual (and prior editions)
US Energy Information Administration (EIA) (2019e) More U.S. coal-fired power plants are decommissioning as retirements continue. Today in Energy, 26 July 2019
US Energy Information Administration (EIA) (2019f) U.S. utility-scale battery storage power capacity to grow substantially by 2023. Today in Energy, 10 July 2019
US Energy Information Administration (EIA) (2020a) Annual Energy Outlook 2019 with projections to 2050
US Energy Information Administration (EIA) (2020b) Monthly Energy Review
US Energy Information Administration (EIA) (2020c) Natural Gas Weekly
US Energy Information Administration (EIA) (2020d) Preliminary Monthly Electric Generator Inventory
US Energy Information Administration (EIA) (2020e) Quarterly Coal Report
US Environmental Protection Agency (US EPA) (2016) Aquatic life ambient water quality criterion for selenium—freshwater 2016. EPA 822-R-16-006. Washington, DC
US Environmental Protection Agency (US EPA) (2019) Disposal of Coal Combustion Residuals from Electric Utilities Rulemakings. https://www.epa.gov/coalash/coal-ash-rule
US Global Change Research Program (US GCRP) (2017) Fourth National Climate Assessment. https://www.globalchange.gov/nca4
Vidic RD, Brantley SL, Vandenbossche JM, Yoxtheimer D, Abad JD (2013) Impact of shale gas development on regional water quality. Science 340:1235009
Wall Street Journal (WSJ) (2019) Banks get tough on shale loans as fracking forecasts flop. 23 December 2019
World Meteorological Organization (WMO) (2019) The global climate in 2015–2019
Zhang X (2016) Emission standards and control of PM2.5 from coal-fired power plants. International Energy Agency, Clean Coal Centre, CCC/267
Zipper CE, Adams MB, Skousen JG (2021a) The Appalachian coalfield in historical context. In: Zipper CE, Skousen JG (eds) Chapter 1, Appalachia's coal-mined landscapes. Springer
Zipper CE, Agouridis CT, Barton CD, Skousen JG (2021b) Conversion options for mining-affected lands and waters in Appalachia. In: Zipper CE, Skousen JG (eds) Chapter 7, Appalachia's coal-mined landscapes. Springer